Time and Space

Time and Space

Barry Dainton

First published in 2001 by Acumen

Acumen Publishing Limited
15A Lewins Yard
East Street
Chesham HP5 1HQ
www.acumenpublishing.co.uk

ISBN: 1-902683-32-3 (hardcover)
ISBN: 1-902683-33-1 (paperback)

British Library Cataloguing-in-Publication Data
A catalogue record for this book is available from the British Library.

Designed and typeset by Kate Williams, Abergavenny.
Printed and bound by Biddles Ltd., Guildford and King's Lynn.

For Gwynneth

Contents

Preface

Is space an ingredient of reality in its own right, or simply nothingness? What does the passage of time amount to? Is only the present real, or is the past real as well? And what of the future? Are space and time finite or infinite? For anyone with an interest in the large-scale composition of the cosmos, no questions are more pressing than these; likewise for those concerned to understand the framework within which we live our lives. But few questions are more challenging. Trying to think clearly about space is not easy. How does one go about thinking about nothing? But time is harder still. While it may seem clear that the past is real in a way the future is not, it also seems clear that the present is real in a way the past is not, so what sort of reality does the past possess? It may seem obvious that time passes, that the present is steadily advancing into the future, but just what does the passage of time involve? These are simple questions about the most fundamental features of our world, yet no obvious answers spring to mind.

Answers to all these questions have been proposed; many of them are fascinating, and many far from obvious. The philosophical literature, both ancient and modern, is both large and wide-ranging, embracing as it does topics as diverse as semantics, causation, modality and phenomenology. The relevant scientific and mathematical literature is no less vast, and the relevant theories are diverse and difficult: the physics of motion; the nature of the continuum; the geometry of flat and curved spaces; relativity theory; and quantum mechanics. This range and diversity make the study of space and time uniquely rewarding. Few other subjects introduce as many unfamiliar and exotic ideas, and perhaps no other subject stretches the imagination so far. But it also makes the subject unusually difficult for anyone approaching it for the first time. The aim of this book is to make this task easier.

I have tried to provide an introduction to the contemporary philosophical debate that presupposes little or nothing by way of prior exposure to the subject, but that will also take the interested and determined reader quite a long way. Anyone completing the book should be in a position to make a start on some of the more advanced philosophical work, as well as coming away with a reasonable (if elementary) understanding of the current state of play in physics. The book is intended primarily for students of philosophy doing courses in metaphysics or philosophy of science (second year and above), but most of it should be intelligible to the interested reader who is new to philosophy: philosophical jargon is used, but sparingly, and often explained *en route*; a good many other terms can be found in the Glossary. Since they will already be familiar with many of the basic themes, avid consumers of popular science should have little difficulty with most of the book, especially those whose appetite for mind-bending ideas from the frontiers of physics is temporarily sated, and who are starting to wonder what the implications of these ideas might be. Much contemporary philosophy of space and time is devoted to precisely this question.

Given the complexity of the subject matter, I have narrowed my focus and concentrated on just two key questions:

- Does time pass?
- Does space exist?

Since these questions cannot be addressed in isolation a good many other topics are discussed as well, but there are some that are not treated at all (e.g. Zeno's paradoxes). These omissions are regrettable, but a more comprehensive treatment would necessarily have been more superficial, and ultimately less interesting, so I had no qualms about making them.

There are three interconnected parts to the book. The first of these, taking up Chapters 2–8, is devoted exclusively to time. Although a good deal of the analytic philosophy of time over the past half century has been devoted to linguistic and semantic issues, I decided not to give prominence to these. The significance of these intricate debates is far from obvious to the uninitiated, and their interest has diminished as the prospect of their having any decisive metaphysical impact has receded. More recently, thanks to the appearance of works such as Tooley's *Time, Tense and Causation* (1997) and Price's *Time's Arrow and Archimedes' Point* (1996), the climate has changed as metaphysical and scientific issues have come more to the fore. I have tried to reflect this development in my treatment here.

Although my approach to time is largely ahistorical, I begin with McTaggart's paradox (1908), which has exerted a significant influence on recent discussions. Although some believe that McTaggart's argument (properly interpreted) reveals the untenability of all dynamic conceptions of time, I argue for a more moderate verdict: some dynamic conceptions perish, but others survive intact. The surviving dynamic conceptions are examined in Chapters 5 and 6 (the latter includes discussions of the "growing universe" model advocated by Broad and Tooley, and "presentism", the doctrine that only what is present is real). Before this, in Chapters 3 and 4, the alternative static conception of time is considered; the (initially astonishing) doctrine that time is essentially like space, in that it does not pass. Of especial concern here is saving the appearances: why is it that time seems to pass if in reality it doesn't? This topic is taken up again in Chapter 7, which is entirely devoted to the phenomenology of time; how time is manifest in our streams of consciousness. This puzzling and important issue is too often treated very superficially or not at all. To round off this opening stage, the possibility (or otherwise) of time travel is examined in Chapter 8.

My approach to space follows a different kind of trajectory. After a general overview in Chapter 9, we take a step back into the past and look in some detail at the sixteenth- and seventeenth-century debates concerning space and motion, the main protagonists being Descartes, Galileo, Leibniz and Newton (with the latter pair having starring roles). This historical detour provides a change of atmosphere after the largely a priori investigation into time, but is also justified on other grounds. Since purely metaphysical investigations into the nature of space do not carry one very far, empirical considerations have to be introduced at an earlier stage, and the Newton–Leibniz controversy provides a useful point at which to start, not to mention the fact that this controversy is still being discussed in the contemporary literature.

Chapter 12 takes the debate about motion and dynamics a step further by examining how the transition to neo-Newtonian spacetime affects the Leibniz–Newton controversy. The move towards the present day is continued in Chapters 13 and 14, where the impact of non-Euclidean geometries is considered, together with Kant's "hand" argument and Poincaré's conventionalism. Chapter 15 is purely metaphysical: the topic is Foster's powerful but complex argument for adopting an anti-realist stance to physical space.

The final part of the book is devoted entirely to contemporary scientific theories and their interpretation. I provide an informal introduction to Einstein's special theory of relativity and Minkowski spacetime in Chapter 16, and consider their metaphysical implications in Chapter 17. It is sometimes assumed that the special theory, by virtue of the way it relativizes simultaneity (and hence the present) establishes the block view of time beyond all doubt, but this is not so. There are dynamic models that are compatible with the special theory, and while these models have some counterintuitive consequences, so, too, does the static alternative, not to mention the special theory itself. Some of the rudiments of Einstein's general theory are expounded in Chapter 18, and its metaphysical implications (mainly concerning substantivalism) are examined in Chapter 19. I close by briefly considering some of the more speculative developments in recent physics.

The book's three parts could be seen as being devoted to *time*, *space* and *spacetime* respectively – for a while I considered separating and labelling them as such – but this would have been misleading. The "block universe" considered in Chapters 3 and 4 is a four-dimensional spacetime continuum, and spacetime returns again (albeit in neo-Newtonian guise) in Chapter 12. On reflection it seemed better to acknowledge the interrelated character of the issues by *not* imposing a formal division into parts.

A problem facing any book such as this concerns the balance between philosophy and science. How much science should be included? How advanced should it be? Where should it go? Needless to say, there is no perfect solution. I have opted for a policy of progressive integration: as the story unfolds the balance tilts from "philosophy + a little physics" to "physics + philosophy of physics", but the latter stage is reached only in the last few chapters, and the level of exposition remains elementary. A more difficult choice was *what* physics to include, and here, too, hard choices had to be made: there is no separate or systematic treatment of quantum theory. This is a deficiency, for it may well be that developments in this area will have a decisive impact in years to come, when quantum gravity theory comes into its own. But we are not there yet, and given the plethora of competing interpretations of quantum theory in its current guise, it is impossible to know what impact it will eventually have. Hence I came to the conclusion that the omission was justifiable; remedying it would have made a long work even longer, and sacrificing relativity was never an option. Quantum theory does, however, make an occasional unsystematic appearance: in §8.6, in the context of backward causation, and in §17.4, where the focus is on the tension between quantum entanglement and special relativity.

Throughout the book I am less concerned with reaching final conclusions than with revealing interconnections. Given the number of important questions in physics and cosmology that have yet to be answered, such conclusions would be premature at best. But, as we shall see, if our understanding of space and time is to improve, progress also needs to be made on certain philosophical problems. Some doubt this is possible, but I am more optimistic. Progress in metaphysics is not impossible, but it does not come easily. When it comes it often involves uncovering patterns of dependency, connections between doctrines not previously suspected, or at least not obvious. By revealing what must stand or fall together, these patterns shed new light on what the coherent metaphysical options might be. I have tried to bring some of these patterns into clearer view.

Acknowledgements

I am indebted to a number of people for ideas and assistance: the Liverpool students who have served as a testing ground for much of this material over the past two years; Stephen Clark, Nicholas Nathan and Howard Robinson for helpful comments on the chapters concerning time; Duncan Cryle; John Earman; Gerard Hurley; Jenny Cordy for material assistance in rendering the third dimension; and Jay Kennedy, who read an earlier draft with great care, and whose many comments assisted me enormously. I am also grateful to Kate Williams, for her design skills and excellent copy-editing, and Steven Gerrard, for much sound advice and his enthusiasm for the project – especially after it became clear quite what it would involve.

1 Preliminaries

By way of setting the stage, this opening chapter introduces some of the most important metaphysical issues concerning time and space. Although many of the questions that can be asked about time can also be asked about space, and vice versa, time raises distinctive issues all of its own, and I will be devoting special attention to these. Some of the distinctive issues concerning space are discussed in Chapter 9.

1.1 Ontology: the existence of space and time

Do space and time exist? This is an obvious question, and one that is frequently addressed, but it can mean different things. Since philosophers raise the question of whether the physical world as a whole exists, it is not surprising that they ask the same question of space and time. However, the question "Do space and time exist?" is usually not asked in the context of a general scepticism. Those who pose this question generally assume that the world is roughly as it seems to be: objects are spatially extended (a planet is bigger than an ant), objects exist at different places and things happen at different times. So what is usually meant by the question is this: "Are space and time entities in their own right, over and above things such as stars, planets, atoms and people?"

The answer is not obvious, and for an obvious reason: both space and time are invisible. We can see things change and move, and by observing changes we can tell that time has elapsed, but of time itself we never catch a glimpse. The same applies to space. We can see that things are spatially extended, and we can observe that two objects are separated by a certain distance, but that is all. Indeed, the very idea that space exists can seem nonsensical.

When two objects are separated by an expanse of empty space there appears to be strictly *nothing* between them (ignoring what appears behind them). If space is just nothing, how can it possibly be *something*? But, of course, whether empty space really is just "nothing" is precisely what is at issue, and invisibility is an unreliable guide to non-existence. Science posits lots of invisible entities (neutrinos, force-fields) that many believe to exist; perhaps space and time fall into the same category. Clearly, the fact that space and time are not directly observable does complicate matters. If we are to believe that space and time exist as entities in their own right, we need to be given compelling reasons; the situation is different with observable things such as rabbits and houses.

Substantivalism
universe = things + space

Relationism
universe = things + relations

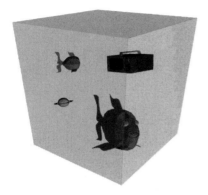
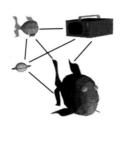

Figure 1.1 The substantival and relational conceptions of space. A pair of space-fish go about their business. The shaded block on the left represents space (or at least a small portion of it) as construed by the substantivalist. On this view, the "void" between material objects is far from empty: it is filled by an extended (if invisible) entity. Relationists reject this entity – and so regard "empty space" to be just that – but they also maintain that material objects (likewise their constituent parts) are connected to one another: by "spatial relations". These relations are usually taken to be basic ingredients of the universe, which connect things directly, i.e. without passing through the intervening void (in this respect the picture on the right is misleading).

There are two opposed views on this ontological issue that have a variety of names, but that for now we can call "substantivalism" and "relationism". Substantivalists maintain that a complete inventory of the universe would mention every material particle and also mention two additional entities: space and time. The relationist denies the existence of these entities. For them the world consists of material objects, spatiotemporal relations and nothing else. Figure 1.1 illustrates the difference between these two competing views; to simplify the picture only a small volume of space is shown, and time is omitted altogether.

It is important to note that in denying the existence of space and time as objects in their own right relationists are not claiming that statements about where things are located or when they happen are false; they readily admit that there are spatial and temporal relations or distances between objects and events. What they claim is that recognizing the existence of spatial and temporal relations between things doesn't require us to subscribe to the view that space and time are entities in their own right. The relationist is, of course, required to give a convincing account of what these "relations" are, and how they manage to perform the roles required of them.

The substantivalist is sometimes said to regard space and time as being akin to containers, within which everything else exists and occurs. In one sense this characterization of substantivalism is accurate, but in another it is misleading. An ordinary container, such as a box, consists of a material shell and nothing else. If all the air within an otherwise empty box is removed, the walls of the box are all that remain. Contemporary substantivalists do not think of space and time in this way; they think of them as continuous and pervasive *mediums* that extend everywhere and "everywhen". The air can be removed from a box, but the space cannot, and if space is substantival, a so-called empty box remains completely filled with substantival space, as does the so-called "empty space" between the stars and galaxies. Roughly speaking, a substantivalist holds that space and time "contain" objects in the way that an ocean contains the solid things that float within it.

It is plain that the verdict we reach on the ontological question concerning space and time has momentous implications for what the universe contains. If the substantivalist is correct, space and time are the *biggest* things there are. The relationist view is much more economical: the vast pervasive mediums posited by the substantivalist simply do not exist.

We will start to explore this issue in Chapter 9, and it will remain centre stage for much of the remainder of the book. Although the initial focus will be on space, the ontological status of time will not be ignored, even though it will not receive separate treatment (in Chapters 2–8, where time occupies centre stage, the substantivalism–relationism controversy will not feature at all). Since any spatial point also has a temporal location, it is more properly viewed as a point in space *and* time, i.e. as a "spacetime point". The substantivalist takes the totality of these spacetime points to be a real entity: "spacetime". Figure 1.2 shows a small portion of a substantival spacetime – the translucent block – over a period of three days (one spatial dimension is suppressed). Clearly, if this conception is correct, time is just as real as space.

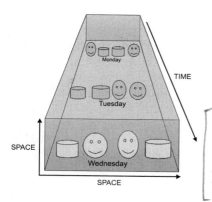

Figure 1.2 Spatiotemporal ("spacetime") substantivalism. Material objects are distributed through an extended medium composed of spacetime points. Since one spatial dimension has been suppressed, each two-dimensional slice of the block represents a three-dimensional volume of space.

1.2 Questions of structure

Irrespective of how the debate between substantivalists and relationists is resolved, there are questions to be addressed concerning what can loosely be called the "structural" characteristics of space and time. Whereas substantivalists take these questions to be about actual entities, relationists take them to be about the ways material things are related in spatial and temporal ways. To get some impression of the flavour of these structural questions consider the following claims:

- Space is infinite in size, three-dimensional, Euclidean (so travellers moving along parallel trajectories will never meet), isotropic (there is no privileged spatial direction) and continuous (infinitely divisible).

- Time is infinite, one-dimensional, linear, anisotropic (there is a privileged temporal direction) and continuous.
- There is just one space, and one time.
- Time and space are *absolute*, in that they are unaffected by the presence or absence (or distribution) of material things.

Some readers may find this characterization quite plausible. Those with some knowledge of ancient and medieval cosmologies will realize that it is similar to one defended by the ancient atomists but soon rejected in favour of the Aristotelian system (which posited, among other things, a finite, non-homogeneous and anisotropic space), and which returned to favour only in the 1500s. Those with some scientific (or science fiction) background will realize that this essentially Newtonian conception was overturned by Einstein's theories of special and general relativity in the early years of the twentieth century. More recently, some quantum theorists have argued that time is branching rather than linear, while others maintain that time does not exist; relativistic cosmologists have argued that there may be a multiplicity of spacetimes (or "baby universes") sprouting from the other end of black holes, and superstring theorists argue for the astonishing view that there are nine (or more) spatial dimensions.

Fascinating as many of these recent theories are, this is not the place for a thorough discussion of them, although their implications are such that they can scarcely be ignored either. Where space is concerned, my main focus will be on the substantivalist–relationist controversy, and I will only be looking at the structural issues that are relevant to it. Since it turns out that a good many structural issues *are* relevant to it, we shall have occasion to enter into at least the shallows of these perplexing waters.

1.3 Physics and metaphysics

Are there any facts about space and time that are necessary (hold in all possible worlds), and that can be established by a priori reasoning? Or are the answers to all of the interesting questions about space and time contingent and a posteriori, and hence to be answered only by the relevant sub-branches of science – physics and cosmology? If we answer "yes" to the latter question, the philosophy of space and time reduces to a branch of the philosophy of science and we must confine ourselves to examining the content of contemporary scientific theories, in particular relativity and quantum theory.

This has something to be said for it for, as we have just seen, scientists have made claims about space and time that provide answers to many of the questions philosophers have struggled with; many (if not all) of the relevant theories are well confirmed and so cannot be ignored. It would also be fair to say that science and mathematics have provided us with new ways of thinking about space and time, ways that philosophers may never have stumbled across if left to their own devices. The following simple but influential argument dates back to ancient Greek times:

> The idea that space is finite can be shown to be absurd. Suppose that space were finite. If you travelled to the outermost boundary of space and tried to

continue on, what would happen? Surely you could move through this supposed barrier. What could stop you, since by hypothesis there is nothing on the other side? In which case, the supposed limit isn't really the end of space. Since this reasoning applies to any supposed edge of space, space must be infinite.

There are replies the metaphysician can make. For example:

> By definition, to move is to change your place in space. Accordingly, it makes no sense at all to suppose you *could* move beyond the edge of space, since by hypothesis there are no places beyond this limit for you to occupy. Consequently, the assumption that space is finite does not result in an absurdity.

This may seem reasonable enough, but there are further moves and countermoves available to the metaphysician. Does the reply tacitly assume a substantivalist view of space? A relationist would reject the very idea that motion consists of moving on to pre-existing spatial locations; if motion consists instead of an object changing its spatial relations to other objects, what is to prevent an object from moving ever further from other objects? And so it proceeds. But as we shall see in Chapter 13, the assumptions underlying this debate were completely transformed in the nineteenth century by the discovery of non-Euclidean geometries and the possibility that a space could be curved. If our space were positively curved – a three-dimensional counterpart of the surface of a sphere – someone could set off on a journey, always travel in a straight line (never deviating to the right or left), and still end up just where they started. Space can thus be finite without possessing any limits, edges or obstacles to movement. This is just one example of how philosophical discussions of space and time have been influenced if not rendered redundant by developments in physics and mathematics; there are others.

However, while the science is important, there are several reasons why it would be a mistake to suppose that *only* science matters, and that considerations of a philosophical sort are dispensable. First of all, physics and metaphysics are to some degree interdependent. When evaluating a theory, when inventing a theory, scientists are themselves influenced, in part, by metaphysical considerations: by what it makes sense to say the world is like. Moreover, the metaphysical implications of a scientific theory – i.e. what implications the mathematical formalism has for how the world is – are often unclear. This is especially true of relativity and quantum theory. Hence there is a need for metaphysical inquiry, even if this inquiry does not take place independently of scientific theorizing.

But metaphysics also has its own distinctive domains and methods of inquiry. While it may well be that to discover the answers to *some* questions about the space and time in our world we will have to listen to the physicist, there are other questions for which this is not so. Some of these questions are about topics that lie altogether outside the domain of physics: for example the role space plays in our ordinary modes of thought; the way time manifests itself in human consciousness; our different attitudes to the past, present and future, and their rationality or otherwise. But even if we confine ourselves to talking about time and space themselves, rather than how we think about or experience them, there are conceptual issues of a distinctively philosophical sort that remain of interest.

Assuming (as most do) that the laws of nature are contingent, there are possible worlds of a recognizably physical sort, worlds that resemble ours in superficial (observable) ways, but where the laws of nature are different. Many of these possible worlds are spatiotemporal (e.g. those where Newton's laws hold, those where Aristotle's laws hold). The fact that these worlds are spatiotemporal means that our *concepts* of space and time are to some degree independent of particular scientific theories, and so we can investigate these concepts independently of any one scientific theory. What must a world be like in order for it to be spatial and temporal? What are the different forms that space and time can take? In their attempts to answer these questions philosophers have introduced many novel and counterintuitive ideas, a good many of which are stranger than anything seriously entertained by scientists engaged in the struggle to ascertain the particular forms that space and time take in this world.

Finally, there is this point. We know that our current fundamental physical theories are imperfect: quantum theory and general relativity have yet to be fully reconciled. It may well be that the theory that emerges from this eventual marriage will have very different implications for the nature of space and time than those of currently accepted theories, so it would be very short-sighted to take *current* scientific theories to be the last word on space and time *in our universe* (let alone other possible universes).

1.4 Time: the great divide

The issues that have been introduced so far, concerning ontology and structure, concern both space and time equally. But there is one issue concerning time that has no spatial counterpart. If asked "How does time differ from space?", after a few moments' reflection you might say "It's not so easy to travel through, it has one dimension rather than three, and it has a direction." I want to focus on the last answer. Why do we say that time has a direction whereas space doesn't? Since there are three dimensions of space and one dimension of time, someone in an awkward frame of mind might point out that space has *more* directions than time, and in one sense they would be right. But what this response omits to mention is that none of these spatial directions is singled out as in any way special or distinctive. Whenever we draw a diagram with an axis representing time we unthinkingly insert a single-headed arrow pointing towards the future. Although we may also insert arrows indicating various spatial directions, no one spatial direction is privileged; up, down, right, left, north, south, east, west, irrespective of what we call them, each direction gets its own individual arrow (see Figure 1.3). This diagrammatic convention has a linguistic counterpart: we sometimes talk of "the arrow of time", but never of "the arrow of space". But what exactly does time's arrow involve?

The pessimist who says "Things always get worse, never better," a sentiment often taken to be confirmed by the second law of thermodynamics, according to which entropy (or, loosely speaking, disorder) will always increase, provides an illustration of one sort of temporal arrow: certain sorts of process favour one temporal direction over the other. But there is, it seems, more to it: the arrow of

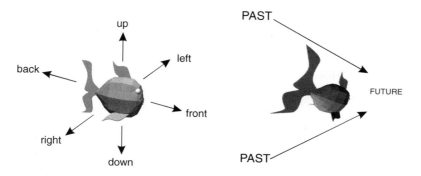

Figure 1.3 A natural way of thinking about space and time. Nothing distinguishes the various spatial directions; in outer space (if not on Earth) we can move equally freely in any direction. Time, however, does possess an inherent direction or directedness, running from past to future.

time runs deeper, there is a directedness that belongs to time *itself*, rather than to anything *in* time. Or so many of us are inclined to think.

We speak of "time passing", "the march of time", "the flow of time", and there are no spatial analogues of these locutions (we don't speak of "space passing", "the march of space", "the flow of space", etc.). As for what we mean when we say that time passes, the following formulation encapsulates the essentials:

What is future will become present; what is present will become past; what is past was once present.

Your birth was once present, but now lies in the past. Your next birthday lies in the future, but it is getting closer every day. Getting closer to what? The present, of course. We can think of the present as advancing into the future, or the future advancing towards the present; either way the distance between the two is always decreasing. What applies to your next birthday applies to all times and all events: they start off in the future, become present and then recede into the past. This process is often called *temporal passage*, and has no obvious spatial counterpart.

This is how we talk, but what is there in reality that corresponds to this talk? It is here that we reach the great divide in the philosophy of time. There are those who believe that temporal passage is a real phenomenon, an objective feature of reality that is independent of the perspective that conscious beings such as ourselves have on things. Anyone who believes this subscribes to a *dynamic* view of time. Some dynamists hold that passage involves a special property of "present-ness" moving along the timeline. Others explain passage in terms of the non-existence of the future: only the past and present are real, but since reality is growing – new times are coming into being as future possibilities crystallize into present actualities – no time remains present for long (see Figure 1.4). Others deny reality to both the past and the future: time consists of a succession of ephemeral presents.

In the opposing camp are the many philosophers who reject the claim that temporal passage is a real and objective feature of the world. These philosophers subscribe to the *static* view of time. According to this doctrine all moments of time (and all events) are equally real, and there is no moving or changing present;

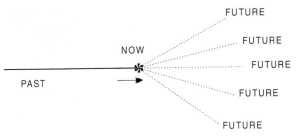

Figure 1.4 Perhaps the most popular of the dynamic views of time: the past is fixed and real; the future unreal and unfixed (there are many ways things may unfold); the present (or now) is real, distinctive and advancing into the future. Despite its appeal, this model of time is not without its difficulties. For example, one might ask: "Just how is the present different from the past?"

nothing becomes present and then ceases to be present. The differences between past, present and future are simply differences of perspective. If someone in Australia says "It's sunny here," they are saying something true if it is sunny at the place where they are speaking; likewise for someone in New Zealand who says "It's raining here." We do not believe that either of these places is special or distinctive simply because they are referred to by the term "here". The same holds, says the static theorist, for "now". Someone in 1801 who says "It's snowing now" is saying something true if it is snowing at the time they are speaking; likewise for someone who says the same thing in 2001. For the static theorist, time and space are ontologically equivalent in this crucial respect: space consists of a three-dimensional arrangement of coexisting locations (places), none of which is special; time consists of a linear one-dimensional series of coexisting locations (times), none of which is special.

Another common label for the static conception of time is the "block view", and for understandable reasons. Think of a long, solid crystalline block embedded within which are small plastic figures in various poses. If the substantivalist view is correct, this is essentially *our* condition, save that the block we are embedded in extends through time as well as space, and so is a four-dimensional rather than three-dimensional entity. All the people frozen in different poses at different places and times in this block are equally real; indeed, some of these people are *you*, as you were yesterday and at still earlier times, and tomorrow and at still later times (see Figure 1.2). Although all the contents of the block are permanent fixtures it is a mistake to suppose that the block is a thing that *endures*. The block does not exist *in* time as we do; it would be more accurate to say that time is *in* the block, for time is simply one of the dimensions that the block possesses. Viewed as a whole, the block is an eternal (or sempiternal) entity (see Figure 1.5).

If, on the other hand, the relationist view is correct, we are not embedded in anything analogous to a block of crystal; the world consists of nothing but material bodies in spatial and temporal relations. But these interrelated things nonetheless constitute a static four-dimensional ensemble; all that is missing is the surrounding medium posited by the substantivalist.

These two views of reality, the static and the dynamic, are as old as Western philosophy. Heraclitus (approx. 540–480BC) held that the world was entirely dynamic, an endless process of change, creation and annihilation. Parmenides (approx. 520–430BC) maintained that the opposite was true: "A single story of a road is left . . . being is ungenerated and undestroyed, whole, of one kind and motionless and balanced . . . nothing is or will be other than what is – since *that* has Fate fettered to be whole and motionless."[1] And the issue is one that still resonates.

Figure 1.5 The static block model of time. As (a) makes clear, to view the block as something that exists within time is quite wrong. Image (b) is an attempt to put flesh on the idea that dynamic features of the universe, such as change and movement, are merely patterns running along the block. Although depictions such as these can be useful, they also highly schematic. The block-universe is really a four-dimensional entity (one dimension of space has been suppressed here). If time were infinite, the block would be infinitely long; if space were infinite in all directions, the block would be infinitely wide and tall. It is only if time were finite and space cube-shaped that the universe would have the shape depicted.

When first encountering the static view many people find it utterly absurd. That we live out our lives in a constantly advancing present, that past and future events are different from those that are present – what could be more obvious? But as static theorists are fond of pointing out, there are plenty of other cases where our everyday experience seems to provide overwhelming support for beliefs that are false. Can you remember how you felt when, as a child, you were told that there are folk in other countries for whom our "down" is their "up"? Even now, are you entirely comfortable with the thought that the Earth is shooting through space at a rate of many miles per second?

Others find the static view more repellent than absurd. Most of us tend to assume that, even if the past is fixed, the future is *open*: what it will bring is as yet undetermined. If the block view is true this is wrong, for the future is just as real, solid and immutable as the past. How our lives will unfold from now until the moment of our deaths is (in a manner of speaking) already laid down. How could it be otherwise if the future stages of our lives are just as real as the past stages? This is not to say that we have no power over the ways our lives will unfold, for we do. We will all make choices, and the choices we make will contribute to the ways our lives will turn out. But if the block view is true, the choices we will make are inscribed in the fabric of reality in precisely the same way as the choices we have already made. While those who find this "temporal determinism" unsettling will be inclined to reject the conception of time from which it derives, it would be a mistake to suppose that the static view *must* be false simply because it has unpalatable consequences. The universe is under no obligation to conform to our preconceptions or preferences.[2]

The static–dynamic issue may impact on how we think of our lives, but it also has consequences of a less parochial sort. A universe in which the sum total of reality is limited to what is (momentarily) present is far *smaller* than a universe in which the sum total of reality includes both the past and future as well, and likewise, albeit to a lesser degree, for a universe in which only the past and present are real. The "smallness" in question concerns temporal rather than spatial extent but, as Figure 1.6 makes clear, it is no less significant for that.

The substantivalism–relationism issue has significant ontological ramifications since, if space is a thing, it is a uniquely *large* thing, extending as it does

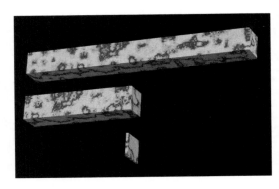

Figure 1.6 Three conceptions of the sum total of reality.

everywhere. But the ontological ramifications of the static–dynamic dispute are vast in comparison.

1.5 Two frameworks

Common sense may tell us that time is dynamic rather than static, but our ordinary ways of talking about time reflect both perspectives. There are two families of temporal terms and concepts that we commonly use, one of which embodies the dynamic view and the other the static view. McTaggart is credited with clearly distinguishing these two ways of thinking. Since the distinction itself and McTaggart's terminology continue to play an important part in contemporary debates, the sooner they are introduced the better.[3]

Imagine finding the following message-in-a-bottle: "The bomb will explode under the Eiffel Tower four days from now." The message is annoyingly uninformative: ignorant as you are of when the message was written, you do not know which time the word "now" refers to, and so you have no way of knowing whether the bomb was intended to explode a hundred years ago, in two days' time, or fifty years in the future. The same would apply if the message read ". . . the day after tomorrow" or even ". . . today". It would be quite different if the message had read: "The bomb will explode under the Eiffel Tower on 1 June 1987."

This simple example provides an illustration of a more general point. We can think of events as being situated at some distance or direction from the present time: for example, "*E* lies in the past (or future)", "*E* is occurring now", "*E* will occur three days from now", or "*E* occurred fifty years ago". Alternatively, we can think of events in terms of when they occurred in relation to other events (or times), without making *any reference to the present at all*: for example, "*E* occurred seven years after the Battle of Trafalgar", "*E* happened three days before the first Moon landing", or "*E* took place in 1977".

McTaggart called the time series that runs from the far past up to the present "now" and then on into the future the "A-series", and the time series that is entirely independent of the present the "B-series". Each is associated with its own range of distinctive temporal characteristics: *pastness, presentness* and *futurity* are the basic A-properties, whereas the basic B-properties are *simultaneous with, earlier than* and *later than*. Subsequent writers also talk of A- and B-concepts, A- and B-sentences, A- and B-statements (what particular A- and B-sentences are used to express on particular occasions of use) and of A- and B-facts. For the sake of convenience I will include this diverse range of items under the rubrics "A-framework" and "B-framework".

The ways of thinking about time enshrined in the A- and B-frameworks are sometimes called the "tensed" and "tenseless" conceptions. Those who adopt this terminology speak of "tensed properties", "tenseless facts" and so forth. Although the concept of tense is usually associated with grammar and linguistics, this more liberal use has a clear rationale. The basic grammatical tenses are past, present and future – the very notions that underlie the A-framework and that are conspicuously absent from the B-framework.

Although the A-series and the B-series involve different ways of thinking about time, it would be a mistake to suppose that the two are wholly independent. The basic B-relationships of being *simultaneous* with, *earlier* than and *later* than, are all to be found within the A-framework: for example, "today" is one day later than "yesterday", and "a week ago" is seven days earlier than today. Moreover, both frameworks order the same events in the same ways. Suppose it is now 1 December. Then clearly "today" and "1 December" refer to the same time, but so do "yesterday" and "30 November", "tomorrow" and "2 December", and likewise for the remainder of the two series. Evidently, although we think of time in terms of the A- and B-frameworks, we don't ordinarily think that there are *two sets of distinct times that run in parallel*; there is just one sequence of times.

The two frameworks may be intertwined, but they are also very different. The B-series is a static ordering of times and events. The first Moon landing in 1969 took place 24 years after the end of the Second World War, and 31 years earlier than the end of the second millennium. These B-facts not only obtain now, they obtain at other times; they will never cease to obtain. If a B-statement is true it remains true. A-statements and A-facts could not be more different. If at the start of December I say "There are now 25 days until Christmas" I am speaking the truth; if I make the same statement a few days later I am not. Why? Because the time referred to by "now" in the two statements is different. The A-series is a dynamic, constantly changing, temporal ordering, and the fact that the present is never still but always advancing into the future renders A-facts transitory and dooms A-statements to changes in truth-value.

1.6 Matters terminological

The following characterizations are rough and ready but they will serve for present purposes.

- The *static view*: all times and events timelessly coexist, and all are equally real; temporal passage is unreal. In short, the block view is true.
- The *dynamic view*: temporal passage is real, and time is not a static ensemble of coexisting times and events. In short, the block view is false.
- The *tensed theory*: A-concepts such as past, present and future have an essential and ineliminable role to play in any metaphysically adequate account of the nature of time.

- The *tenseless theory:* B-concepts such as earlier than, later than and simultaneous with are all we need for a metaphysically adequate account of time.

The static–dynamic dispute is over the nature of time, how time is in and of itself;
the tensed–tenseless dispute is over the terms or concepts needed to describe time
and its contents. These are distinct issues, and we should not at this point make any
assumptions about how they will be resolved. Consequently, we need to be careful
how we use and understand terms like "A-theorist" and "B-theorist". As I will be
using the terms, a B-theorist is someone who subscribes to the static block view,
and an A-theorist is someone who doesn't.

2 McTaggart on time's unreality

2.1 Could time be unreal?

In an attempt to get us to take seriously the idea that reality might be very different from how it seems, Descartes introduced an all-powerful demon. This demon, we are to suppose, is causing us to hallucinate, to have experiences that do not correspond to anything real. The experiences in question are the perceptual experiences that we are having at the moment and those we remember having in the past: the experiences that we unthinkingly take to be revealing the world to us. The point of this scenario is to lead us to appreciate that our experience does not guarantee that the world is as we normally believe it to be. Let us vary the scenario. The all-powerful demon informs us that it has not been providing us with hallucinatory experiences – the world really is pretty much as it seems – but it has instead taken the decision to abolish time.

What would you make of this announcement? Should you care? What difference *would* it make? Perhaps the world would cease to exist: when the demon acts the present winks out of existence, and the past along with it. Or perhaps everything would come to a stop. People and things wouldn't cease to exist, they would simply stop changing. They would be frozen in mid-sentence, in mid-air, and (timelessly) remain in this condition. These suggestions should not be taken too seriously, but most of us would surely agree that the absence of time would make a very significant difference.

McTaggart had no truck with demons, but he did maintain that in actual fact time is unreal. His claim is all the more remarkable (and hard to believe) because he did not also claim that we are being provided with delusory perceptions. McTaggart maintains that time would be unreal even if the world were exactly like it seems to us to be (except for the fact that time is unreal). You may think you have never heard of anything more ridiculous, but by way of a defence for his claim McTaggart deploys a metaphysical argument to the effect that time *could not possibly* exist. This intriguing and (at times infuriating) argument continues to be discussed and taken seriously, and many hold that, even if it does not establish what McTaggart thought it did, it still tells us something important about the nature of time.

2.2 Change as the essence of time

McTaggart's argument revolves around the A-series (which runs from the past, on to the present and into the future) and the B-series (which runs from earlier to later), which we encountered in §1.5. He suggests that "we never *observe* time except as forming both these series" (1908: 458), and that anyone who believes in the reality of time should believe that both series are real. His overall argument for the unreality of time has this form:

1. If the A-series does not exist, time does not exist.
2. The A-series is metaphysically incoherent, and therefore does not exist.
3. Therefore time does not exist.

As is plain, if 1 and 2 are both true, his conclusion 3 follows. We will be looking at his arguments for 2 in the next section; for now I want to focus on 1.

McTaggart's argument for 1 involves establishing two claims. The first is (on the face of it at least) uncontroversial: time is the dimension of change. This isn't to say that in a temporal world there is always something changing; there are powerful arguments to the effect that "temporal vacua" – periods of time during which nothing changes – are conceptually speaking possible.[1] Even if temporal vacua are (conceptually) possible, it still seems true that time is the dimension in which changes do and can occur. A world without time would certainly be a world without change. McTaggart's distinctive (and undeniably plausible) claim is that a world in which change *cannot possibly* occur is a world without time.

The second claim is far more contentious: change requires the A-series. Why believe this? McTaggart asks us to consider a poker that is red hot at one end and cool at the other. The poker possesses different properties at different places; it is spatially extended and one part of it is hot whereas another part is cool. Does this variation in properties over space amount to a change? Clearly not. A hot poker that cools down over a period of a few minutes involves change, involving as it does a change in properties over time, or so common sense suggests. Now just suppose that the B-theorist is right and that time could exist without an A-series, and consider a poker that is hot at t_1 and cold at t_2. The B-theorist will say that this is an instance of change, but McTaggart objects: what is it that *changes* here? Everything about the B-series is permanent so it is *always* true that the poker is hot at t_1 and cold at t_2. These B-facts obtain at all times. Variations in properties across space at a given moment of time do not amount to change because all the places in question coexist. The B-theorist who holds that all moments of time are equally real is, in effect, viewing time as we ordinarily view space: as an extended dimension of coexistent locations. As a result there are no essential differences between spatial variation *at* a time and the B-theorist's account of change *over* time: in both cases we have an object possessing different properties at different locations in a static dimension. It is quite different if we introduce the A-series into the equation. We still have the permanent B-facts – the poker is hot at t_1 and cold at t_2 – but in addition we have ever-changing A-facts. The poker's being hot at t_1 is first in the distant future, then the near future, then the present and thereafter the ever more distant past.

Hence McTaggart's conclusion: a world without ever-changing A-facts, a static B-world, is a world where change does not and cannot occur, and so is a world without time. McTaggart does not disagree with those who hold that a poker that is hot at one time and cold at another time has changed; his claim is that in the absence of changing A-facts there are no *times* at all, and so change cannot occur.

McTaggart's line of argument here may fail to convince fully, but it is also true that there is *something* to what he is saying. B-theorists undoubtedly do regard time as being more closely akin to space than we are accustomed to supposing it to be, and to this extent the B-theory account of change runs counter to common sense. Let us turn now to McTaggart's reasons for thinking that the A-series cannot be real.

2.3 McTaggart's A-paradox

The objection to the A-series for which McTaggart is best known begins with a statement of the obvious: past, present and future are incompatible attributes. If an event *M* is present, for example, it cannot also be either past or future. However, since everything starts off as being future, then becomes present before sinking into the past, it follows that *M* does possess all three A-properties, and the same applies to all events (or times). Since the A-properties are incompatible, this is impossible; but if the A-series were real, it would be the case. The A-series is thus paradoxical, and cannot exist.

This reasoning may well seem suspect, and McTaggart is not blind to the obvious reply. The A-theorist will insist, quite properly, that while it is true that nothing can possess incompatible properties *at the same time*, this is not how A-properties behave:

> It is never true, the answer will run, that *M is* present, past and future. It *is* present, *will be* past, and *has been* future. Or it *is* past, and *has been* future and present, or again, *is* future and *will be* present and past. The characteristics are only incompatible when they are simultaneous, and there is no contradiction to this in the fact that each term has all of them successively.
> (McTaggart 1908: 468, original emphasis)

However, McTaggart maintains that this plausible reply fails. He starts by claiming that the reply involves a vicious circle: confronted with an apparent paradox involving the A-series, the A-theorist is employing the A-series itself to get around the problem. However, he again anticipates a plausible response: "Our ground for rejecting time, it may be said, is that time cannot be explained without assuming time. But may this not prove – not that time is invalid, but rather that time is ultimate?" (1908: 470). It may well be impossible to explain notions such as goodness or truth in completely different terms, but this does not lead us to reject the notions in question; perhaps time is also a primitive concept, one that cannot be explicated in other terms. But McTaggart has a further objection up his sleeve. He maintains that the proposed solution introduces a vicious infinite regress.

The solution that McTaggart has in mind runs thus. An event *M* doesn't have the incompatible characteristics of being past, present and future at the same time, it has them successively; for example, an event *is* present, *will be* past and *has been* future. McTaggart says that this, in effect, introduces a *second A-series*, and that this A-series will suffer from the same difficulty as the first. What does he mean? In the 1908 article he doesn't explain. In his 1927 restatement he goes into more detail. In what follows I expand still further.

McTaggart starts by providing a way of reformulating A-statements such as:

(A) *M* is present, will be past and has been future.

He suggests that when we say "*X* has been *F*", we are asserting *X* to be *F* at a moment of past time; when we say that "*X* will be *F*", we are asserting *X* to be *F* at a moment of future time; when we say that "*X* is *F*", we are asserting *X* to be *F* at a moment of present time.[2] We can now restate (A) thus:

(A)′ *M* is present at a moment of present time, past at a moment of future time and future at a moment of past time.

Or, more succinctly:

(A)″ *M* is present in the present, past in the future and future in the past.

In using these formulations we are introducing *compound* or second-level temporal predicates. The availability of these second-level predicates may seem to provide an escape route from the initial paradox, but this is an illusion generated by an incomplete grasp of the real situation. There are in fact *nine* second-level predicates, not just three:

First-level	Second-level	
	is past in the past	[1]
is past	is past in the present	[2]
	is past in the future	[3]
	is present in the past	[4]
is present	is present in the present	[5]
	is present in the future	[6]
	is future in the past	[7]
is future	is future in the present	[8]
	is future in the future	[9]

Assuming that time passes, the position of an event in the A-series is constantly changing, so every event falls under each and every one of these second-level predicates. Take an event such as the death of Kennedy and consider predicate 1, "is past in the past". This was true yesterday, since yesterday is now past, and Kennedy's death occurred before it. Take predicate 2, "is past in the present". This is true *now*, while predicate 3, "is past in the future", will be true tomorrow; predicate 4, "is present in the past", is true at the date Kennedy died, as is predicate 5, "is present in the present". Predicate 6, "is present in the future", applies to the event *before* it happened, as do predicates 7, "is future in the past", 8, "is future in the present" and 9, "is future in the future".

Now, some of these second-level temporal predicates are compatible with one another; for example, predicates 3, 5 and 7, which were introduced to avoid the initial paradox. But some are not compatible; for example, predicates 2, 5 and 8, which together state that *M* is past in the present, present in the present and future in the present. McTaggart can thus reasonably claim that introducing second-order tenses fails to eradicate the initial paradox since precisely the same paradox exists at the higher level of temporal predication.

In reply, the A-theorist might insist that events don't have all nine second-level properties simultaneously, rather they have them successively. In describing this succession *third-level* temporal predicates are required. But once again, the passage of time ensures that each and every third-level predicate applies to everything, and not all of them are compatible, so we again have a contradiction. So we have to move to fourth-level predicates, but the same applies here, and so we have an infinite regress. It is easy to see that the contradiction is not removed by ascending in the hierarchy, no matter how many levels we add. The three predicates:

past
present   in the present . . . in the present . . . in the present . . .
future

are equivalent to the first-level predicates "past", "present" and "future", so if there is a contradiction connected with the first level, the contradiction is also present at each and every level in the (infinite) hierarchy.

This regress is undeniably vicious. We begin with a contradiction, and to remove it we appeal to a higher-level of temporal predicates, but since each successive level contains a paradox, the contradiction is never removed. McTaggart thus concludes that the A-series is inherently contradictory, and so cannot correspond with anything in reality. It doesn't make sense to describe things as present, past or future, if we understand these terms as the tensed theorist does, as properties that all events eventually possess. Given that he also maintains that without A-change the relationships of "earlier than" and "later than" cannot be real, he concludes that time does not exist. What *does* exist, according to McTaggart, is an entirely atemporal "C-series": a static block of events, whose contents are ordered but not in a temporal way.

2.4 Other routes to the same place

What are we to make of McTaggart's A-paradox? If you find yourself thinking "Interesting, but I'm not convinced; there's something fishy going on, but I'm not quite sure what!" you would not be alone. Broad observed of the A-paradox: "I suppose that every reader must have felt about it as any healthy-minded person feels about the Ontological Argument for the existence of God, viz., that it is obviously wrong somewhere, but that it may not be easy to say precisely what is wrong with it" (1998: 77). But then again, there are contemporary philosophers of time who disagree: Horwich, Le Poidevin, Mellor, Oaklander and Schlesinger have all argued that McTaggart's paradox is genuine.

For those inclined to resist McTaggart's reasoning, one plausible diagnosis is to deny that the alleged vicious regress ever gets off the ground. McTaggart simply assumes that a tensed theorist who believes that time passes is committed to the claim that, for any event E, "E is past *and* present *and* future". Are tensed theorists really obliged to use this unnatural and obviously problematic formulation in describing what is involved in the passing of time? Smith, himself an advocate of the tensed theory, rejects this formulation in favour of the following: for any event E,

E will be past, is now present, and was future; *or* E is now past, was present, and was (still earlier) future; *or* E is now future, will be present, and will (still later) be past.   (1993: 171, original emphasis)

Not only is this disjunctive formulation more natural than McTaggart's somewhat strained constructions, but it seems free from paradox. (See §5.2 for further discussion of this disjunctive approach.)

However, the fact that McTaggart's argument is flawed does not mean that his conclusion is false: as we are about to see, the particular dynamic model of time he was seeking to discredit *is* deeply problematic, even if not for the reason he proposed.

McTaggart's target is a theory of time that could aptly be called the "transitory tensed property" view. In its simplest guise, this theory posits a static time-line (the B-series), and a moving present, with the latter construed as a distinctive property, "presentness", which is successively instantiated at different times. The theory faces two problems. First, what is the nature of this property? Secondly, does it really make sense to suppose that this property is transitory? I will deal with these in turn.

2.5 The nature of A-properties

Properties can be divided into those that are intrinsic and those that are relational. A sugar cube's shape, size and composition are among its intrinsic properties. Its relational properties include being smaller than Mount Everest, being to the left of a teacup, being heavier than a flea and being above the floor but beneath the ceiling. The cube has many of its relational properties in virtue of having the intrinsic properties it does (if it were much, much bigger, it would be larger than Mount Everest), but the two sorts of property are nonetheless quite different. Having a relational property involves being related in some way or other to one or more other objects. An object's intrinsic properties are a matter of how the object itself is, irrespective of its relationships with other objects.

The doctrine that A-properties are intrinsic (or "qualities" in McTaggart's terms) can take different forms. There could be just one intrinsic A-property of *presentness*; pastness and futurity would then be defined relationally, in terms of changing B-distances from the present. Or there could be three intrinsic A-properties: *pastness*, *presentness* and *futurity*. Or, alternatively, just two, with the third defined relationally. The important point is that, on this view, an object that possesses an intrinsic A-property has an additional feature, akin to size, shape and mass, that does not depend on its relationships with other things. If, on the other hand, A-properties are all relational, then possessing any given A-property makes no difference to how an object is in itself; it is merely a matter of being related in some way to something else.

McTaggart suggests that A-properties are more likely to be relational rather than intrinsic (1908: 467). He does not explain why, but it is certainly not easy to see how they could be intrinsic, and to this extent at least his position is understandable. Suppose we try to explain the meanings of the A-predicates in the way that McTaggart himself suggests: "Here is a hand. The hand you see before you is present, your dinner this evening is future, your breakfast this morning is past." What intrinsic feature does the hand you are now looking at have that a hand in the past lacks? Does the hand situated in the past lack colour, mass, shape or size? If so, why is this ghostly residue a *hand* at all? But if the past hand possesses colour, mass, shape and size, what property does it lack that the present hand possesses? Saying "the property of presentness" is of no assistance for, even if we suppose this is true, we have no idea what difference possession of this property makes to an object.

So should an A-theorist say that A-properties are relational, the option McTaggart thought most plausible? Almost certainly not. We know that all B-type relations within the B-series are permanent; since events are always changing their A-properties, if the latter are relational it follows that an event's A-properties must involve a relationship with something *external* to the B-series – call it X – and hence external to the time-series itself. But what could this X be? As McTaggart observes, "it is difficult to say" (1908: 468). But even if we could say, the relationist is not out of the woods. We are now supposing that the passage of time consists of X moving along the B-series, momentarily bestowing presence on the times it touches: first X confers presentness on t_1, then on t_2, then on t_3, and so on. The trouble is, since we are also assuming that being present does not involve any alteration in a thing's intrinsic properties, coming to be related to X in the presence-conferring way makes no difference to how a thing is in itself. Consequently, if X were moving in the "wrong" direction – towards what we call the past – we would be none the wiser. The passage of time, thus construed, is simply not worth having: any serious A-theorist will want to insist that time's passing makes a significant difference of *some* sort to the things in time.

In which case, the A-theorist is obliged to return to the first option, and hold that A-properties are intrinsic. As for the problem of the character of the A-properties, one option is to say something like this:

> It is a mistake to think that presentness is a single intrinsic property, an ingredient to be found in every presently existing thing; there is no such property. The movement of the present consists of a distinctive and transitory *alteration* in at least some of the intrinsic properties objects possess.
>
> By way of an analogy, suppose a reel of a black and white movie film is unwound and glued to a long glass wall. As you walk along the wall, looking at the film, you see a long series of frames or "stills", each of which depicts the action in the film at a different instant. This static series of images corresponds to the B-series. Suppose you now see, far to your right, a single image shining brightly in full colour. An instant later the image has returned to black and white, but now the next image is illuminated, and then the next. This fleeting change in the intrinsic properties of individual images corresponds to the present moving along the static B-series. Not ll the intrinsic properties of the image change (e.g. shapes do not), but others do.
>
> Our own experience confirms this general picture. We know perfectly well what *some* intrinsic properties are like when they are present: the properties we are aware of in our own conscious experience. We all know what pain

feels like. A pain that exists in the past is not being felt, and so lacks this phenomenal quality, and the same applies to sounds, colours and textures as they feature in experience. Of course, we do not know what the intrinsic character of a past experience is like, simply because as soon as an experience ceases to be present it is no longer experienced. But we can be confident that, whatever this intrinsic character is like, it is different from that of present experiences. As for material things, although we know quite a lot about their causal and structural properties (e.g. the shape, size and mass of an electron, how an electron reacts with other particles), we have no knowledge of their non-structural intrinsic properties (e.g. what the intrinsic nature of an electron is like in itself). Given this ignorance, we certainly cannot rule out the possibility that the non-structural intrinsic properties of material things undergo changes when they become present.

I am not suggesting that this is a viable position, but it may be the best available. In any event, even if we *can* make sense of A-properties as intrinsic, a problem remains.

2.6 The overdetermination problem

The difficulty I have in mind arises when the following three claims are combined.

1. All actual events have permanent locations somewhere along the static B-series.
2. A-properties are intrinsic properties of events.
3. A-properties are transitory: every event changes from being future to present and then past.

Here (as elsewhere) I use the term "event" in a broad way, to refer to what obtains at a given time, or over an interval of time, irrespective of whether the objects involved undergo changes. To simplify, let us return to the idea that there is a single intrinsic property of "presentness" which objects possess only when they are present. This simplification will not affect the main point.

Consider some event E at t. On the assumption that A-properties are both intrinsic and transitory, E gains and then loses presentness. An enduring object, such as a lump of clay, can gain and lose intrinsic properties: one and the same lump could be spherical at noon and non-spherical at midnight, or red all over on Monday, black all over on Tuesday and green all over on Wednesday (see Figure 2.1). But this sort of change consists in possessing different properties at different times. The sort of change we are envisaging E undergoing is not like this: E both possesses and lacks presentness at precisely the same time, t, for the only time that E exists at is the time at which it occurs: t. This seems quite absurd; just as absurd as supposing that our lump of clay could be both spherical and not spherical at precisely the same time, or be differently coloured (all over) at the same time (see Figure 2.2).

It is true that some parts of an extended event can be present when others are past and future, but the problem remains. Just focus on any single instantaneous (or very

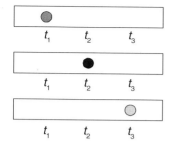

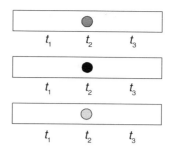

Figure 2.1 Ordinary change. The same object possessing different (and incompatible) properties at different times.

Figure 2.2 McTaggart-style change. The same object possessing different (and incompatible) properties at one and the same time.

brief) *part* of this event; this part, let us suppose, occurs at t_1. Can this event-slice both possess and lack presentness at t_1? It is hard to see how it could if presentness is an intrinsic property: to suppose otherwise would be akin to thinking that the same material object could be both spherical and cubic at the same time. So-called "substantial change" occurs when an object loses an essential property and so ceases to exist (e.g. an ice cube melts into liquid water). But we cannot think of E's gaining and losing presentness in this way, for E does not cease to exist: it exists at the same location in the static B-series when it is both future and past.

I will call this the "overdetermination problem". It may seem insuperable, but in fact there are responses to it available to the A-theorist. Whether these responses really help is another matter.

First response: enduring times

Our problem is that, while one and the same object can have incompatible intrinsic properties by having them at different times, it seems incoherent to suppose that *a single time* (or the events at that time) can have incompatible intrinsic properties at that very time. Posing the problem in this way suggests a solution: why not say that a single time can possess incompatible properties in just the same way as an enduring object; that is, by possessing them at *different* times? For this to be the case there must exist an additional dimension of time, *meta-time*, which is such that ordinary moments of time endure along this extra dimension. There is now no longer any difficulty in supposing that the same (ordinary) times can change their properties, as is clear from Figure 2.3.

The bold arrow indicates the location of the present, which moves from t_1 to t_2 and then to t_3. This movement is rendered possible by the fact that our universe as a whole (and so "ordinary" time) is a persisting entity: it endures through meta-time, as indicated by the vertical line on the left. The present is at t_1 (in ordinary time) at $m\text{-}t_1$ (in meta-time); at $m\text{-}t_2$, t_1 is no longer present, but t_2 is, and at $m\text{-}t_3$, t_2 is no longer present, but t_3 is. We are now supposing that times (and the events that occur at them) are enduring entities; they endure through the additional dimension of meta-time, and one and the same (ordinary) time or event can have different intrinsic properties at different meta-times, just as an ordinary material object can have different properties at different (ordinary) times.

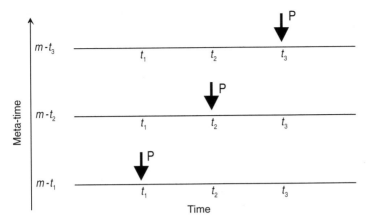

Figure 2.3 The "two-dimensionalist" solution to the overdetermination problem. Ordinary times (such as t_1, t_2 and t_3) persist through a "meta-time", and the property of "presentness" is possessed by different ordinary times at different moments of meta-time. The claim that one and the same time (or event) can possess different and incompatible properties is thus rendered intelligible, but the cost is high. The universe acquires an additional dimension, filled with innumerable copies of the same sequence of events; all that distinguishes these repeated "histories" is the location of the present, which is different in each.

But if one problem is solved, others remain. From an ontological perspective, introducing meta-time is an extravagance: an additional dimension of reality is being posited. Isn't this a blatant case of multiplying entities beyond necessity? To this obvious objection the A-theorist might reply:

> On the contrary. It is perfectly true that if we want to record when an event occurred, a one-dimensional time-line suffices, but what we do not usually recognize is that this one-dimensional representation is drastically incomplete: it provides no information at all about the movement of the present. We know from our ordinary experience that there is a privileged moving present. A one-dimensional representation of time is thus inadequate; we need meta-time.

Introducing unobservable entities in order to explain observational data is a perfectly respectable scientific strategy. But even if we grant the A-theorist the relevant data – that our ordinary experience supports the doctrine that there is a unique and moving present – it is not clear that the proposed meta-time does the required job, as the following objection makes clear:

> Although we are supposing that different ordinary times possess presentness at different meta-times, since all meta-times are equally real, doesn't it remain the case that *every* moment of ordinary time possesses presentness? In which case, the uniqueness of the "now" is lost. Moreover, this model of time is not in the least way dynamic: the posited *two-dimensional* system is entirely static. The proposed second temporal dimension fails to explain what it is claimed to explain, and so is entirely without justification.

This objection has considerable force. The appeal of the meta-time model lies in its purported ability to explain how presentness can be gained and lost; but in fact the model yields the result that *every* moment permanently possesses presentness, albeit at different moments of meta-time. Since the theory fails to explain what it sets out to explain, there is no reason to accept it.

To meet this objection the defender of meta-time might say:

> Ah, but you are making the mistake of forgetting that meta-time is itself a *temporal* dimension, and as such it has its own unique present: the *meta-now*. The ordinary "now" is that time which possesses presentness in the meta-now. Since there is just one meta-now, and it is always moving, there is a single privileged ordinary "now", and this too is always moving.

However, this clearly solves nothing. To make sense of a unique moving meta-now we would have to introduce a further dimension of time, *meta-meta-time*, so that different meta-times could be present at different meta-meta-times. But we are now launched on a regress that is as absurd as it is infinite: to secure a moving present in meta-meta-time an additional temporal dimension is required; this meta-meta-meta-time then requires its own moving "now", and hence yet a further temporal dimension, and so on, endlessly.

Second response: other worlds

Instead of positing additional dimensions of time, the A-theorist might seek to overcome the overdetermination problem by positing additional worlds. Bigelow has proposed a model of this sort, observing:

> Not only can contrary properties attach to the same object at different times, but also in different possible worlds. Something may be small at one time and large in another; but similarly, it may be small at one possible world and large at another. (Which is just to say that, although it may actually be small, it could have been large.) Possible worlds can be used to keep contrary properties out of each other's hair.
> (1991: 5)

If temporal passage is real, then statements such as "What is now past was present" are true. As Bigelow notes, if something *was* present, then it is clearly *possible* for this thing to be present, which means that it *is* present in some possible world. Which possible world? Not the actual world, where it is past, so some other possible world.

More generally, we can interpret the claim "What is present was future, and will be past" in the following modal way: "What is actually present could have been future and could have been past." Or, in terms of possible worlds, something that has the property of presentness in the actual world also exists in other possible worlds where it lacks this property; in some of these other worlds it possesses futurity, in others pastness. Since the thing has these contrary properties in different worlds, there is no problem with overdetermination.

Figure 2.4 illustrates Bigelow's model. Each of $W(orld)_D$–$W(orld)_H$ represents the entire history of a possible world, each of which is exactly the same in all non-temporal respects; "a", "b", etc., refer to the entire state of a world at a given moment of time (same letter, same sort of state); pastness is represented by bold

World$_H$	a b c d e f g H *i j k l*
World$_G$	a b c d e f G *h i j k l*
World$_F$	a b c d e F *g h i j k l*
World$_E$	a b c d E *f g h i j k l*
World$_D$	a b c D *e f g h i j k l*

Figure 2.4 Bigelow's modal solution to the overdetermination problem. These possible worlds ("possible universe" is a less misleading term) are exactly the same in all respects save one: the properties of **pastness**, PRESENTNESS and *futurity* are differently distributed in each.

type, futurity by italics and presentness by capitals. Although each of these worlds contains the same sorts of object, organized in the same ways, the worlds are nonetheless distinct, since the temporal properties (presentness, etc.) are distributed over these objects in different ways in each world.

These various worlds can be ordered with respect to one another in terms of pastness and futurity. A world such as W_G is "in the past" of a world such as W_H, since everything that is past or present in W_G is past in W_H; a world such as W_E is "in the future" of a world such as W_D, since everything that is present or future in W_E is future in W_D. We can now say the following: something was true in a world W when it *is* true in some world in W's past; something *will* be true when it *is* true in some world in W's future. The claim that "what is present was future" is true in a world W just when everything that is present in W is future in some world in W's past. The claim "what is past was present" is true in a world W just when everything that is past in W is present in some world in W's past. Similarly for other formulations that express the passage of time. Bigelow goes on to define a general earlier/later relation in terms of pastness, presentness and futurity, and explains how his basic scheme can be modified to accommodate different conceptions of time (e.g. a branching future, a non-existent past). These and other refinements need not concern us here.

Before we can assess this modal model we need to be clear about its precise content. Bigelow concludes his discussion thus: "Time is real; or at least, time is no less real than possible worlds" (1991: 18), a remark that points to a question for which he does not venture an answer: what is the status of these other possible worlds? There are two main positions on this question: *modal realism* and *modal actualism*. According to the modal realist, all possible worlds are equally real, all exist in exactly the same way; although the inhabitants of each world regard their own world as actual and the others as merely possible, each claim to actuality is equally valid. The modal actualist, by contrast, maintains that only one world is actual, only one world exists, the other possible worlds are *merely* possible. These different views of possible worlds, when combined with Bigelow's modal model of passage, yield very different conceptions of time. Unfortunately, neither conception is in the least appealing, or dynamic.

If realism is true, there are as many B-series as there are moments of time, and in each of these series a different moment is present. In this form, Bigelow's theory delivers a model that is, in effect, indistinguishable from the two-dimensional meta-time that we considered earlier, and is problematic for the same reasons. Our lives extend through an extra dimension: we are not confined to one world, or one spacetime, we are scattered beings whose lives are spread across an infinite number of four-dimensional planes, but the world we are spread across is a wholly *static* world. As Bigelow himself notes "the theory [is] one which can be described in ways which make it seem like a 'static' structure, and hence in ways which make it seem to miss the point of the passage of time as a 'dynamic' process" (1991: 4).

Modal realism is an extreme and highly controversial view of modality. Let us suppose, as many do, that modal actualism is true. Is the situation improved? Far from it. If only one possible world is actual, there is just one B-series and *just one time that is present*. This is an extremely odd position, which has odd consequences, however it is interpreted. The A-theorist could say "Although there is just one present, since your current experience is clearly present, you can be sure that you are enjoying this privileged moment." But since this entails that we only ever enjoy a single moment of experience, it is not a position that holds great appeal. But if the A-theorist adopts the alternative position, and says that the unique present may be located at *any* point in history, their account is doubly problematic. Not only is there no passage (since the present is "stuck" at a single moment), but presentness becomes an utter irrelevance: most people (most likely all) will never live through the present moment, but their lives are none the worse for it. Either way, the modal theory does not provide us with a truly dynamic account of time, nor does it explain why time *seems* dynamic, so there is no reason to accept it.

2.7 Consequences

Since none of the responses to the overdetermination problem hold the slightest appeal, it seems that we must reject those conceptions of time that are vulnerable to this problem. Versions of the tensed theory (concerning how time should be characterized) that take tensed predicates to refer to intrinsic tensed properties should be rejected. So, too, must dynamic conceptions of time that posit intrinsic and transitory A-properties. Does this mean that *all* dynamic conceptions of time have been refuted? Not in the least, for there are dynamic theories that do not posit transitory tensed intrinsics: there is the "growing block" view, according to which an event is present not by virtue of possessing a distinctive (but transitory) intrinsic property, but by virtue of being the most recent addition to the universe; there is the doctrine that *only* the present exists, a view sometimes combined with the claim that there is a succession of presents. We will be considering the intelligibility or otherwise of models such as these in Chapter 6.

It is interesting to note that McTaggart himself realized that there are models of time that are clearly dynamic, but also quite different from the particular form of dynamic model he believed (rightly, albeit for the wrong reason) to be paradoxical. At an early stage during the exposition of his argument he writes:

> Could we say that, in a time which formed a *B* series but not an *A* series, the change consisted in the fact that an event ceased to be an event, while another event began to be an event? If this were the case, we should certainly have got a change.
>
> But this is impossible. An event can never cease to be an event. It can never get out of any time series in which it once is . . .
>
> Since, therefore, what occurs in time never begins or ceases to be, or to be in itself, and since, again, if there is to be change it must be change of what

occurs in time (for the timeless never changes), I submit that only one alternative remains. Changes must happen to the events of such a nature that the
occurrence of these changes does not hinder the events from being events,
and the same events, both before and after the change.   (1908: 459–60)

He goes on to claim that the only properties an event can come to possess and then
lose while remaining the same event are A-properties such as pastness, presentness
and futurity.

What is interesting here is that McTaggart clearly appreciated that someone
might argue that the universe is dynamic by virtue of the fact that events come into
existence or depart from existence. Yet he rejects this view out of hand with little or
nothing in the way of supporting argument. Why? Because he took it for a given
that our ordinary conception of time comprises both the A-series and the B-series,
and the B-series consists of a sequence of events that are all equally real and
permanent.

But this is moving far too quickly. While there is no denying that history books
and memories provide us with information about what occurred in the past, and
calendars and diaries mention future times, facts such as these do not in themselves
entail that the past and future are just as real as the present, nor that most people
believe as much. Indeed, most people believe future events have no existence at all,
and many believe that past events cease to exist when they cease to be present.
These beliefs may turn out to be incoherent, but they are beliefs that many hold. By
simply assuming that the B-series is real, and hence that times and events cannot be
created or annihilated, McTaggart is guilty of begging the question against those
dynamic conceptions of time that maintain the opposite. In effect, his starting
premise is that the world is fundamentally Parmenidean. He then considers
whether such a world can *also* possess a dynamic Heraclitean character, while
retaining its essentially timeless Parmenidean nature, and, not surprisingly, he
concludes that it cannot. Strangely, or perhaps not so strangely, contemporary
Parmenideans, such as Mellor, who argue that McTaggart's A-paradox deals a
lethal blow to all dynamic theories, do not draw attention to this point.

3 Static time

3.1 Static theories of time

Many contemporary philosophers are convinced that McTaggart was essentially correct: our world is a static four-dimensional ensemble, lacking a moving present, wherein all times and events are equally real. But few (if any) of these philosophers follow McTaggart by holding that time is unreal. According to these contemporary B-theorists, what we call "time" is simply that dimension which exists in addition to the three of space and in virtue of which the properties of things can change; a thing "changes" by having different properties at different locations in this additional dimension. Since not everything happens at once and things do change, there clearly is such a dimension, and so time is undeniably real. McTaggart concluded that time was unreal because he believed that time essentially involves a moving present, and that no such thing could exist. Modern B-theorists take a different tack, and while agreeing with McTaggart that there cannot be a moving present, they argue that this amounts to a discovery about the real nature of time. We are accustomed to advances in knowledge overturning commonly held beliefs about how things are. The Earth does not seem to move, but it turns out that it does; time seems to pass, but it turns out that it does not. Where is the difference?

This chapter and the next are given over to an examination of the static view of time. I am not assuming that this theory is true. As we saw in the preceding chapter, while there are powerful arguments against the view that A-properties are transitory intrinsics, there are also dynamic models of time that take a different form, and we cannot conclude that the B-theory is true unless and until these alternative accounts all prove untenable. However, there are good reasons for taking a closer look at the B-theory before considering these wider issues.

Of all the models of time that we will be considering, the B-theory is currently the most popular among philosophers and physicists, but in many respects it is also the strangest and most counterintuitive.[1] Most people believe that time passes; most people believe that there is an objective difference between past, present and future. B-theorists maintain that these beliefs are false. Like anybody seeking to defend a view that runs counter to common sense, the B-theorist is faced with two tasks, one positive and one negative. The negative task is to provide reasons for thinking that the common-sense beliefs in question are false. McTaggart's A-paradox provides one such reason, but we shall see that other B-theorists are more inclined to lean on scientific considerations. The positive task is more complex. B-theorists must provide us with a clear account of the nature of time and change, but they must also convince us that this account is one that *could* be true. Most people believe that time passes because it *seems* to: the belief is not groundless; it is not sustained merely by custom or tradition; there are features of our experience, features of the world itself, that provide seemingly good reasons for holding the belief. Hence the B-theorist must explain why it is that so many of us have false beliefs about the nature of time, and show that, surprising as it may seem, nothing in our everyday experience requires us to hold that there is an objective difference between past, present and future. This is no mean task, and although it will occupy us for this chapter and the next, we will be returning to it in Chapter 7.

Just as different B-theorists take different approaches to the negative task, they also approach the more complex positive task in different ways. In several parts of this chapter I will be drawing on Mellor (1998), a revised edition of his classic *Real Time* (1981). But although Mellor's writings on this topic have been influential, not all B-theorists agree with all aspects of his position, and on occasion I will indicate some of these divergences.

3.2 Passage and experience

Common sense tells us something like this: only the present is fully real; the future is wholly non-existent and as yet unfixed (in the sense that it is not yet determined what will happen), whereas the past is fixed but not real in the same way as the present; the present is constantly advancing. This picture of time is imprecise and on reflection puzzling (what, precisely, is the status of the past?), but it is also very appealing. Call it the "natural view".

It is not surprising that we find this view appealing, for it seems firmly grounded in the character of our everyday experience. Someone trying to describe the most general temporal features of everyday experience might well come up with something like the following:

> We have a good deal of highly detailed knowledge of what happened in the past, but little or none about what will happen in the future (except for the case of highly stable and regular systems whose behaviour can be accurately predicted). We can anticipate experiences that we will have in the future, and we can remember experiences that we had in the past, but only present experiences are fully and properly real, only sensations that are occurring now are actually felt. The passage of time is something we can directly observe in the world: we can see one event happening after another (e.g. we can observe a car moving along the street passing one house, then another, then another). Deliberation and choice are oriented towards the future. I am now (as always) in the position of having to decide what I *will* do, never what I *was* doing. We cannot change or affect the past, but by choosing to do one thing rather than another we can help determine what happens in the future.

If this is how we are inclined to describe our experience, the appeal of the natural view of time is readily understandable: the two fit together perfectly. The idea that only present events are fully real is supported by the confinement of experience to the "now". The fact that we can perceive things changing, and find the experienced present to be continually advancing into the future, supports the idea that the present is continually advancing into the future. That we have epistemic access to the past of a sort that we do not have to the future supports the fixed past–unfixed future doctrine.

That everyday experience makes the appeal of the natural view readily understandable is not in question. What *is* in question is whether there is anything here that is truly incompatible with the competing B-conception of time. B-theorists argue that there is not. I will start by considering what the B-theorist can say about two aspects of the natural view that concern experience.

The presence of experience

Consider first the idea that we can only experience what is present. If we take this to mean "we can only perceive what is present" then the claim is simply false. Light is fast, but not infinitely fast. When you look at the Sun you are seeing it as it was eight minutes or so ago; when you look at a star you may well be seeing it as it was millions of years ago. As Mellor points out, whether a thing is past, present or future doesn't make any difference to how it looks: "We cannot for example refute someone who claims to see the future in a crystal ball by pointing to the visible pastness of the image: there is no such thing. Our reasons for thinking we cannot see the future rest not on observation but on theory" (1998: 16). This truth (obvious when it is pointed out) points the way to a more promising formulation of this "confinement doctrine": the objects we perceive may not be present but the sensory experiences we have when we perceive them surely are. Conscious experiences are confined to the present.

But are they? Let us suppose (with the B-theorist) that all events are equally real. It follows that all experiences are equally real: your pain five days ago is just as real as your pain today, and your pain next week. From this perspective, it is clearly not true that experience is "confined to the present". What is true is this straightforward tautology: an experience happens when it happens. Every experience occurs at some time, and is "restricted" to the time at which it occurs; but this is true of everything that is located in time. It doesn't follow from this tautology that the *only* experiences are the experiences occurring at the time in question. This would only follow if earlier and later events are unreal. So it seems that the doctrine that experience is confined to the present doesn't support or confirm the doctrine that past and future are unreal; it presupposes this doctrine. Even so, there is something that needs explaining. Why do we feel that only present experience is really felt, and hence that experience occurs only in the present?

Imagine waking up to find yourself in a strange place. You are sitting in a field of grass, next to a lamp that illuminates the surrounding area. There is complete silence. As you look around, you can see nothing whatsoever. Apart from the small patch of grass illuminated by the lamp there is darkness everywhere. Not surprisingly, you conclude that you are alone. You could not be more wrong. A few yards to your right there is another lamp, and another person waking up to find themselves surrounded by total darkness; likewise to your left. In fact, you are in a line of people stretching for many miles in either direction, all of whom are sitting

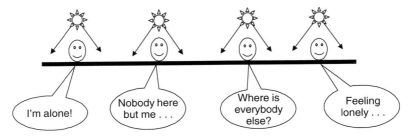

Figure 3.1 Illusory spatial solitude. Since it is dark, and light emitted by the lamps fades away after travelling only a couple of feet, no one can see their neighbours, so everyone concludes they are alone in the field. But they are wrong.

in their own small pools of light, all of whom are alarmed to find themselves alone in a strange place.

Why is it that nobody can see anyone else? The answer lies with the strange form of light emitted by the lamps, which only extends a few feet before dying away. According to the B-theorist, we find ourselves in an analogous position in time. What stretches only a short distance is not light through space but consciousness over time: the temporal span of direct awareness is very brief. And as in the analogous spatial case, the fact that at any given time we are not aware of experiences occurring at other times does not mean that these experiences are not there.

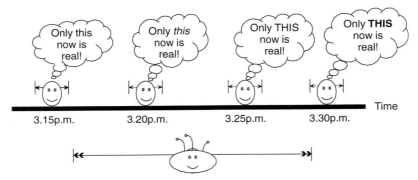

Figure 3.2 Illusory temporal solitude. According to the block theorist, the impression we have that reality is confined to a very brief present is (in part) due to the fact that the span of our direct awareness (or "specious present") is very brief. The "lived present" of a being with a wider span of awareness (such as the alien depicted here) would include times we would regard as past and future.

Experiencing passage

How can the B-theorist account for the experience we have of time passing if time doesn't really pass? According to many B-theorists, Mellor included (1998: 122), the solution lies in the ways our memories accumulate. Suppose you are in a dentist's waiting room. Every few minutes you take an anxious glance at the clock. Although the minute hand of this clock looks stationary, it is in a different position every time you look at it. Your first glance shows it to be 10.15. When you next look you see it is 10.25, a while later you see that it is 10.29. Is this all you experience? Very probably not: in addition to what you perceive, you also have various memories of what you have recently been doing. So, for example, when you see that it is 10.25 you also remember seeing the clock showing it to be 10.15; when you see the clock showing 10.29, you remember observing it showing 10.25 *while also remembering having seen it showing 10.15*. According to the B-theorist, our sense that time is passing is due to the fact that our perceptions and memories combine in this sort of way, not only over the span of a few seconds or minutes, but over days, months and years.

On the face of it this account seems quite plausible. Vary the scenario, and suppose you are suffering from Korsakoff syndrome: you forget everything that happens to you within a minute or so. When you see the clock showing 10.25 not only do you have no memory of having seen it showing 10.15, but you have no recollection of what you were doing two minutes earlier. Likewise when you see the clock showing 10.29. In this memory-impoverished condition, what sense would you have that time is passing?

Memory also has a part to play in explaining why we feel that our lives unfold in the direction of the future. Even if our future experiences are just as real as our past ones, as the B-theorist maintains, at any given point during our lives we have detailed knowledge only of the earlier parts of our life; indeed, for all we know we might be about to drop dead, and so not have a future at all. It is thus easy to believe that our lives don't extend beyond the present, even if in fact they do. This belief leads to another: that any continuation of our lives involves the addition of something new to the already completed past. Since we remember having these same beliefs at past stages of our lives, and anticipate having them in the future, we always have the impression of living at the outermost limit of an ongoing future-directed process.

Whether this memory-based account of experiencing passage is entirely adequate – especially over the very short term – is something we will be looking into in more detail in Chapter 7. It is enough for now to recognize that the B-theorist has at least the beginnings of a plausible account, one that does not require a moving present or the non-existence of times other than the present. All that it requires is for people to have different memories at different stages of their lives.

3.3 A-truth in a B-world

There is a further reason why it is natural to believe that there must be an objective difference between past, present and future, one that we have not yet considered. Quite generally, when a claim about the world is true, there is something about the world, a fact that obtains, that makes the claim true. This fact is the claim's "truthmaker". If I say "Canada is larger than France," the relevant truthmaker is the fact that Canada extends over a greater area of the Earth's surface than France does. Suppose I say "The battle of Hastings is in the past." I am making an A-statement, and this statement is clearly true, but what is its truthmaker? The obvious candidate is the tensed A-fact that Hastings took place in the past. Likewise for other A-statements. Not all A-statements are true, but many are. If those that are true have tensed A-facts as truthmakers, it seems that we live in an A-world after all.

How can the B-theorist respond to this? Simply denying that there are any true A-statements is bold, heroic even, but hopelessly implausible. One response, which dates back to Russell (1915), is to hold that A-statements do not require A-facts as truthmakers because they are equivalent in meaning to B-statements. If this were the case, the absence of any A-facts would not matter, since there are B-facts, and these could serve as truthmakers for both sorts of statement.

This approach is nowadays widely known as the "old tenseless theory", and is generally agreed to fail: A-statements cannot be translated into B-statements without loss of meaning. Even though this is generally agreed to be true, it is worth pausing to consider some of the difficulties involved in such translations. It seems clear that two statements cannot have the same meaning if one can be true when the other one is false. Consider:

S_1: Event E is past.

S_2: Event E is earlier than t.

S_1 is a typical A-statement; S_2 is a B-statement. It is easy to see that these can have different truth-values: if S_2 is true it is true at all times, whereas S_1 is only true when E is in the past. A-statements have *time-varying* truth-values, they are true at some times but not at others, whereas B-statements have *time-invariant* truth-values. This alone seems to show that A- and B-statements differ in meaning. But there are further difficulties. Suppose the following is uttered at 3.25p.m.:

S_3: Bill is happy now.

We might try to translate S_3 by the B-sentence:

S_4: Bill is happy at 3.25p.m.

There are two reasons for rejecting the claim that S_3 and S_4 have the same meanings. First, the two sentences employ different concepts. Since someone might understand "now" but have no grasp at all of what "occurs at 3.25p.m." means, how could S_3 and S_4 have the same meaning? Secondly, someone could understand both S_3 and S_4, but nonetheless believe S_3 to be true while believing S_4 to be false. How is this possible? Simple. The person in question sees that Bill has a broad grin on his face, but (mistakenly) believes that it is 12 noon.

Advocates of the "new tenseless theory" of time, such as Smart (1980), Mellor (1981, 1998) and many others, agree that A-claims cannot be translated into B-claims, but argue that this doesn't matter: A-claims can nonetheless have tenseless B-truthmakers (or truth-conditions). Here is how Mellor has recently stated the case.

Mellor distinguishes between beliefs, statements of beliefs, the sentences that are used in stating beliefs and the contents of beliefs, which he calls "propositions", and takes to be what sentences expressing beliefs mean. Thus when Jack says "Bill is happy" and Jacques says "Bill est heureux", they are using different sentences to make the same claim or statement; that is they are using different sentences to express the same proposition, and have beliefs with the same contents. On the face of it, it is not easy to see how an A-belief could have a B-truthmaker. A-beliefs are true at some times but not at others, whereas B-beliefs always have the same truth-value. For example:

B_1 = Jim races tomorrow.

B_2 = Jim races on 2 June.

B_2 is always true but, assuming that Jim does indeed race on 2 June, B_1 is only true on 1 June. Clearly, since the truth-value of any given B-belief is constant across time, whereas an A-belief such as "Jim races tomorrow" is true at some times and false at others, no *single* B-fact can make such an A-belief true. How many B-beliefs does it take? The answer, it turns out, is a great many: (roughly speaking) a different B-fact is required for each different time at which a given A-belief can be true or false. How can this be? What are the relevant B-facts? How can a single A-belief have many different B-truthmakers?

The solution relies on the type-token distinction. A token of a type is a particular instance, specimen or exemplar of it. For example, Mary is a token of the type *humanity*; "the" and "the" are two tokens of the same type of word, *the*. Just as the type-token distinction can be applied to words, it can be applied to beliefs, sentences and statements.

Our problem was that a given A-belief *type* can have different truth-values at different times, so no A-belief type can have a single B-truthmaker. However, when we switch our attention to *tokens* of A-beliefs a way forward opens up. A token of an A-belief is a particular person holding a particular belief at a particular time. The fact that a person holds a given belief at a certain time is a *B-fact*, and hence holds at all times. Consequently, all we need is a way of stating the conditions under which a token A-belief is true in tenseless B-terms. There are two general strategies available, which Mellor sketches in these terms.

- The *token-reflexive account*: For every A-proposition P about any event E, any token of P is true if and only if it is as much earlier or later than E as P says the present is than E.
- The *date account*: For every A-proposition P about any event E, any token of P is made true at any t by t's being as much earlier or later than E as P says the present is than E.

Both accounts seem to give the right results. Consider the A-proposition "E occurred yesterday". My saying "E occurred yesterday" is a token of this proposition. The first account tells us that this A-statement is true if and only if E does indeed occur a day earlier than the token itself. (It is because this account makes reference to the token itself that it is *token-reflexive*.) The second account says that, if I make this statement on 2 June, then what I am saying is true if and only if E occurs on 1 June. Analogous considerations apply for the A-belief "E is occurring now". According to the token-reflexive account, a token of this belief is true if and only if it occurs when E occurs (i.e. the token itself is neither earlier nor later than E). The date account yields the result that, if the belief is tokened on 2 June, it is true if and only if E also occurs on 2 June. Underlying these differences is a crucial similarity: both accounts state the conditions under which tokens of A-propositions (and hence A-beliefs, A-statements, etc.) are true in entirely B-terms.

In *Real Time* Mellor preferred the token-reflexive account; in *Real Time II*, he prefers the date theory – to handle certain tricky cases that critics have developed in the intervening years. We don't need to go into these details here. The important point is that the B-theorist has a plausible strategy for explaining how A-statements and beliefs can be true even if there are no A-facts to make them true.

3.4 Another A-paradox

Mellor takes a further step in arguing that, if we try to supply A-sentences (statements, beliefs, etc.) with *tensed* truthmakers, we end up in a paradoxical situation, and consequently we have no option but to accept that A-statements have tenseless truthmakers. More specifically, Mellor argues that, if tensed sentences have tensed truth-conditions, the statements that we make using these sentences will undergo changes in their truth-value, changes that *do not in fact occur*.

By way of an example, suppose that on Monday I see a man with short hair and I mistakenly believe that this man is John, who has worn his hair long for many years. I exclaim, "So, John has short hair now." Call this claim C. In actual fact, at this time John has long hair, so C is false. Suppose that it is now a month later and John has just taken the plunge, visited the barber and emerged with short hair. Is C now true? Of course not. The claim that I made a month ago remains false. If I were *now* to claim "John has short hair now", my claim would be true, but the claim I made a month ago, *that very claim*, was false then and is false now. Mellor's allegation is that, *if* tensed statements have tensed truthmakers, then statements such as C *would* change their truth-value. The argument runs thus.

Suppose a tensed sentence S = "E is happening now" has a tensed truthmaker, the obvious candidate being *the fact that E is now happening*. The tensed theorist will thus say:

> Any token of S = "E is happening now" is true if and only if E is happening now.

In other words, all tokens of the tensed sentence S, irrespective of when they occur, are made true by this tensed fact, and hence, if this fact does obtain, then all tokens of S are true, irrespective of when they occur. Suppose that on Monday I say "It is now Tuesday." Call this statement S_T. Applying the tensed account, this statement is true if and only if it is now Tuesday. Since it is now Monday, the statement is false. However, a day later, on Tuesday, the truthmaker for S_T now obtains, so the statement in question is now true: S_T has undergone a change of truth-value!

Mellor is certainly right in maintaining that we do not normally suppose that truth-value changes of this sort occur. Moreover they *don't* occur if the truthmakers of tensed statements are of the tenseless sort considered above: for example, if a token of "It is now Tuesday" is true if and only if the token in question is uttered on a Tuesday, then a token of this sentence uttered on Monday remains false even on Tuesday. So the idea that tensed claims are made true by tensed facts does look to be in difficulty. Whether this difficulty is insuperable is another matter. In §5.1 we will encounter a proposal for a cure.

3.5 The indispensability of the A-framework

If there are no A-facts why do we constantly think in A-terms? Mellor argues that the A-framework, in particular A-beliefs, are of vital practical importance even though there are no A-facts.

Beliefs contribute to our actions; what we do is a product of what we *desire* or *want* and what we *believe*. If I want an ice-cream, and believe that there is ice-cream in the fridge, then I will go to the fridge to get some ice-cream; my belief helps direct my action in the way appropriate to satisfy my wants. Of course, my wants will only be satisfied if there really *is* ice cream in the fridge: successful action requires true belief.

Now, it is very often the case that the success of an action depends on when it is carried out. If I want to buy some food from the shops, then my visit to the shops will only be successful if I go when they are open. I have to act at the right time. Suppose that I believe (truly) that the shops are open between 8a.m. and 8p.m. Since this B-belief is always true, it cannot successfully guide my actions: success depends on my going to the shops at some times but not others. Clearly, the belief that will successfully guide me must be true at some times but not others; it must be an *A-belief*. An example of the right sort of belief would be "It is *now* 3p.m." The belief "If it were 3p.m. then the shops would be open", a B-belief, will not do, for although this belief is true, it is always true, and so it cannot guide timely action.

To act at the appropriate times we must have A-beliefs that are constantly updated to reflect the different times at which they are held. I should believe "The shops are open now" between 8a.m. and 8p.m. but not at other times. I should believe "The banks are closed today" only on days when the banks are indeed closed. How do we update our A-beliefs? Usually by using our senses. To put it crudely, we believe what we see, and (cases of known time-lag apart) we believe that what we see is happening *as we see it*; happening *now*. If I look at my watch, and it says "4p.m." then I believe "It is now 4p.m." and act accordingly.

Again, there is a good deal more to be said about these topics, but the important point is that we can understand why the A-framework is indispensable for practical purposes without assuming that there are any A-facts. The account just sketched appeals only to B-facts and B-relationships. A-*beliefs* play an ineliminable role in successful timely actions, but, as we have seen, B-facts can serve as truthmakers for A-beliefs.

3.6 Questions of attitude

Experiences that we are currently undergoing are of more concern to us while we are undergoing them than experiences we have yet to have, and experiences we have already had. This is understandable: if I am now in pain, this present pain is real to me (now) in a way that my past and future pains are not. But we also have differing attitudes to past and future experiences. Although I suffered horribly at the dentist last week, this fact doesn't much bother me at all. If someone were to say "Give me £10 and I will change the past, and make it so that pain never occurred" I would not be inclined to pay, even if I believed that the past *could* be changed in this way. But if someone were to say "Give me £100 and I will ensure that you feel no pain when you visit the dentist next week," I might well be tempted by the offer, especially if this were the only way of avoiding the pain in question. Quite generally, our future experiences are of far greater concern to us than our past experiences (especially where the experiences in question are of the painful and pleasurable variety). Perhaps the most dramatic example of our bias towards the future is Lucretius's observation that while death (i.e. our own future non-existence) is something we deeply dread, the vast periods of personal non-existence preceding our births is of no concern to us whatsoever.

Some A-theorists have held that this asymmetry in our attitudes is problematic for the B-theory. The best-known of these arguments originated with Prior (1959), and runs thus. After several long hours of torment, a fierce headache at last fades away, and you think to yourself "Thank goodness that's over!" What is it that you are expressing relief about? Could it be the tenseless fact that this token thought occurs later than the headache in question? Although according to the B-theorist it is this tenseless fact that makes the thought true, this answer seems plainly false:

> the trouble is that this was as much a fact before and during the headache as it is now the headache is over. It always was and always will be a fact that this particular token of "That's over" occurs later than the headache it refers to. What is more, that fact could have been recognized as such in advance. In particular, you could have decided in advance to say "That's over" after the headache and known about the fact in that way. So if that were the fact that you were thanking goodness for, you could just have well thanked goodness for it before or during the headache as afterwards. Which, of course, is non-sense. So it seems that you must be thanking goodness for some *other* fact, something that was not a fact at all until the headache ceased. The tensed conclusion appears irresistible; the pastness of the headache, for which you are thanking goodness, must be an extra fact over and above the tenseless fact that makes "That's over" true. (Mellor 1994: 296)

A straightforward response to this objection is to agree with Prior that the fact that you are thanking goodness for is indeed a tensed rather than a tenseless fact (Mellor himself recognizes that there are tensed facts, if only in the trivial sense that every true statement states a fact), but also to hold that the relevant tensed fact is identical to (or refers to the same state of affairs as) a tenseless fact. As for why we thank goodness for the tensed fact but not the tenseless fact, there is a simple explanation. Lois Lane believes that Superman is a superhero, but she does not believe that Clark Kent is a superhero, yet Superman and Clark Kent are one and the same person. Our beliefs are attitudes to objects (or states of affairs) that are represented in a certain way, and the same object (or state of affairs) can be represented differently. The same holds, it seems reasonable to suppose, for attitudes such as dread and relief. Trapped in a burning building, Lois Lane would feel relieved if she thought Superman were about to arrive; the thought that Clark Kent were about to arrive would leave her indifferent.[2]

However, recognizing that the same object (or state of affairs) can be repre-sented differently only solves part of the problem. It is obvious why Lois Lane has the attitudes she does: she believes Superman to be a superhero, and Clark Kent to be a mild-mannered reporter. It is not obvious why we tend to care more about states of affairs when they are characterized in tensed rather than tenseless terms. Why is it that we have such different attitudes to the past and future?

Do we care more about our future experiences because we can still do something about them, whereas the past is beyond our control? No. We dread future pains that we believe are *inevitable* in just the same way as we dread future pains that we believe are avoidable (imagine yourself in prison, facing inescapable torture this afternoon). As Parfit says, "Our ground for concern about such [inevitable] future pains is not that, unlike past pains, we can affect them. We know that we *cannot* affect them. We are concerned about these future pains simply because they are not yet in the past" (1984: 168, original emphasis).

That we care more about future pains just because they *are* future may seem to pose a problem for the block theorist. If there is no ontological difference between past and future – if all experiences are equally real – what could justify our bias towards the future? However, while it is far from obvious how this bias could be justified in the context of the block view, this is only a problem for the B-theorist if the bias in question *is* justified. Perhaps it is rooted in nothing more than irrational instincts implanted in us by natural selection. It is easy to sketch the outlines of a suitable evolutionary story: no doubt natural selection has ensured that we are motivated to act in ways that will enhance our survival prospects; since we can only act in ways that influence our futures, we have an overriding concern with the future compared with the past.

But there is a second point. Irrespective of how successful the B-theorist's attempts to explain the bias are, it is by no means clear that adopting a tensed or dynamic conception of time changes the situation. Why should we care less about pains that *were* present than pains that will *become* present? What is it about *pastness* that makes experiences less worthy of concern?

B-theorists can also take encouragement from the fact that closer examination reveals our "bias towards the future" to be rather less straightforward than it initially appears. As Gallois points out (1994), our bias is not towards the future *per se*. Were you to travel back in time to the last ice age, the thought of enduring the terrible cold might well fill you with dread; but the experiences that you are concerned about in this case lie in the *past* and not the future; or rather, they lie in the world's past but your *personal future*, a reference frame determined by the ordering of the stages of your life (see §8.2 for more on the distinction between personal and world times that time travel generates).

Parfit's case of the "Past or future suffering of those we love" reveals an ambiguity of a different sort:

> I am an exile from some country, where I have left my widowed mother. Though I am deeply concerned about her, I very seldom get news. I have known for some time that she is fatally ill, and cannot live long. I am now told something new. My mother's illness has become very painful, in a way that drugs cannot relieve. For the next few months, before she dies, she faces a terrible ordeal. That she will soon die, I already knew. But I am deeply distressed to learn of the suffering that she must endure.

A day later I am told that I had been partly misinformed. The facts were right, but not the timing. My mother did have many months of suffering, but she is now dead.   (Parfit 1984: 181)

If past pains are of no significance I should be relieved. But am I? Obviously not. The tragedy for my mother is just as great; she suffered just as much from her illness, and I am just as upset on her account. So it seems that where *other people* are concerned, especially other people whom we love, the timing of any pains they suffer is of little significance.

The existence of this striking self–other asymmetry poses a problem for anyone seeking to justify the bias towards the future in the first-person case. Might it not be that the pattern of concern that we display towards other people is *more* rational than our self-oriented concern? More rational because less influenced by the instinctive drive for self-preservation instilled by natural selection? If so, it may well be unreasonable for us to care greatly about our own future experiences while being largely indifferent to our past experiences: the rational pattern of self-concern (if such there be) may well be entirely in tune with the block view of time.[3]

3.7 B-theories of change

In its crudest form, the B-theory of change is easy to state: an object changes by virtue of having different properties at different times. McTaggart objected that a poker's being hot at t_1 and cold at t_2 doesn't amount to change because B-facts such as these always obtain. In response, the B-theorist can ask, "Why can't some changeless facts *be* changes? If the poker is hot at t_1 and cold at t_2, why can't these facts not constitute a change in the poker? Does it matter that these facts are and always will be facts?" McTaggart's response was to deny that the B-theorist has any right to talk of *times* at all in the absence of A-change. Since this is precisely the assumption that B-theorists reject, this response does not trouble them at all. However, while agreeing on these fundamentals, B-theorists find much else to disagree on when it comes to making sense of change.

Objects (trees, people, houses, etc.) are often thought to be different from events (concert performances, conversations, battles, etc.) because of the kind of *parts* that each has. Since a battle is typically spread over a fairly large area it has a good many spatial parts (i.e. what is going on at different places at a given time), but it also has temporal parts. To put it crudely, it has a beginning, a middle and an end, and these different phases are as much parts of the battle as any of its spatial components. Ordinary material objects seem to be different. While a typical person such as Alice has various spatial parts (her hands, feet, bones, organs, etc.), it seems wrong to think she has temporal parts in the same way as a battle. Alice exists now, and she existed yesterday, but it seems that Alice *in her entirety* exists at both these times. If you met Alice you didn't meet just part of her, you met the whole thing, Alice herself. Alice-yesterday and Alice-today are not two parts of a person, they are each a whole person in their own right; moreover, they are the very same person: Alice-yesterday and Alice-today are *numerically identical* (the very same object) even though they are very likely *qualitatively different* (they don't have exactly the same properties).

This distinction between objects (or substances, continuants) and events comes very naturally to us, and some B-theorists, such as Mellor, not only subscribe to it, but they deploy it in defending themselves against McTaggart's charge that the B-theory reduces change to mere spatial variation. Mellor points out that change is something that a single persisting thing undergoes. Events cannot change in this way. Suppose a performance of a symphony is loud at t_1 and quiet at t_2. What we have here are two distinct entities, two temporal parts of the symphony, possessing different properties. Likewise, Jim's being happy yesterday and Alice's being sad today doesn't amount to a change; it is a case of two distinct objects having different properties. Alice's being happy yesterday and sad today *is* a change. Entities such as Alice can change precisely because they do not possess temporal parts. Now consider a typical case of spatial variation. If a stick is red at one end and white at the other, this isn't change; rather than one and the same entity (the whole stick) possessing different properties, we have two distinct parts of the stick possessing different properties – that is, two numerically different entities. If we were to say that *the stick as a whole* is red and white we would be saying something false, since there are parts of the stick that aren't red and parts that aren't white. Hence we have a clear difference between change and spatial variation: the latter involves numerically distinct entities possessing different properties, whereas the former involves numerically identical entities possessing different properties.

Mellor takes a further step. He argues that persistence involves causation. Alice's present condition is (in part) a causal consequence of her preceding condition. More generally, if O at t_2 is numerically identical with O at t_1, then O at t_2 must be causally dependent upon O at t_1, directly or indirectly. Since causation operates over time, but not space (Mellor argues against the possibility of causes being simultaneous with their effects), we have a further respect in which genuine change differs from spatial variation. Change requires persistence, persistence requires causation; since time alone is the dimension of causation, time alone is the dimension of change, and hence the allegation that the B-theory obliterates the distinction between time and space (or change and mere variation) is successfully rebutted.

Mellor's account of change aims to be non-revisionary. In effect he is saying something along these lines: "The B-theory of time may strike you as very odd when you first encounter it, but it is not *that* odd, because it turns out that you can adopt it without having to undertake a radical overhaul of your current conceptual scheme." Other B-theorists take a quite different line. They start off with the assumption that if our ordinary ways of thinking about time are mistaken it may well be that other parts of our ordinary conceptual scheme are mistaken too. They then argue that if the B-theory is true some natural assumptions about what persistence involves turn out to be false, and as a result a revisionary account of change starts to look very appealing. Their reasoning runs as follows.

Think of a spatially extended physical object, such as a worm. Suppose that this worm has ten segments; some of these segments may be very similar to each other, but nonetheless each is a distinct object (a worm-segment) in its own right. If we think of a persisting object from the four-dimensional perspective of the B-theory a similar picture emerges. The space–time diagram in Figure 3.3 represents Alice over a period of time.

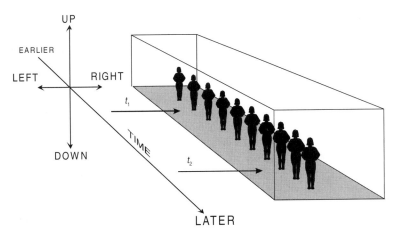

Figure 3.3 The four-dimensionalist view of persistence. A brief phase of Alice's life is shown. On the "traditional" view of persistence, Alice-at-t_1 and Alice-at-t_2 are numerically identical. The four-dimensionalist views matters differently: like all persisting objects, Alice has temporal parts, and these parts are numerically distinct objects; Alice is identical with the sum of these parts. Consequently, if you were to meet Alice at t_1, you would not be meeting Alice-as-a-whole, but merely one of her parts.

The diagram depicts Alice as an extended four-dimensional object. This object is composed of parts. Some of these parts are of the familiar spatial sort: at t_1 Alice can be divided into a head, two arms, two legs and a torso. But isn't it clear that Alice also has *other* parts too? To be sure, Alice exists at t_1; but since she also exists at t_2 (and other times too) it is surely wrong to say that *Alice-in-her-entirety* exists at t_1. If you see Alice at t_1 you aren't seeing the whole Alice, you are simply seeing the part of her that exists at that time (and place). She has plenty of other parts, existing at other times, and these you don't see. The common-sense assumption that when you see a person on a given occasion you are seeing the whole person, and hence that the person is wholly and entirely present before you, would be true if only the present time were real, and many of us do commonly assume this. But if we adopt the B-theory, this assumption is false. Since all times are equally real, people and other persisting things do not exist only in the present; they extend into the past and future too.

If persisting objects *do* have temporal parts, the difference between change and spatial variation that Mellor stresses evaporates. If Alice-at-t_1 and Alice-at-t_2 have different properties, these properties are possessed not by one and the same object, but by two numerically distinct objects (person-stages). These objects are indeed parts of a single thing (the whole temporally extended Alice), but then so, too, are the red and white ends of a stick, or the front-end and the rear-end of a worm.

This reasoning can be resisted, but I won't enter further into the twists and turns of this particular debate, which is primarily of importance in the metaphysics of identity and object-hood. I want to mention instead a different (and arguably more profound) way in which the B-theory can be revisionary to greater or lesser extents.

3.8 Emergent time

If we choose to think of ordinary objects as four-dimensional entities possessing temporal parts, change becomes more akin to spatial variation, but does it *become* spatial variation? Not if the temporal dimension of the four-dimensional block universe is different in some way from the three spatial dimensions, different in a way that renders the distinction between time and space well founded. Many B-theorists assume that there is such a difference. Some (like Mellor) believe that causality runs only along the temporal axis. Others, sceptical of causality as a time-distinguishing feature (see §4.5 and §4.6), look instead to physics. Although relativity theory posits a four-dimensional manifold (or continuum of points), this manifold is not a four-dimensional space, and the distinction between space and time is maintained (albeit in an altered form).[4]

But this may change. Relativistic gravitational theory and quantum theory have yet to be combined in an entirely adequate way, and the search for a satisfactory theory of "quantum gravity" continues. Some physicists believe that the distinction between space and time will not survive in quantum gravity theory, for the simple reason that this theory will not mention time at all.[5] Rovelli writes, "In quantum gravity, I see no reason to expect a fundamental notion of time to play any role . . . The time 'along which' things happen is a notion which makes sense only for describing a limited regime of reality . . . At the fundamental level, we should, simply, forget time" (1999: 13). This is only well-informed speculation – there are plenty of physicists who take a different view – but suppose that Rovelli turns out to be right. Belot and Earman suggest the metaphysical consequences would be considerable:

> there is no change [in such a world] since no physically real quantity evolves in time . . . change is illusory . . . This is a very radical thesis – it is, for instance, much stronger than the doctrine that there is no room for temporal becoming in a relativistic world, since proponents of the tenseless theory of time are confident that they can account for the existence of change.
>
> (1999: 178)

If change is illusory, then so, too, is time, or so McTaggart would argue, and it would be hard to disagree.

But does it follow from the as yet only hypothetical disappearance of time as an independent variable from the equations of fundamental physics that change *is* unreal? Only if we take fundamental physics to be our sole guide in deciding which (if any) dimension of the universe can legitimately be called "time". Not all B-theorists would want to take this step, as this passage from D. C. Williams's well-known exposition of the block theory clearly displays:

> it is plain that things, persons and events, as a matter of natural fact, are strung along with respect to the time axis in patterns notably different from those in which they are deployed in space. The very concept of "things" or "individual substances" derives from a peculiar kind of coherence and elongation of clumps of events in the time direction. Living bodies in particular have a special organized trend timewise, a *conatus sese conservandi*, which nothing

has in spatial section. Characteristic themes of causation run in the same direction, and paralleling all of these, and accounting for their importance and obviousness to us, is the pattern of mental events, the stream of consciousness, with its mnemic cumulation and that sad anxiety to *keep going* futureward which contrasts strangely with our comparative indifference to our spatial girth. An easy interpretation would be that the world content is uniquely organized in the time direction because the time direction itself is aboriginally unique. Modern philosophical wisdom, however, consists mostly of trying the cart before the horse, and I find myself more than half convinced by the oddly repellent hypothesis that the peculiarity of the time dimension is not thus primitive but is wholly a resultant of those differences in the mere *de facto* run and order of the world's filling. It is then conceivable, though doubtless physically impossible, that one four-dimensional area of the manifold be slewed around at right angles to the rest, so that the time order of that area, as composed by its interior lines of strain and structure, runs parallel with a spatial order in its environment. It is conceivable, indeed, that a single whole human life should lie thwartwise of the manifold, with its belly plump in time, its birth at the east and its death in the west, and its conscious stream perhaps running alongside somebody's garden path.

(1951: 468, original emphasis)

The speculation that ends this passage is striking, and for sheer *oddness* it has few rivals in the literature of philosophy.[6] And this is why it serves Williams's purposes so well. The observable (and introspectible) contents of the world are not spread through the four-dimensional continuum in an entirely haphazard fashion. There are distinctive patterns of development and organization that are only found along the temporal axis, so much so that when asked to envisage one of these patterns – a whole human life, from birth to death – being distributed across space in the way that it is usually distributed across time, we simply don't know where to start. There may be universes where this does not hold – where it is common to find people's streams of consciousness running alongside roads and garden paths – and in such universes there might well be no rationale for maintaining or imposing a general (as opposed to a person- or object-specific) distinction between time and space, but our universe is not like this. Given the asymmetrical ways in which the contents of our universe are distributed, the distinction between the temporal and spatial dimensions is manifestly well founded, even if the distinction between time and space is not "primitive" or "aboriginal" (and so not registered by fundamental physics), but depends entirely on the "*de facto* run and order of the world's filling".

In saying that the distinction between space and time is "well founded" I mean that it is grounded in objective facts concerning the asymmetrical patterning and distribution of events (we will be taking a close look at these asymmetries in the Chapter 4). This is undeniably so – at least at the macroscopic (and introspectible) levels in this region of our universe – and future developments in physics are simply irrelevant to the distinction drawn in this way. To this extent time is real rather than illusory, and the same applies to change.

This said, if it does turn out that time is absent from the equations of quantum gravity theory (or whatever theory in physics turns out to be fundamental), this will be an interesting and significant result. There will be a clear sense in which time is not a fundamental feature of the universe, especially if it turns out that the basic ingredients of the physical world are quite capable of existing without giving rise to the asymmetrical formations that ground the space–time distinction. If this should prove to be the case, time will have the status of an *emergent* phenomenon, in somewhat the same way as macroscopic material things such as planets and animals (which "emerge" only when elementary particles are arranged in certain ways). But even if the status of time is thus diminished, its *existence* is scarcely threatened, for at the macroscopic level the distinction between spatial and temporal dimensions remains perfectly real. Cats are emergent entities; are they illusory?

According to Belot and Earman, the doctrine that time is not physically fundamental is stronger and more radical than the doctrine that "temporal becoming" is unreal. This is questionable. As far as departures from our ordinary ways of thinking are concerned, it is passage (or "becoming") that matters: the (metaphysical) thesis that time does *not* pass, that all moments are equally real, is truly astonishing when first encountered; almost impossible to believe. The idea that the space–time distinction, although real enough at the macro-level, is emergent relative to the most fundamental level of physical reality, is far less radical in this respect. If time can survive in the metaphysical framework of the block-universe, it can certainly survive discoveries from the frontiers of physics.

4 Asymmetries within time

4.1 The direction of time

A French nursery rhyme runs "L'eau est comme le temps: il coule, il coule, il coule." If time really did flow, then, like a flowing river, it would also have a direction. But if the B-theory is true, time is not going anywhere: there is no moving "now", no slippage into the past, no crystallization from mere possibility into present actuality. Every time and every event is equally real. But why does time seem to have a direction if it doesn't? The B-theorist's explanation will take this general form: time does not flow or pass, but there are nonetheless asymmetries to be found among the *contents* of time, the material and mental processes we know to exist, and it is these asymmetries that we mistake for temporal flow and that provide time with its apparent direction (cf. the passage from Williams in §3.8).

Whether or not these content-asymmetries suffice to give time *itself* a direction depends on how we view time. If the substantivalist is correct, and time is an entity in its own right over and above any contents it might have, then there is a sense in which time might be perfectly symmetrical in itself, and so lack any "directedness", even if its contents do not. By way of an analogy, imagine looking at a glass cylinder that gradually changes in colour from blue to red; you assume that the glass itself has an in-built colour asymmetry, but you are wrong: the glass is perfectly clear throughout (and so entirely symmetrical), but embedded within, a short way beneath the surface, is an asymmetrically coloured sheet of paper. This option is not available if we follow the relationist in identifying times with collections of simultaneous events: asymmetries in content now amount to asymmetries within time itself.

But there are two further complications to bear in mind. Time as a whole can be asymmetrical even if its contents are not. For example, rather than being infinite in both directions, it could be infinite in only one; or it could be infinite in both directions, but be linear in one and branching in another; or it could branch in both directions, but in different ways. Secondly, if we ignore these large-scale questions and assume a linear time, there could be content-asymmetries that are only local, and so not of a kind that can provide a direction to time as a whole. To return to our earlier analogy, suppose that our cylinder is blue at both ends, turning to red in the middle, with exactly similar gradations in shading in both directions. If the content-asymmetries within time were of this overall form, then although most people would find time in their neighbourhood to be directed, time as a whole would lack a direction. We could call one end of this (finite) universe "the start" and the other "the end", but since the universe is perfectly symmetrical about its centre, the opposite choice would be equally justified.

We will be encountering a universe of this last sort later in the chapter but, this case aside, we will not be concerned with the more esoteric possibilities just mentioned. Our focus will be on the observable content-asymmetries to be found in everyday life. Thus far we have looked only very briefly at just one family of content-asymmetries: those that contribute to what is sometimes called "the psychological arrow". We do have the impression that we live our lives in the present, and that the present is constantly advancing into the future. As we have seen, B-theorists can appeal to the factors such as changing A-beliefs and patterns of memory-accumulation to explain why it is that we have this impression. However, there are other aspects of the psychological arrow, and further asymmetries within time that we have not yet tried to explain.

If all conscious beings were to vanish tomorrow there would be nobody left around to have the subjective impression that time is passing, but there are good reasons for believing that plenty of asymmetries within time would remain, even if no one were around to notice them. Using time-lapse photography, the events of weeks or years can be squeezed on to a film lasting only a few minutes. Suppose that someone has made a time-lapse film of what has been going on in your garden over the past few months. Would you have any difficulty in knowing if the film were running forwards or backwards? Of course not. If the film were running backwards you would see large plants shrinking rapidly down to small seedlings, smudges of slush concentrating themselves into ripe plums that suddenly leap upwards and attach themselves to the branch of a tree, paw-prints gradually appearing in mud prior to the arrival of a cat walking backwards, rings of water converging into the centre of the pond from which a ball rises and flies over the fence into the next garden, and so forth. A film running backwards shows things occurring in the wrong order; later events are shown as occurring before earlier events.

Given that we can so easily tell whether a film registering everyday sequences of events is running backwards or not, there must be certain sequences of events that are always (or typically) ordered in the same way with respect to earlier and later. This is why, for a world such as our own, it seems reasonable to think that there is an objective difference between earlier times and later times, a difference that does not depend on the psychological peculiarities of conscious agents such as ourselves. In this chapter we will be examining what B-theorists have to say about asymmetries of this sort. While all B-theorists agree that there are *some* objective asymmetries, there is a significant difference of opinion as to precisely what these are. As we shall see, some B-theorists defend the provocative view that many of the content-asymmetries we are inclined to think of as objective are not: they are products of the distinctive perspectives that we bring to bear on a world that is more symmetrical, in temporal respects, than commonly supposed.

4.2 Content-asymmetries: a fuller picture

Before starting to consider what the B-theorist can say about the various content-asymmetries, it will help to have more of them clearly in view. Some of the asymmetries listed below have already been mentioned; others we have yet to encounter. The brief characterizations provided are only intended to give a rough idea of what each asymmetry involves. They should not be taken to be definitive characterizations (nor should the list be taken to be complete).[1]

- *Entropic asymmetry*: the second law of thermodynamics states that entropy (or "disorder") increases over time.
- *Causal asymmetry*: some events are causes of others; causes usually occur earlier than their effects.
- *Fork asymmetry*: it is very common to find many later events that are correlated with (or caused by) a single earlier event; for example, everyone in a coach party who had fish at lunch falls ill a few hours later. The "inverse fork" (one later event correlated with or causing many earlier events) is much rarer.
- *Explanatory asymmetry*: we tend to explain later events by reference to earlier events (Jim died because he drank poison, he didn't drink poison because he died).
- *Knowledge asymmetry*: although we can often make accurate predictions about what will occur, we have detailed and reliable knowledge of the past of a kind that is not available for the future.
- *Action asymmetry*: our deliberations are oriented towards the future, not the past (no one spends time wondering about what to do yesterday); what we do can influence the future, but not the past.
- *Experience asymmetry*: we experience our lives unfolding in a present that inexorably moves forwards in the direction of the future.

As will immediately be clear, these asymmetries are not wholly independent; at least some (perhaps all) are interrelated. But quite how they are related is not so obvious, and we will see that different B-theorists have different views not only about the precise pattern of interrelationships, but about which are fundamental and which depend on others. Before proceeding, it is worth pausing. What exactly is it that the B-theorist *has* to explain? A B-theorist might say:

> I don't *have* to explain any of this. Perhaps it is just a brute fact that the contents of time are distributed and related in asymmetrical ways. In any case, what does the A-theorist have to offer by way of an explanation of these asymmetries? If temporal passage could exist, it could exist in worlds organized very differently from our own, in worlds where there are no content-asymmetries at all, or where there are asymmetries, but they are differently oriented with respect to past and future. Positing an objective difference between past, present and future doesn't explain anything; nor does temporal passage.

This response is by no means absurd – it may well be true that there are some basic facts that cannot be explained in terms of other facts – but the B-theorist should only adopt it as a last resort. First, if the B-theorist *could* explain some or all of the content-asymmetries, the B-theory itself would be more believable. Secondly, and more importantly, A-theorists can provide, at least in rough outline, explanations of at least some important asymmetries.

Finding the result of last week's lottery is easy: just pick up a newspaper. Finding the result of *next* week's lottery is not so easy. Past events leave their marks on subsequent presents, but future events do not leave their marks on earlier presents; this is why our knowledge of the past is far more detailed than our knowledge of the future. If, as some A-theorists maintain, the future is unreal, this asymmetry is easily explained: since future events do not exist, they clearly cannot causally influence or leave traces on anything; past events can and do. This is why we do not find instances of backward causation, or inverse forks. It is why our actions can influence the future, which is not yet fixed, but not the past, which is. It is why we explain later events by reference to earlier events: when an event is present, there are no future events to influence it.

Can the B-theorist offer anything remotely as plausible by way of explanations of these phenomena?

4.3 Entropy

It has often been claimed that a crucial key in understanding the direction of time lies in thermodynamics, and more specifically with the second law of thermodynamics, which dictates that entropy tends to increase over time. Some go as far as to claim that the entropic asymmetry is the ultimate key: it alone is responsible for *all* the other content-asymmetries. Atkins writes:

> We have looked through the window onto the world provided by the Second Law, and have seen the naked purposelessness of nature. The deep structure of change is decay; the spring of change in all its forms is the corruption of the quality of energy as it spreads chaotically, irreversibly, and purposelessly in time. All change, and time's arrow, point in the direction of corruption. The experience of time is the gearing of the electrochemical processes in our brains to the purposeless drift into chaos as we sink into equilibrium and the grave.
> (1986: 98)

To be in a position to evaluate this claim we need to take a closer look at what the second law of thermodynamics actually amounts to. Since the complexities here are considerable, I have been obliged to simplify somewhat.

Entropy is a technical concept that can be defined in different ways, but the general idea is easily grasped. Roughly speaking, entropy is a measure of the extent to which the energy in a system has spread out in a disorderly fashion through the space available to it, and hence of how close a system is to its equilibrium state. If a system's energy has largely dissipated (i.e. it is more or less evenly spread out) it is in a high entropy state; if, by contrast, its energy is concentrated in just a few places, it is in a low entropy state. The second law of thermodynamics thus states that in any closed system (i.e. no outside influences and no leakage) the energy within that system will tend to spread out, until finally it is evenly distributed through the available space, and having reached this condition of equilibrium it will remain there. In our universe, energy is concentrated in stars separated by vast extents of (very cold) space, so the universe is currently in a low-entropy condition. But since the stars are pouring their energy into the surrounding space at an impressive rate, and thus spreading it around, the entropy of the universe is increasing, in accordance with the second law of thermodynamics.

The entropic account of the earlier–later distinction is in many respects analogous to the gravitational account of the up–down distinction:

> For Aristotle, the down direction is a primitive notion. He probably believed that at all points of space a downward direction existed, and that at all of these "downs" were in the same spatial direction. But now we realize that "down" is just pointing in the direction in which the local gravitational force is pointing. We now understand that there are regions of the universe in which no direction is down and none is up, and we accept without difficulty that the downward direction for someone in Australia is not parallel to that for someone in New York. That is how it is with the future–past distinction, Boltzmann claims. Where there is no local entropic asymmetry, there is no future–past of time.   (Sklar 1992: 149)

As Sklar goes on to note, the gravitational account of "down" is plausible because it explains all the relevant facts about the downward direction: if you drop an egg it falls down; gas filled balloons float upwards (due to air pressure created by gravity); we can *feel* the difference (due to gravitational forces acting on mechanisms within our inner ears), and so on. Does the entropic account of earlier and later similarly explain the relevant facts?

It can certainly explain why some of the processes that allow us to tell whether a film is running backwards or not unfold as they do. If you see a film showing leaves blowing about in the wind and ending up in a neat pile in the middle of the lawn, you can be reasonably certain that the film is running backwards: since this course of events is highly improbable, it is much more likely that the leaves were initially piled up by a human gardener, as human gardeners commonly do such things. What is far less clear is whether any of the other asymmetries listed in §4.2 can be explained in the same sort of way. Would a reversal in entropy also reverse the direction of causation and explanation? Would machines (such as clocks) run in the opposite direction? Would we have detailed knowledge of (what are now) future happenings? Would the various components of the psychological arrow reverse their direction? It is certainly not obvious that they would; far from it.

The first point to note is that thermodynamic processes are not in fact irreversible. In the course of pioneering work in the nineteenth century, Boltzmann attempted to reconcile the time-asymmetric second law of thermodynamics with the time-symmetric laws of classical mechanics. He showed that the tendency for entropy to rise could be explained in terms of the statistics governing particle interactions. However, as critics were quick to point out, his explanation also applies in reverse: even if it is overwhelmingly *probable* that the entropy of a typical system of particles will always increase, does it not follow that there is some probability, no matter how small, that the entropy of such a system will at certain times decrease? This objection first came to Boltzmann's attention in the 1870s, and was reinforced in 1889 by Poincaré's "recurrence theorem", according to which the particles in any energetically isolated state will, given enough time, return to a state that is arbitrarily similar to their initial state, and do so infinitely often. Simplifying a complex story, Boltzmann took the reversibility objection on board and accepted that so-called irreversible processes are not truly irreversible after all. Energy in a system does tend to spread out (this is by far the most probable way for a low-entropy system to evolve), but on occasion (very rarely) concentrations of matter and energy will occur, and entropy will decrease. But only for a short while; these improbable concentrations will soon break down, and equilibrium will be restored. More generally, although there is a tendency for disorder to increase, there will be exceptions; very rarely, bizarre things will happen (e.g. milk stirred into tea will separate out again, leaves will be blown by the wind into neat piles). More generally still, any system that at a given time is in a high (and highly improbable) state of order is likely to be in a more disordered (and more probable) condition at later times *and* earlier times.

These considerations do not preclude our defining the direction of time in terms of entropy, but they do complicate matters. Boltzmann took the view that what we call "the universe" was quite likely to be only a tiny part of a far larger whole, most of which is at thermodynamic equilibrium, or very close to it. From the vantage point of the universe as a whole, the observable stars amount to nothing more than a tiny and localized eruption of massively improbable (but statistically inevitable) order in a vast sea of disorder. As the stars fling their energy into space, our region of the universe is in the process of returning to equilibrium. Hence, over our region taken as a whole, entropy is increasing, despite occasional and improbable small-scale reversals. Boltzmann suggested that this regional entropic increase provides time with its direction (*earlier* = lower entropy, *later* = higher entropy). Since the universe as a whole is already at equilibrium point (or very close to it), its average entropy is neither increasing nor decreasing, and time is without a direction; it is only at the regional scale (e.g. clusters of galaxies), where significant entropy gradients are to be found, that time has a direction. In recent years, with the advent (and widespread acceptance) of the "big bang" cosmology, most scientists take the view that our universe started off in a low-entropy condition, and that, with the exception of the occasional small-scale local reversal, its entropy has been increasing ever since.[2]

Leaving these complications to one side, the entropic account of the direction of time faces another serious problem. What exactly is the connection between the tendency for entropy to rise and the other observable content-asymmetries? What is the link between the laws of statistical thermodynamics (concerning particle interactions and energy distributions) and asymmetries involving causation, knowledge and explanation?

As Earman points out (1974), applying the thermodynamic concept to systems as diverse as decks of playing cards, human beings, printed records, and so on, involves a considerable extension of the concept beyond its established range of legitimacy, and there is no guarantee whatsoever that the extension *is* legitimate. Moreover, if all the other content-asymmetries were the direct result of local entropy increase, then we would expect a local *decrease* in entropy (which is perfectly possible don't forget) to reverse the other asymmetries. But there is no reason at all to believe that this would occur. There are many ways in which the entropy of a region can decrease, and most of these ways seem very unlikely to have any significant effects on any macroscopic structures (such as people and mountains) located within the region in question.

For example, suppose that (due to a freakish sequence of molecular motions) the water in the eastern half of the Atlantic Ocean were to become, on average, 20°C warmer than the western half, a process that takes two weeks. Throughout this period the entropy in the Atlantic would be decreasing. What effect would there be on its occupants? Well, no doubt a great deal of plant and animal life would be significantly affected by the dramatic and unseasonal changes in temperature; there would be serious meteorological consequences too. But there is no reason to believe that anything else of interest or significance would happen. Life on board ship might become more difficult due to extreme weather conditions, but it would not start to run in reverse: radio and video transmissions from the shore would seem perfectly normal (rather than running backwards), clocks would continue to turn in the same direction, and so on. The same considerations seem to apply quite generally. One way for entropy in a particular region of outer space to increase is for the particles that were previously distributed in a uniform fashion (e.g. one atom per cubic metre) to become concentrated in just one half of it, leaving the other half completely empty. As the latter region gradually empties, entropy within it is decreasing. If you were to fly a spaceship into this region, is there any reason to suppose you would find yourself living in a time-reversed fashion? Again, there is no reason to think so. Since you are entering a region of space that is almost devoid of matter and energy, and you may well not interact with anything at all, why should anything on your ship change in any way whatsoever?

Or consider a more specific case: the knowledge asymmetry. Since a footprint on a sandy shore will invariably belong to an earlier rather than a later walker, we reasonably take it as a reliable indicator of what has happened in the recent past. Can we explain this in terms of the tendency of entropy to increase? Reichenbach argued that we could. A high entropy beach (i.e. nearly flat, no distinct imprints) is much more probable than a low entropy beach (one featuring some distinctive patterns, such as a footprint). So if we encounter a footprint in the sand, it is reasonable to conclude that this is not due to a random fluctuation from the high entropy norm, but rather the product of a previous interaction with some external influence, such as a human foot. This sounds reasonable enough, but why assume (as Reichenbach does) that the external influence occurred in the past rather than in the future? This is a perfectly reasonable assumption, since experience has taught us to expect that any footprints we happen to observe are the product of earlier rather than later strollings along beaches. But if we are seeking to explain the direction of time in terms of entropy increase, we cannot simply *assume* that it is more probable that the external influence occurred earlier rather than later. The entropic account must explain why there is this asymmetry. But it seems unable to do so.[3]

These considerations suggest at least this much: that there is no direct or obvious connection between entropy increase and many of the asymmetries that we normally associate with the direction of time. Of course, if a defender of the entropic account could provide a detailed and plausible account of how, despite appearances to the contrary, local entropy increase does in fact explain the other asymmetries, the situation would be quite different. Some B-theorists are optimistic about the prospects for developing an account along these lines (we will be encountering a variant of this approach shortly). Others do not share this optimism, and adopt a very different approach.

4.4 The causal route

Of the various ways of explaining the content-asymmetries available to the B-theorist, there is one that stands out by virtue of its straightforward simplicity: the causal approach. Consider again the footstep we see on the beach. Why do we take this as evidence of an earlier rather than a later occurrence? An appealing line of reasoning runs thus. We know that the footprint is unlikely to have occurred spontaneously, which means that it must have been created by something else. To put it another way, the footprint is the *effect* of some other event, which was its *cause*. Since we know that causes precede their effects, we conclude that the event responsible for the footprint occurred at an earlier (rather than a later) time. Likewise for all the traces and records that provide us with detailed knowledge of past occurrences: these are all effects, and since effects always occur later than their causes, there can be no records or traces of future events. By appealing to causation we can easily explain an asymmetry that seems highly problematic for the entropic account.

We can explain much else too. Why do we tend to explain later events in terms of earlier events? Because explaining an event often consists in citing its cause, and causes come earlier than their effects. We can perceive past occurrences but not future ones because to perceive something is to be causally affected by it, and causes occur earlier than their effects. Memories too can plausibly be viewed as causal imprints, so the fact that causes precede their effects explains why we have memories of the past but not the future. Actions have consequences; effective actions have desired consequences; if actions are the causes of their consequences, then since an action's effects must occur later than the action itself there is no mystery in why our actions can influence the future but not the past, and why our deliberations are future-directed rather than the reverse.

So far so good: it seems that most (perhaps all) of the content-asymmetries can easily be explained in causal terms. We can now take an additional step and explicitly define the direction of time in causal terms. Roughly speaking, we can hold that if E_1 causes E_2 then E_2 occurs *later than* E_1 precisely because it is an *effect* of E_1. We cannot simply say "for any two events, E_x precedes E_y if and only if E_x causes E_y", since many events are not effects of the events that precede them. All events are ordered in time but relatively few are causally related to one another. Mellor, who himself defends a causal account of the direction of time, suggests a solution along these lines. Take two times, t_1 and t_2. A lot is going on at these two times, and many events at t_1 are not causally related to many events at t_2, yet many events at t_1 *are* causally related to events at t_2, so we can say that t_1 is *earlier than* t_2 provided that *some* events at t_1 are causes of events at t_2. Given the complexity of the universe, this will always be the case. In short, the temporal order of all events can follow from the causal order of only some.[4]

4.5 Causation in question

This is all wonderfully simple and straightforward, so much so that it may come as a surprise to discover that many B-theorists (perhaps a majority) find the causal account untenable. To appreciate why, we need to probe more deeply into what defenders of the causal account need to establish for their account to be viable.

The claim that for E_1 to be *earlier* than E_2 it is sufficient that E_1 *causes* E_2 would be undermined if it turned out that backward causation were a possibility.[5] In cases of backward causation, the effect occurs earlier than the cause. Hence, if we also say that a cause occurs earlier than its effects, we have a situation in which one event is both earlier and later than another. At the very least, admitting the possibility of backward causation considerably complicates the causal theorist's task. Whether backward causation *is* possible (or actual) is a controversial issue, but many B-theorists are unwilling to dismiss it (not least because the most obvious objection to it – future events do not exist, and so cannot cause anything – is unavailable to them).

A second problem is no less serious. Unless causal theorists can distinguish between the events that are causes and those that are effects without appealing to the earlier–later distinction, their account of the direction of time will be circular, and so trivial. It will not do to say something along these lines:

> Causes and effects are events of kinds whose instances regularly occur in succession, and do so by virtue of the laws of nature rather than mere coincidence; of these pairings, the *cause* is the earlier of the two, the *effect* the later of the two.

According to Hume, causes *precede* their effects only because there is a linguistic convention of using the word "cause" to refer to the earlier instances of event-types that are known to be regularly paired in the appropriate ways. To avoid the charge of triviality, the causal theorist must reject the Humean view, and provide a substantive account of *causal* priority (i.e. how causes differ from their effects) that does not appeal to *temporal* priority (i.e. it will not do to say that causes differ from their effects by occurring earlier than them).

It might seem that there is no real difficulty here. Surely causes *make their effects happen*; effects don't make their causes happen. Hence effects depend on their causes, in a way that causes don't depend on their effects. Since this notion of dependency is asymmetrical and (seemingly) non-temporal, we have all we need.

However, on closer examination it turns out to be far from obvious that an asymmetrical dependency relationship of the required sort really exists. As noted earlier, it may well be that the laws of physics are time-symmetrical; if so, any type of event that can be a cause can also be an effect, and vice versa. Moreover, in any given concrete situation where a cause c produces an effect e, if we say that in the given circumstances, c is both necessary and sufficient for e, then since the relationship "necessary and sufficient" is symmetrical, we are clearly committed to holding that e is necessary and sufficient for c. For example, suppose a spark ignites a fire. In the circumstances (oxygen present, water absent, no other flames, etc.), the fire would not have occurred if the spark had been absent, so *the spark was necessary for the fire*. It is also true that, in the circumstances, *the spark was sufficient for the fire* (nothing else was needed). Now look at the situation in another way. In these circumstances, for the fire to ignite the spark had to occur, so the fact the fire *did* ignite guarantees that the spark occurred: *the fire is sufficient for the spark*. Also, since the occurrence of the spark guarantees the fire will ignite, the spark couldn't occur without the fire, so *the fire is necessary for the spark*. The two events, the cause and the effect, are thus related to one another in a perfectly symmetrical way. If this applies generally, what could possibly distinguish any cause *c* from an effect *e* other than the fact that *c* occurs *earlier* than *e*?

Causation is a complex topic; it may well come in different forms, and there are many competing accounts of it. The only thing that is clear and relatively uncontroversial is that finding an objective and non-temporal difference between cause and effect is a far harder task than one might have imagined, and so a good many philosophers have concluded that there is no such difference. In which case, the causal asymmetry is not something that exists in reality, but something our modes of thought project *on to* reality, in somewhat the same way as (many would argue) we project colour on to the physical things we perceive, things that in themselves do not possess colour-as-we-perceive-it.

If the causal relation is inherently symmetrical, causal accounts of the direction of time (of the earlier–later relation) cannot succeed; or rather, they are at best circular and so trivial. However, as we saw earlier, the causal theorist can provide simple and appealing explanations of the other content-asymmetries (e.g. knowledge). What are we to make of these explanations? As we shall now see, even if we assume that causation *is* an inherently asymmetrical relation, its explanatory potential is suspect.

4.6 Time in reverse

The universe is currently expanding, and it may continue to do so for ever, but equally it may not. If there is enough mass in the universe to halt the current expansion (recent evidence suggests not, but the issue remains open), at some point in the future gravitational forces will cause the universe to fall back on itself. The big bang will be followed by a big crunch.

Big bang–big crunch models fall into two classes. There are models where entropy is low at the big bang and keeps on increasing right up to the big crunch (which is thus far less ordered than the big bang), but there are also models where entropy increases only until contraction kicks in, at which point it gradually decreases, and the universe becomes more orderly as it contracts. It is quite possible for a universe of the latter sort to be highly symmetrical: at either end is a singularity (a dimensionless point of infinite density), and the sequence of events running from each singularity to the mid-point are very similar. Highly symmetrical universes of the second type are sometimes called "Gold universes", after the physicist Thomas Gold, who raised the profile of such models in the 1960s.[6]

Let us imagine that we live in a Gold universe. If we could observe the post-contraction phase it would be like looking at a film running backwards; we would see light and energy pouring into stars rather than out of them, people would die before

Figure 4.1 A time-symmetrical "Gold" universe. A universe that is bounded at both ends, and that is broadly (not necessarily exactly) symmetrical about its temporal mid-point. The inhabitants of the left-hand side regard A (their big bang) as lying in their past, and B (their big crunch) as lying in their future; the inhabitants of the right-hand side regard B (their big bang) as lying in their past, and A (their big crunch) as lying in their future.

they were born, and trees would start off as rotting logs, rise up to towering heights and then dwindle into seedlings. This might seem utterly impossible: how could the laws of nature permit such bizarre happenings? Very easily. As noted earlier, many believe that the laws of fundamental physics are time-reversible, in which case any physical process can run in reverse. But in addition we must take care not to assume that *our* way of looking at things is in any way privileged. The inhabitants of the "other side" will regard *their* half of universe as expanding and *ours* as contracting; what we call the end of the universe, they would call the beginning; what we regard as inflationary expansion they would regard as gravitational implosion. If they were to peer over into our side, it would seem to them that they were looking at a film running backwards: from their perspective it would seem that time in our half of the universe is running in reverse. They, too, would wonder how such a thing could be possible, and their perspective is just as valid as our own.

This scenario opens up some intriguing topics. Could we causally interact with our (seemingly) time-reversed neighbours? Could we communicate with them? It might seem not. Suppose we transmitted a sequence of messages M_1, M_2, M_3 and M_4. From the perspective of our neighbours, M_4 would be the earliest message and M_1 the latest; consequently, if they respond to M_4, depending on the mode of communication adopted, we may well receive the response before we have transmitted the message itself! However, as MacBeath (1983) shows, with sufficient ingenuity, these hurdles can be overcome (I won't go into details but will provide a clue: the solution involves programming a computer to transmit one's initial messages at later times). In a later paper MacBeath suggests that if we say "that a space has *n* dimensions if there are *n* respects in which its occupants can, *qua* occupants of that space, vary continuously but independently" (1993: 186), then time-reversal of the sort we have been considering actually involves time itself possessing *two* dimensions rather than one. But interesting as they are, these issues cannot be pursued here.

Suppose that the causal account of the direction of time were true. Would the time in the "other half" of a Gold universe have a direction opposite to what it has in ours? It would if the arrow of causation pointed in a different direction in the other half, but there is no reason to believe that this would be the case: if (as we are currently assuming) causation is an intrinsically asymmetrical relation between events, why should its direction be reversed just because the universe starts to contract? So let us assume that causation runs in just one direction from one end of the universe to the other: from our big bang to the big crunch. Something rather disturbing now emerges. It is possible for the *apparent* direction of time to become completely disconnected from its *real* direction. Not only are the inhabitants of the "other half" of our universe very similar to us in all respects, but so, too, is the world they inhabit. The various content-asymmetries (with the exception of causation) all run in the same direction: away from *their* big bang, in the direction of increasing entropy.

I will call the apparent direction of time the "world-arrow". Our time-reversed counterparts exist in a situation where their world-arrow points in the opposite direction to the causal-arrow. Of course, they would be completely oblivious to this fact, since, as is now very obvious, the direction of the causal-arrow has no observable consequences whatsoever. More disturbing still, for all we know *our* world-arrow might be pointing in the opposite direction to the causal-arrow! If it were we would be none the wiser, nor any the worse off.

This thought-experiment does not refute the causal theory; it does not prove that events are not in fact related by an asymmetrical dependency relationship. But it does a good deal to undermine whatever intuitive support the causal theory in this form has, for it brings into sharp focus how *little* the theory explains: the apparent asymmetries within time, the asymmetries that determine the direction of our world-time, may well be entirely independent of the causal-arrow. If so, we must look elsewhere for an explanation of the temporal asymmetries that we find in the observable world.

We can draw a further lesson from the thought-experiment. The causal theorist's explanations of the content-asymmetries may be suspect, but the fact that these explanations are plausible (at least until examined more closely) suggests that there is indeed an intimate connection between our ordinary concept of causation and these other asymmetries: the causal-arrow and the world-arrow point in the same directions, and it is hard to believe that this alignment is contingent. Wouldn't our time-reversed counterparts be perfectly justified in thinking that the direction of causation in their part of the universe runs in the direction they call earlier-to-later? If this is right, if the causal- and the world-arrow must point in the same direction, we can ask *why*? One answer would be that the causal-arrow determines the direction of the world-arrow, but since this answer is suspect, we must consider the other option: perhaps it is the world-arrow, or some ingredient of it, that determines the causal-arrow.

4.7 Fundamental forks

In his *Asymmetries in Time* (1987), Paul Horwich embarks on the ambitious task of providing an account of the relationships between all the familiar temporal asymmetries within the context of a block universe. What is striking about the account he defends is the very lowly status causation occupies within it: see Figure 4.2 (a slightly modified version of Horwich 1987: 201). Rather than being the asymmetry that explains all others, causation comes towards the very end of the hierarchy of explanatory priority.

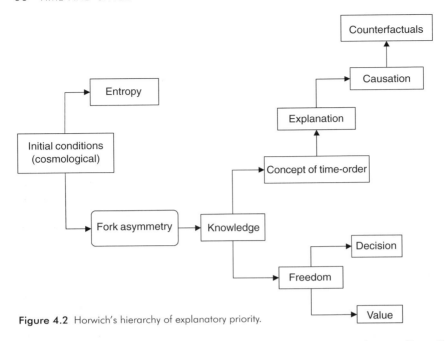

Figure 4.2 Horwich's hierarchy of explanatory priority.

For Horwich, the knowledge asymmetry is the key to understanding all those other asymmetries depicted further to the right; he argues that the knowledge asymmetry itself is (largely) a product of the fork asymmetry, whereas the latter exists because of the special conditions that obtained in the early universe.

To bring this chapter to a close I will provide an outline of some of the main elements in Horwich's explanatory scheme. The account, as Horwich himself concedes, is not entirely adequate as it stands. In addition to certain speculative components (inevitable, given the current state of physics), a good deal of detail requires filling in, and some parts are more persuasive than others. Nonetheless, by virtue of its treatment of causation, Horwich's approach to the content-asymmetries is quite different from those that we have considered thus far, and this alone makes it worth examining.[7]

Forks

Since the fork asymmetry plays such an important role in Horwich's scheme we had best start here. He describes the asymmetry thus: "Highly correlated event types are invariably preceded by some unified common cause, but they need not have joint effects. We never find a pattern as depicted [on the right of Figure 4.3] in which the correlated events are linked only by a joint effect" (1987: 73).

If X and Y are types of event whose occurrences are linked, that is they frequently occur together (and so are "highly correlated"), we will usually find that there is a certain type of prior event Z that causes both. The reverse of this – two highly correlated events having a typical subsequent effect but no typical prior cause – never occurs (or only rarely). So, for example, thunder and lightning are strongly correlated types of event (they often occur together), and they have a common prior

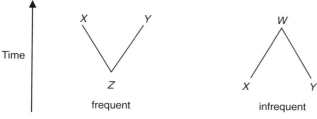

Figure 4.3 The fork asymmetry.

cause (high atmospheric static charge), but there is no single type of effect that invariably follows a lightning discharge (e.g. sometimes a tree is destroyed, sometimes a house is hit, sometimes a field, etc.). While this all sounds plausible enough, we cannot characterize this fork asymmetry in terms of causes and effects if we hope to use this asymmetry to explain causation. Horwich thus provides an alternative cause-free formulation along these lines: it is common to find strongly correlated events at t statistically associated with a single earlier event; it is rare to find events at t similarly associated with a single later event.

It is worth pointing out that there is a more general kind of fork asymmetry, which is not restricted to strongly correlated events. Detonating a given type of bomb will have very different consequences in different environments (outer space, within buildings, under water); the effects of detonations are thus not strongly correlated, in Horwich's sense. Yet in each particular case the effects can be traced back to a single cause. In such cases there is an asymmetry of *over-determination*. Owens characterizes it thus:

> For any event, the number of events at any given point in the future for which it is necessary is greater than the number of events at any given point in the past for which it is necessary. (Or equivalently: For any given event, the number of events at any given point in the future which are sufficient for it will be greater than the number of events at any given point in the past which are sufficient for it.)
> (1992: 89)

Any given detonation will typically have a single sufficient prior condition, such as the pushing of a switch. The detonation is thus necessary for just one immediately preceding event. But it is necessary for each of its many subsequent effects, which would not have occurred in its absence, and hence each of these effects is, in the circumstances, sufficient for the detonation. In such cases, which are very common, events are massively overdetermined (in Owens's sense) by later events but not by earlier events. These overdetermining later events may be strongly correlated, but they need not be. Bearing this point in mind, let us move on.

Causation

Horwich agrees with Hume that causation is in itself a symmetrical relationship, and that causes precede effects only because of a convention about how these words should be used. However, he also maintains that the relevant convention is not arbitrary. The stipulation that causes come before their effects meshes with, and is supported by, a cluster of other beliefs about causation that we have, namely:

1. Causes explain their effects.
2. Causes are (in some sense) ontologically more basic than their effects.
3. Correlated events are causally connected.
4. Causes are the means by which we bring about the effects that are our ends.

Given these beliefs about what a cause is, and the temporal asymmetry of knowledge, explanation, and so on, the idea that causes occur earlier than their effects is entirely natural. Or so Horwich goes on to argue.

Both 2 and 4 are linked to the knowledge asymmetry. The vague idea that causes are "more real" than their effects is rooted in the widespread belief that the past is more real than the future, and this belief is in turn rooted in the fact that we have detailed knowledge of the past but not of the future. As for 4, Horwich suggests that causation is (in part) defined by its association with our experience of deliberation and control: "we define causation as that general relation between events that is exemplified when an event is deliberately brought about by free choice" (1987: 143). If, after due deliberation, I decide to put poison in Jim's drink, and Jim subsequently dies, I have caused Jim's death. More generally, we would all agree that if C is used by someone to produce a desired end E, then C causes E. As for why our deliberation and intentions are future-oriented, consider this: could you seriously deliberate about performing a certain action A if you know for certain that you will not do A? Of course not. We can only deliberate and form genuine intentions to perform actions that we (believe) we are *free* to perform. Horwich argues that:

> the feeling of freedom is associated with our ignorance, prior to deliberation, of what we are going to do. If we knew as much about the future as we do about the past (and with the same certainty that is provided by memory) this feeling could not be sustained. We can then explain, in terms of the sense of free choice, why rational decisions are not made for the sake of past events and why we care about the future more than the past.   (1987: 204)

The knowledge asymmetry also plays a role in the temporal orientation of explanation. As we have just seen, the belief that the past is real and fixed, whereas the future is not, is (according to Horwich) grounded in the knowledge asymmetry, and this same belief helps sustain the practice of explaining later events in terms of earlier ones, since for anyone who holds this belief it will seem natural (even if it is not really the case) to suppose that later events depend on earlier events in a way that earlier events cannot depend on later events. But the contribution of the fork asymmetries to the asymmetry of explanation is even more marked. As Owens stresses, a genuine coincidence consists of a conjunction of events *that has no explanation* (1992: Ch. 1). Such a conjunction has no explanation if its component events each have necessary and/or sufficient conditions that are independent of one another. If, on the other hand, there is some *single* event that is necessary and/or sufficient for each component of the conjunction, we have an explanation, and the conjunction in question is no coincidence. Ten people on a coach party falling ill is a coincidence if they all ate different meals; it isn't if they all ate the fish for lunch. In virtue of the fact that future-directed forks predominate, and given the asymmetry of overdetermination, non-coincidental conjunctions of events are nearly always explained by reference to earlier events. It is common to find a single earlier event that is both necessary and sufficient for several subsequent events, and that thus reveals the later conjunction to be non-coincidental, but it is rare to find the same in reverse. Generally speaking, when (as often happens) a given earlier event is necessary for several subsequent events, none of the latter is individually necessary for the earlier event, since each is individually sufficient for it. Since the direction of explanation runs from earlier to later, and since causes are events which explain their effects, so, too, does the direction of causation.

Knowledge

Horwich argues that the knowledge asymmetry can be explained in terms of the fork asymmetry, a conclusion that emerges from a highly abstract discussion of the nature of recording devices (1987: 84–90). I will try to convey the gist of the basic idea in a more informal way.

An expanse of wet concrete constitutes an efficient recording system: if someone steps in it, an accurate imprint of their footwear (or foot) will be made, and is likely to endure for a good while, since once concrete is set it is physically robust. Anyone who subsequently sees the imprint in the concrete will have a good chance of being right if they believe that someone with just that shape and size of foot stepped in the concrete a short while before it hardened. Any foot-shaped imprint in concrete is a particular macro-state consisting of trillions of particles existing in a particular configuration. As is clear, these same particles could exist in trillions of different configurations, and since concrete slabs generally start off in a randomly mixed condition, it is highly unlikely that any given foot-shaped configuration would develop through chance alone. It is even more unlikely that all the very many (but slightly different) foot-shaped imprints to be found in different concrete surfaces would arise by chance alone. Viewed from the micro-level, the creations of these imprints are strongly correlated events: particles in many different concrete surfaces all coming to be arranged in highly improbable foot-shaped configurations. These correlated events require an antecedent event of a highly specific sort (e.g. a foot pressing downwards). We can be reasonably certain that the event that explains the correlation occurred earlier than the correlation itself because we know that, whereas future-directed forks of this kind are common, inverse forks are rare.

Records come in diverse forms; for example, fossils, newspaper reports and memories. But despite this diversity there is also similarity:

- in each case we have a macroscopic system whose constituent parts have a tendency to enter into highly specific (and improbable) configurations;
- these configurations are strongly correlated with *some other type of event* (e.g. foot-shaped imprints are correlated with downwards pressing feet);
- this other type of event almost always occurs *earlier* than the configuration in question.

This last fact obtains by virtue of the fork asymmetry. If instances of regularly occurring improbable conjunctions of event-types were commonly associated (statistically) with a single type of subsequent event, there might be many reliable records of future events; but since they aren't, there isn't.

The source of the fork asymmetry

Since all the other asymmetries lead back to the fork asymmetry, directly or indirectly, we are left with one outstanding question: why do normal forks predominate over inverse forks? Why does the fork asymmetry exist at all in a universe governed by time-symmetrical laws?

This is a crucial question, and Horwich can offer no more than a speculative cosmological conjecture. He makes the assumption that the early universe was in a state that combined order with disorder. The disorder obtained at the micro-level: elementary particles existed in a state of random flux. Within this *micro-chaos* there was a limited degree of *macro-order*: mass-energy was not distributed uniformly through space, but concentrated in the structures that eventually gave birth to galaxies, stars and planets. These assumptions are compatible with many current theories of the early universe. Now, suppose events of types X and Y are strongly correlated with one another – they co-occur at many times and places (e.g. heat and smoke). It is unlikely that the sequences of events leading up to these frequent conjunctions of X and Y will consist of two independent processes, for this would require a vast and repeated conspiracy at the micro-scale, a conspiracy that is ruled out by the assumption of initial micro-chaos. Hence we would expect to find that there is some single type of event Z (e.g. fire), the occurrence of which is strongly correlated with (or nomologically sufficient for) the subsequent occurrence of X- and Y-type events. This, of course, is precisely what we do tend to find: it is the normal fork pattern. The question that needs an answer is why do *inverse* forks not occur? Why are strongly correlated simultaneous events not regularly followed by a single *later* event of a specific type?' The answer, suggests Horwich, lies with the *implicit microscopic order* that exists at later times.

Let us suppose that the universe does not collapse under its own gravitation, but its expansion gradually slows. From the second law of thermodynamics, we know that the energy that is currently concentrated in stars will eventually disperse, and so the later stages of the universe will not contain any recognizable order: there will be a thin homogeneous spread of low-energy particles and fields. Now, if we consider the evolution of the universe from the direction of earlier-to-later, we see a progression from an initial state that combines macro-order and micro-chaos to later states that consist only of micro-chaos, an asymmetric progression that at the micro-scale is governed by time-symmetric laws. But now imagine viewing this same process unfold in the other direction, from later-to-earlier. What would you see? On the large-scale, you would see the cold thin soup of particles gradually coming together to form dim stars, which gradually grow brighter as more and more energy flows into them; occasionally, you would see vast numbers of particles converging to a single star (a super-nova in reverse). On the much smaller scale of our Earth, you would see huge numbers of astonishing coincidences, such as wind blowing leaves into neat heaps in the middle of lawns, dirty clothes scattered around a room ending up washed and neatly folded in cupboards and drawers, and so forth. Adopting the time-reversed view makes one thing very clear: although the thin cold particle soup that exists at the later stages of the universe might look like a random distribution, it isn't. Given the distributions of particles that exist at these times, and the time-symmetrical laws governing them, they will evolve into the highly organized systems that exist at the present time. The apparently random distribution is in fact highly ordered; this order is all but impossible to discern, it is *implicit*, but it is there nonetheless.

The presence of this implicit order explains why inverse forks are rare. Even if the scenario sketched above is implausible, and the implicit order dissipates over long periods of time, it certainly exists over the short term ("short" by cosmological standards). So given the hypothesis of initial micro-chaos, strongly correlated events will generally only occur when there is a single earlier event that is nomologically sufficient for them all. The implicit order in the events that occur *later* than strongly correlated occurrences (even a short while later), means that there is no need for correlated events to be followed by some single type of event that is nomologically sufficient for them all: even if the subsequent events are many, varied and apparently random, in each specific case and context, these later events are nomologically sufficient for the earlier events. And this is what we find. Thunder and lightning are always preceded by a build up of atmospheric static; there is no one *succeeding* condition of remotely comparable significance.

If we explain the content-asymmetries along the general lines sketched out by Horwich, we cannot rule out the possibility of time-reversed universes. Since the initial conditions at each end of a Gold universe are very similar – both exhibiting micro-chaos – it is to be expected that each half will develop in much the same way. In which case, not only will both halves of the universe be characterized by fork asymmetries, but the forks will point in different directions. If, as Horwich maintains, all the other content-asymmetries depend on the fork asymmetry, these, too, will be differently oriented, causation included. In which case, the expanding and contracting phases of the universe will be time-reversed with respect to one another. Time's arrow may thus be less securely fixed than we tend to think. Or at least, this would be the case if we lived in a static B-world, a world where temporal passage is illusory, where *time itself* possesses no inherent direction. Whether this is so or not has yet to be established.

As Horwich himself stresses, his account of why the fork asymmetry exists rests on largely speculative premises (e.g. a low-entropy big bang), and certainly needs further development before it can be considered fully adequate. He points out that all that matters for his account of the other content-asymmetries is the existence of the fork asymmetry itself (1987: 76), but even here there are questions of detail to be addressed, and there are those who doubt whether this asymmetry is up to the job that Horwich assigns to it.[8] But for our purposes what matters is this: even if the B-theorist cannot appeal to causation to explain the asymmetries that we find in time, there are alternative strategies that are available, strategies that may well prove viable. If Horwich's proposal proves problematic, there may well be variants of it that will prove more successful.

However, in the absence of any fully worked out proposal along these lines, and given the difficulties encountered by the entropic and causal accounts, B-theorists cannot yet be said to have provided a fully satisfactory account of the various content-asymmetries that we have considered. If it proves impossible to formulate such an account, B-theorists would have no option but to regard the asymmetries (or at least, those that remain unexplained) as a matter of brute fact: it just so happens that the contents of our world are distributed through a four-dimensional continuum in certain asymmetrical patterns. This is not ideal, but neither is it necessarily fatal, since in any theoretical system there is always something that is not explained in other terms. Of course, adopting this stance *would* prove damaging if there were viable dynamic conceptions of time possessing greater explanatory resources. Whether there are such alternatives to the static conception is the issue we shall be examining next.

5 Tensed time

5.1 Tense versus dynamism

In §1.6 I noted that we should not assume that the static–dynamic dispute maps neatly onto the tenseless–tensed dispute. The latter revolves around the sorts of concepts that are needed to describe the world in a metaphysically adequate way, whereas the former concerns the reality or otherwise of temporal passage. The B-theory we have just been considering combines two claims:

- The world is static, all times and events are equally real and coexist.
- The world can be fully described in tenseless terms.

In Chapter 6 I will be looking at ways of taking issue with the first claim; in this chapter I will consider the implications of rejecting the second.

In a sequence of papers spread over some years, E. J. Lowe has sought to defend the A-framework against its critics (such as McTaggart and Mellor), and thus establish the viability of the tensed conception of time.[1] Since many B-theorists have argued that the tensed conception is incoherent, his arguments on this point are of interest in their own right. In Chapter 2 I introduced McTaggart's A-paradox, but since it proved peripheral to the main point that McTaggart was trying to make, I did not consider it in any detail; a remedy was suggested but not examined closely. Similarly, in Chapter 3 we encountered Mellor's argument that tensed statements have tenseless truth-conditions, but we have not yet considered counterarguments. But there is another reason why Lowe's position is worth looking at. Although he is a staunch defender of the tensed conception, Lowe does not engage in grandiose metaphysical speculations of the sort that we will be considering in Chapter 6. Does this mean that such speculations are pointless and unnecessary? Can someone who finds the B-theory implausible and/or repugnant find all the consolation they need in the tensed conception of time? Or is something more needed? Let us find out.

5.2 Taking tense seriously

Lowe distinguishes the tensed and tenseless views thus:

> the tensed view takes McTaggart's "A-series" terminology to be indispensable for the metaphysics – not just the epistemology and semantics – of time,

whereas the tenseless view takes "B-series" terminology alone to be indispensable for the metaphysics of time.   (1998: 43)

A mark of Lowe's commitment to the A-framework is his firm stance on the question of the proper use of tenseless sentences. He rejects the idea that properties can be tenselessly predicated of concrete historical events. The only objects that can tenselessly possess properties are abstract entities (such as numbers) that do not exist in space or time; anything that *does* exist in time possesses properties in a tensed way: "What it *is* to exist in time is to be a potential subject of tensed predications, in my view. That is, for a thing to exist in time is for some property to *be now* exemplified by that thing, or to *have been* exemplified by it, or to *be going to be* exemplified by it" (1998: 44, original emphasis).

An advocate of the tenseless view would take a sentence such as "There is rain in Durham on 12 March 2099" to be a tenseless sentence akin to "2 + 2 = 4", but for Lowe it can only be properly understood in a tensed way. Hence it is equivalent to the sentence "There is now rain in Durham on 12 March 2099" (which is false since it is not now 2099 or 12 March), or it is a grammatically improper way of trying to say "There will be rain in Durham on 12 March 2099", or, finally, it is an abbreviated way of saying "There was, is now, or will be rain in Durham on 12 March 2099".

This point may seem rather pedantic, but it turns out to have far-reaching implications. For a start, if Lowe is right, *all* facts about concrete reality are tensed, and so the idea that a tenseless description of the world could be complete is a non-starter. But what of the B-theorist's claim that tensed statements have tenseless truthmakers? Lowe has a response to this too.

Mellor argues that A-theorists face an impossible dilemma: they must accept that either (i) tensed sentences are made true by tensed facts that are *non-token-reflexive*, or (ii) tensed sentences are made true by token-reflexive facts that are *tenseless*. Adopting (ii) is tantamount to rejecting tensed facts and thus the metaphysical significance of tense: surely if there were tensed facts one of the things they would do is make tensed sentences true! But (i) is clearly incoherent. The tensed sentence "World War Two is now over" is true if tokened after 1945, and false if tokened earlier; but how can this be if the sentence is made true by a tensed fact that is independent of when the sentence is tokened?

Lowe accepts that the truth-conditions of tensed sentences must be token-reflexive, but rejects Mellor's claim that such truth-conditions must be tenseless. He suggests that the truth-conditions of (tokens) of a sentence such as "It is now raining in Durham" can be stated in either of the following ways:

A token u of the sentence-type "It is now raining in Durham" [is] true if and only if u [is] uttered at a time t such that it [is] raining in Durham at t.

A token u of the sentence-type "It is now raining in Durham" [is] true if and only if u [is] uttered simultaneously with an occurrence of rain in Durham.

Why are the occurrences of "is" in brackets? Because Lowe wants to make it absolutely clear that these statements of truth-conditions are tensed. Since sentence-*types* are abstract entities they can have properties predicated of them tenselessly. But the formulations above concern sentence-*tokens*, and concrete time-bound entities such as these can only be the subject of tensed predication, or so Lowe maintains. Accordingly, when he says that "a sentence-token *is* true if and only if such-and-such *is* the case" he means either that it is *now* true, or else what he is saying is a potentially misleading abbreviation of the *disjunctive tensed predication* "was, is now, or will be true". It is the latter he has in mind in the above statements of truth-conditions (to make them fully general rather than present-tensed). Hence "[is] true" means "was, is now, or will be true".

5.3 McTaggart revisited

This approach provides Lowe with a response to McTaggart's A-paradox. Like C. D. Broad before him, Lowe believes that McTaggart is guilty of a simple mistake: the alleged paradox doesn't even get off the ground. As we have already seen in Chapter 2, McTaggart argued that if we use A-terminology we are committed to saying that *every* event is past, present and future, despite the fact that these are incompatible attributes. Lowe responds (quite plausibly) by maintaining that at most we are committed to saying something along these lines:

> For any event *e*, (i) it either was, is now, or will be true to say that "*e* has happened", and (ii) it either was, is now, or will be true to say that "*e* is happening now", and (iii) it either was, is now, or will be true to say that "*e* will happen".
> (1998: 46)

As is clear, there is nothing paradoxical here at all. There *was* a time when tokens of the sentence "Napoleon is now alive and well" were true, but tokens of this sentence were not true before Napoleon was born, and they are not true now, and (barring miracles) they won't be true in the future either. Likewise, tokens of "The year 2010 is in the past," are not true now (or in the past), but they will be one day. And so on.

As Lowe concedes, careful formulations such as the one above can be something of a mouthful, so we might easily find ourselves falling into misleading ways of talking. The mistake to avoid is to suppose there is no alternative to the misleading forms of words. So, for example, we might be tempted to express the fact that there was a time when tokens of the sentence "Napoleon's death lies in the future" were true by saying that Napoleon's death *was future*, or even *has the property of being future in the past*. McTaggart himself employed these compound tenses when expounding the A-paradox (as does Smith's disjunctive formulation quoted in §2.4). But as Lowe points out, interpreted literally, formulations such as these are simply incoherent: to say that an event *E* is future in the past is like saying that an event *E* is happening *here over there*. If someone insists "But *E* is happening *here* over there, you know what I mean!" what they really mean is something like this: an utterance *over there* of the sentence "*E* is happening here" is true (Lowe 1987).

5.4 Is tense enough?

These metalinguistic constructions provide Lowe with a plausible way of refuting the charge that the A-framework is inherently contradictory. Provided we take the necessary care, we can talk in tensed terms without fear of talking nonsense. The tensed theory of time cannot be refuted on the grounds that it is internally incoherent. But what does Lowe's tensed theory of time actually amount to? What, on his view, is the metaphysical difference between past, present and future?

As noted earlier, he says that the tensed view of time is distinguished by the claim that the notions of past, present and future are indispensable ingredients in our metaphysical understanding of what time is. If he is right in claiming that for a thing to exist *in* time is for some property to be *now* exemplified by it, or to *have been* exemplified by it, or to be *going to be* exemplified by it, he may well be right that the notions of past, present and future *are* indispensable ingredients in our metaphysical understanding of time. But what does talking in tensed terms really signify? Wellington's victory at Waterloo lies in the past. How does this event differ from an event that lies in the present or future? Do present events have a distinctive property, *presentness*, which past and future events lack? Do past events possess a distinctive property of *pastness*, which present and future events lack? Lowe rejects this idea, citing overdetermination difficulties of the sort we encountered in §2.6:

> There is one very good reason why I use this sort of metalinguistic construc-
> tion, and this is that I do not want to suggest that "presentness" is some sort
> of *property* of events on the grounds that if presentness (and likewise
> pastness and futurity) is conceived as being a property of events, then it is
> difficult to see how it can *either* be a property which an event has always had
> and always will have, *or* be a property which an event now has, has had, or
> will have only temporarily – and no other option seems available.
>
> (1998: 47, original emphasis)

This is a very plausible position, but if the difference between being past, present and future is not a matter of possessing different properties, what does the difference consist in?

One compelling answer is that whereas present events actually exist, past and future events do not, and Lowe would go along with this, up to a point. He is happy to concede that events that lie in the past do not exist *now*, and likewise for events that lie in the future. Of course while events such as the Battle of Waterloo do not now exist, they *did* exist, and future events *will* exist. Is there anything more to be said? Someone might insist:

> What you are saying sounds reasonable, but I want to know more. Are you
> saying that past and future events are just as real as present events, that all
> events are on the same ontological plane, so to speak, as the B-theorist
> maintains? Or are you saying that future and/or past events do not form parts
> of the sum total of reality?

Lowe does not address quite this question, although he does address one somewhat similar to it, so the following reply goes some way beyond what Lowe actually says, but not by much.

> I do not say that there is no ontological distinction between past, present and
> future. I am happy to concede that some future-tensed statements may be
> neither true nor false (e.g. it *is now* neither true nor false to say "It will rain in

Durham on 12 March 2099") (Lowe 1998: 50). But neither do I say that past and future events are unreal, or that they do not form parts of the "sum total of reality". What distinguishes past events is precisely that they *are past*, not "unreal"; likewise for future and present events. Moreover, the "sum total of reality" is not restricted to the present, since some events are (now) past and others (now) future (even though we do not as yet know what these will be).

While there is no denying that some events are present whereas others are past, it is not clear what this difference amounts to on Lowe's account, given that he rejects the idea that pastness and presentness are properties possessed by events. His response is to say that tensed concepts such as "past" are

> so fundamental that they are semantically irreducible . . . Do we not take the same view of an operator such as *negation*? "It is not the case that" cannot be given a non-circular semantic explanation, yet of course we grasp its meaning as clearly as we grasp the meaning of anything.
>
> (1998: 49, original emphasis)

Again, this is a perfectly respectable reply: some concepts are so basic that they can't be explained in terms of other (independent) concepts. Perhaps *past*, *present* and *future* fall into this category. But this reply, although hard to refute, does not entirely satisfy.

Lowe does not explain why some future-tensed statements can lack a truth-value when (presumably) their past counterparts (e.g. "It rained in Durham on 12 March 2099BC") do not. He might well say that this is simply an aspect of what is involved in being past and future. But is it plausible to suppose that this is *all* there is to be said? Suppose someone says,

> The future is unreal, not just in the sense of "not existing now", but in the sense of *not existing at all*. If we could compile a complete inventory of what exists, not just now but at all times, past and present events would be included, but not future events.

This claim, if intelligible, would explain why statements about the future contingencies lack truth-values. By eschewing claim such as these, Lowe is unable to provide an explanation of comparable power.

Lowe's position amounts to a *deflationary dynamism*.[2] On one level the account offers opponents of the static view what they want. There is a difference between past, present and future; time does pass, at least in the sense that an event that at one time can truly be characterized as being present can also, at different times, be truly characterized as past and future. But on another level the account leaves something to be desired. It turns out that the only difference between being past, present and future is that events can legitimately be *called* "past" after they have occurred, "present" when they are occurring, and "future" before they occur. Since tensed predicates do not denote genuine properties, it is hard not to feel that the difference between past, present and future is merely verbal. Is this all this difference boils down to? It is hard not to feel that there is something more to be said. The models of time we will be looking at next try to say more.

6 Dynamic time

6.1 The growing block

In *Scientific Thought*, after a clear statement of the problems posed by supposing that there are transitory tensed properties, Broad proposes a very different dynamic theory of time. Focusing on his current state, he says that this:

> is just the last thin slice that has joined up to my life-history. When it ceases to be present and becomes past this does not mean that it has changed its relations to anything to which it was related when it was present. . . . When an event, which was present, becomes past, it does not change or lose any of the relations which it had before; it simply acquires in addition new relations which it *could* not have before, because the terms to which it now has these relations were then simply non-entities.
>
> It will be observed that such a theory as this accepts the reality of the present and the past, but holds that the future is simply nothing at all. Nothing has happened to the present by becoming past except that fresh slices of existence have been added to the total history of the world. The past is thus as real as the present. On the other hand, the essence of a present event is, not that it precedes future events, but that there is quite literally *nothing* to which it has the relation of precedence. The sum total of existence is always increasing, and it is this which gives the time-series a sense as well as an order. A moment t is later than a moment t^* if the sum total of existence at t includes the sum total of existence at t^* together with something more.
>
> The relation between existence and becoming . . . is very intimate. Whatever is has become, and the sum total of existence is continually augmented by becoming. There is no such thing as *ceasing* to exist; what has become exists henceforth for ever. When we say that something has ceased to *exist* we only mean that it has ceased to be *present*; and this only means that the sum total of existence has increased since any part of the history of the thing became. (Broad 1923: 87–9)

The process that Broad subsequently calls "absolute becoming" is the coming-into-being of events. Time passes, the universe is dynamic, because new events are continually being created. When an event comes into existence it remains in existence. Past events are perfectly real; just as real as present events. Future events have no existence at all. As is very clear, Broad has no truck with transitory tensed intrinsics: present events do not possess a distinctive property of *presentness*, which past events lack. An event is present by virtue of belonging to the most recent reality-slice, that is all. It does not change its intrinsic properties as it sinks into the past; that is, as additional slices of reality are added to the universe. Broad's model of time is exactly the same as the B-theorist's block view with one crucial difference: if Broad is right, the block is not static but *growing*.

By what increments? How large are the new slices of reality? In principle, there seems little reason why the increments could not be huge hunks of time – a universe of ten billion years' total duration could come into being in just two stages – but for obvious reasons this idea lacks appeal. If we define "later than" in Broad's way, these billions of years would contain only two successive times! More importantly, the appeal of the growing block model rests on its conformity with the common-sense idea that the present is a brief (or momentary) interface between a real past and a non-existent future; this is lost if we take the present to be the most recent addition to reality, and allow these additions to be of sizeable duration. Hence growing block theorists have two plausible options: they can say that each reality slice is instantaneous, or they can say that each reality slice is of some finite but minimal duration. Anyone who is sceptical of the idea that even an infinite number of strictly durationless "slices" can ever add up to anything of finite physical duration will prefer the second option.

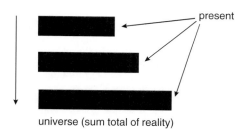

present

universe (sum total of reality)

Figure 6.1 A growing block universe. As new presents are created, the sum total of reality increases. The growth in question is temporal. Although our universe (apparently) originated in a big bang singularity and has been expanding (spatially) ever since, other possible universes may start large and shrink in size over time – in such cases each new present will have a smaller spatial extent than its predecessor.

This model of time has considerable appeal to anyone reluctant to abandon the natural view. The past is fixed and determinate, the future is unreal and the present is both constantly changing and distinctive, even if only by virtue of being reality's most recent accretion. Moreover, the model provides an intuitively appealing explanation for some important content-asymmetries. We have trace evidence (and hence detailed knowledge) of the past but not the future because the former is real and the latter is not, and only what is real can leave traces of itself. The absence of (easily discernible) backward causation can be explained in the same way: non-existent events cannot cause anything; real events can.

But is the model truly coherent? The closer examination that follows will yield this tentative verdict: there are versions of the model that may well be coherent, but in certain respects the model is not the friend of the natural view that it initially seems.

6.2 Overdetermination

By not positing intrinsic tensed properties, the growing block theorist avoids the overdetermination problem in one form, but the same problem in other guises lies in wait unless appropriate measures are taken. Consider the following versions of the model:

A Past, present and future moments of time are equally real and permanent, but may be empty. As events are created, time is gradually filled with concrete events.

B Moments of time come into being with things and events; there are present and past moments, but no future moments.

Version A succumbs to the overdetermination problem that we encountered in §2.6. Does it make sense to suppose that a given moment of time could be both empty of events and not empty of events? Could the universe at some point in 2010 be both entirely devoid of matter and also at the very same time contain billions of galaxies? Of course not: just as an object cannot be in two different and incompatible conditions at a single time, the universe cannot be in two different and incompatible conditions at a single time. To avoid this problem the growing block theorist must subscribe to version B. If events and the times at which they occur are created together, times are not overdetermined.

time

Figure 6.2 The wrong way for a universe to grow.
If we envisage the material universe growing through pre-existing substantival time, we encounter the overdetermination problem: there will be times which possess the incompatible attributes of being empty of events, and not being empty of events.

But has the overdetermination problem really gone away? Consider the universe at an earlier time t_1 and at a later time t_2. Let us call the sum totals of reality that exist at these times S_1 and S_2. S_1 includes everything that happens at t_1 along with everything that has happened earlier than t_1. S_2 includes everything that happens at t_2, along with everything that has happened earlier, and so includes S_1. Suppose that E is an event that happens at t_1. It seems that we can say the following about E:

1. The sum total of reality to which E at t_1 belongs consists of S_1.
2. The sum total of reality to which E at t_1 belongs consists of S_2.

These two claims are inconsistent; they cannot both be true. A rock cannot, at a given time, be in a pond containing two different total volumes of water. How can one and the same event be a part of two different sum totals of reality at the time it occurs?

There is a solution to this problem. We are familiar with the distinction between states of affairs that are real, or that exist, or that are *actual*, and those that are not. It is plausible to suppose that the notions of *existence* and *actuality* are primitive. As Tooley plausibly argues at the outset of his impressively wide-ranging defence of a growing block model (see Tooley 1997), in a dynamic world the actual–non-actual distinction is insufficient, since the states of affairs that are actual may be different at different times. What we need, accordingly, is the concept of existing or of being actual *as of* a given time. What is actual as of 1066 is quite a bit less than what is actual as of 1966. Tooley proposes that we take this notion of *actual as of a particular time* as a primitive. Corresponding to this ontological notion there is the semantic notion of *truth as of a time*. Since the totality of facts can vary from time to time, the concept of truth appropriate to a growing reality is temporally relativized: as reality grows, so, too, does the totality of truths.

From this two consequences flow: the same proposition can have different truth-values at different times, and there are three truth-values: true, false and neither-true-nor-false. Consider:

3. There are (tenselessly) no dinosaurs.
4. There are (tenselessly) unicorns in the year 2010.

If we suppose that it is now 2001, 3 is not true: the Jurassic Period is actual as of 2001, and there are dinosaurs in the Jurassic Period. In 2001, 4 is not true either, but for a different reason: the year 2010 is not actual as of this time, so the statement is neither true nor false. In 2010 it will become true if there are unicorns then, and false if not. So the truth-value of 4 will change from indeterminate to either true or false.

The notion of truth as of a time provides the solution to the problem of the same event belonging to different sum totals of reality. As of t_1, 1 is true and 2 is indeterminate (since some parts of the sum total of reality S_2 are not actual as of t_1). As of t_2, 1 is false and 2 is true.[1]

I suggested earlier that a growing block theorist should hold that a universe grows by durationless or very brief stages. A natural follow-up to the question "By what increments does a universe grow?" is "How *fast* does a universe grow?" However, if times and their occupants come into being together, this question has no real sense. The question "How much time passes between the addition of one reality-slice and the next?" would be perfectly meaningful if a block-universe existed in an external time dimension (for example, if a meta-time axis ran vertically past the horizontal blocks in Fig. 6.2). But since the block theorist recognizes (or should recognize) no such external time, all we can say is that successive phases of reality come into being at (and along with) the times when they occur. That a specific speed cannot be produced in response to the "How fast?" question is sometimes taken to be disastrous (or at best a serious embarrassment) for the growing block theory (cf. Nahin 2001: 199); but since the question in this context has no sense, the error is with those who press the question *expecting* an answer of this sort.

6.3 Dynamism without tense

As I pointed out in §1.6, the issue of whether time is tensed or tenseless is distinct from the issue of whether time is static or dynamic; consequently, it may be possible to combine a dynamic view of time with a tenseless view. Tooley develops his account of the growing block model along these lines. A proponent of the tenseless view subscribes to the claim that the world can be fully described in tenseless terms. Tooley not only subscribes to this claim, but he rejects the idea that there are genuine tensed properties, and provides an exhaustive analysis of tensed concepts in tenseless terms. Space precludes a full exposition of Tooley's account, but I provide some examples of his approach, which in some respects is novel.

Tooley takes a *tensed sentence* to be any sentence that employs tensed concepts, such as "present" or "past". Most tensed sentences are indexical: a sentence such as "It is hot now" specifies which time is present not by explicitly stating a date, but by having a meaning such that the time at which the utterance occurs indicates the time of the present. Are there tensed sentences that are non-indexical? Tooley claims that there are:

Event E lies (tenselessly) in the present at time *t*.

This is tensed (since it mentions the present), but it specifies the time that is present in a non-indexical way, by mentioning a date. Tooley offers a tenseless analysis of this sentence:

E is an instantaneous state of affairs, E is actual as of time *t*, and no state of affairs that is later than E is actual as of time *t*.

What of non-indexical sentences about the past and future? Tooley suggests the following analyses (1997: 196):

E *lies in the past* at time *t*: Event E is earlier than time *t*, and *t* is an instantaneous state of affairs, *t* is actual as of time *t*, and no state of affairs that is later than *t* is actual as of *t*.

E *lies in the future* at time *t*: Event E is later than time *t*, and *t* is an instantaneous state of affairs, *t* is actual of time *t*, and no state of affairs that is later than *t* is actual as of *t*.

As for indexical sentences, drawing on the above, Tooley suggests analyses such as these:

Any utterance or inscription, at time *t**, of the sentence "Event E is now occurring" is true (false, indeterminate) at time *t* if and only if it is true (false, indeterminate) at time *t* that E lies in the present at *t**.

Any utterance or inscription, at time *t**, of the sentence "Event E has occurred" is true (false, indeterminate) at time *t* if and only if it is true (false, indeterminate) at time *t* that E lies in the past at *t**.

I will not pursue Tooley's analyses any further. I introduce them solely because they provide a concrete illustration of how a dynamic view of time can be combined with the tenseless view. There are no doubt other ways in which the combination can be worked. To cite one instance, Tooley defends a causal account of the earlier–later relation, and (cf. Horwich and Price) not all tenseless theorists would go along with this.

6.4 The thinning tree

Before moving on to assess the growing block model, there is another dynamic model – in some respects similar to the growing block – that deserves a mention.

The account in question has been developed and elaborated over a number of years by Storrs McCall.[2] For reasons that will shortly be obvious, I will call it the "thinning tree".

The growing block model conforms with the widely held view that whereas the past is real and fixed and the present is real and transitory, the future is (as yet) unreal, and so consists at most of an open realm of possibilities. In Figure 1.4 these possibilities are represented schematically, as a fan of dotted straight lines; a more accurate representation would be a collection of spreading and dividing *branches*. Tomorrow I may stay at home or go to work, I haven't yet decided. If I stay at home, I will either do some gardening or read a novel; if I go to work I will either spend the day in my office catching up on paperwork, or attend a lengthy meeting. The same pattern applies quite generally: each of the initial possible courses of events at any given time for a given object or system (of which there are usually many more than just two) is associated with numerous further possibilities, and each of these sub-branches itself divides into further possible courses of events (if I spend the day in the garden, I will either repair the lawn, dig the border, prune the apple tree and so on). With the passage of time (the creation of new reality-slices) a few of these possibilities are realized. The remainder – the vast bulk – are systematically eliminated.

This process of "branch attrition" is depicted in Figure 6.3. As the present approaches the first node (n_1) the possible futures represented by b_1 and b_2 (each with their attendant sub-branches) still exist; but as the present moves past n_1, branch b_1 in its entirety is eliminated (I decided to go to work), leaving b_2 and its attendant sub-branches (the first two of which are b_3 and b_4). As the present passes n_2, b_3 is actualized and b_4 is entirely eliminated; at this point b_5 and b_6 both remain possible, but as the present passes n_3 this will change: one branch will be actualized, and the other eliminated. And so it continues. In reality, of course, there are many more branches (perhaps infinitely many) than are depicted here. Quite how many depends on what we decide to include as "possibilities". There are more logically possible futures than there are physically possible futures (i.e. ways things might develop that are compatible with the laws of nature). Indeed, in a universe where nomological determinism holds, there is only a single physically possible future.

So we see that a growing block universe is also a "thinning tree" universe: every gain for actuality (every creation of a new reality-slice) is accompanied by a loss of possibilities; as the present advances, the branching tree of possible futures is progressively thinned.

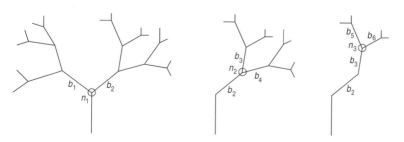

Figure 6.3 The thinning tree.

McCall's dynamic conception of time is essentially the model just described – the branches of his "universe-tree" are restricted to physically possible futures – but with just one crucial difference. McCall subscribes to a restricted version of "modal realism" (see §2.6), the doctrine that possible states of affairs are just as real as actual states of affairs. Consequently, he holds that all the future branches on the universe-tree are just as real as the "trunk" – the portion representing the past and present:

> At each stage, as the present (the first branch-point) moves up the tree, it is a purely random matter which branch survives to become part of the trunk. There is no "preferred" branch, no branch which is singled out ahead of time as the one which will become actual. Instead, all branches are on a par. All are equally real and, together with the trunk, constitute the highly ramified entity I shall call the "universe". Since it is never exactly the same at any two times the universe is a dynamic not a static thing. Nevertheless it is the same universe throughout, just as a child can look very different at different times and yet remain the same person the whole of its life.   (McCall 1994: 4)

For McCall, "actual" does not mean *real*, since every branch of the universe-tree is equally real. An event (or time) becomes actual when, due to the process of progressive branch-attrition, it stands alone and without competitors. Whereas the growing block model posits an *increase* in the sum total of reality, in McCall's model there is a progressive *decrease*: with every advance of the present untold billions of perfectly real stars, planets and people wink out of existence.[3]

Intriguing though it is, I will not discuss McCall's conception further. If the growing block model is metaphysically coherent, then so, too, I suspect, is McCall's version of the thinning tree. If a linear block universe can be of different sizes as of different times, it is hard to see why a *branching* block universe cannot be of different sizes as of different times in the way that McCall suggests – by having fewer branches rather than by having different temporal extents. But even if it is a coherent conception, McCall's model is problematic on several counts. As a conception of time it is radically counterintuitive; indeed, with its multiplicity of real futures it is far more counterintuitive than the static block. The same applies to McCall's version of modal realism, which abolishes the distinction (as usually understood) between possibility and actuality. But in addition, and perhaps most seriously, the model is ontologically profligate, and so falls to Occam's razor. Accepting the enormous expansion in the sum total of reality (as of certain times if not all) that McCall's model requires would be warranted if there were no other way to make sense of the world as we find it. But it is far from clear that this is the case.[4]

6.5 How can a block grow?

An important aspect of the growing block's appeal lies in the clear and unambiguous way that the model provides time with a direction. Thus far I have assumed, as it is natural to do, that there is a genuine sense in which a universe whose contents as of one time are greater than as of another time is *growing*. Further reflection suggests that the situation is not so clear-cut.

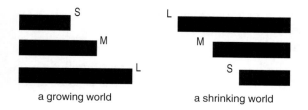

Figure 6.4 Growth v. Shrinkage How does a world that grows via "'absolute becoming" differ from a world that shrinks via "absolute annihilation"?

In Figure 6.4, S, M and L (standing for small, medium and large, respectively) refer to three distinct times. At first view, the "growing world" seems very different from the "shrinking world"; whereas the former starts small and becomes larger, the latter starts big and becomes smaller.

However, if we focus solely on the facts represented in these diagrams, the difference begins to seem less obvious. The following descriptions capture all the facts on display, and yet apply equally to *both* worlds:

- The universe as of S is smaller than it is as of M.
- The universe as of M is smaller than it is as of L.

Where is the "inherent directedness" here? All we have is a *dynamic* block, a universe whose temporal extent is different as of different times.

Of course, I have made no mention of "becoming", and it does seem natural to suppose that a universe in which facts "come into being" is growing rather than shrinking, but consider another simple universe of finite temporal extent. The universe in question consists of three "slices", S_1, S_2 and S_3, occurring at three times t_1, t_2 and t_3. Each slice contains the same rubber ball. In S_1 the ball is blue, in S_2 it is yellow and in S_3 it is green. Now consider these two ways of describing this universe.

1. As of t_1 the sum total of reality consists only of S_1. S_2 and t_2 come into existence by the process of absolute becoming, and the sum total of reality as of t_2 consists of S_1 and S_2. S_3 and t_3 are created, and as of t_3 the sum total of reality consists of S_1, S_2 and S_3.
2. As of t_3, the sum total of reality consists of S_3, S_2 and S_1. By the process of *absolute annihilation*, S_3 and t_3 depart from existence, leaving t_2, as of which time the sum total of reality consists of S_2 and S_1. S_2 and t_2 undergo annihilation, leaving t_1, as of which time the sum total of reality consists of S_1.

It is natural to interpret 1 as a description of a growing universe, and 2 as a description of a shrinking universe. The scenarios are described sequentially, and it is easy to adopt the tacit assumption that the sequence is temporal, and so unfolds in the earlier-to-later direction, where the latter terms have all the usual associations (we can remember earlier events but not later events, etc.). Interpreted in this way, 1 and 2 describe two different universes: one that starts small and increases in size as time passes, and another that starts large and decreases in size as time passes. But if we read 1 and 2 without making any assumptions about how the sequences temporally unfold, the differences between them all but vanish. As in our initial case, the objective facts about this universe appear to be these:

- the sum total of reality is not constant, there are reality-slices that exist as of some times but not as of others;
- the sum total is smallest as of t_1, largest as of t_3 and intermediate as of t_2.

In saying this I am assuming that "absolute becoming" and "absolute annihilation" are alternative ways of describing the fact that reality-slices exist as of some times but not others. Unless this is wrong, and there is some reason for preferring one term over the other, it seems that 2 is just as accurate a description of our simple universe as 1. Given that this is so, isn't it just as accurate to describe this universe as *shrinking by incremental annihilation* as it is to describe it as *growing by incremental creations*?

Not if we define our terms in the way Broad does. Since Broad stipulates that times as of when the block is larger are *later* than times as of when it is smaller, it is true by definition that the universe is growing; that is, that the block is larger at later times. But there is an alternative to Broad's definition. Consider:

D_1 t_2 is later than t_1 if and only if the sum total of reality is *greater* as of t_2 than it is as of t_1.

D_2 t_2 is later than t_1 if and only if the sum total of reality is *smaller* as of t_2 than it is as of t_1.

Broad opts for D_1, but why is Broad's definition of "later" preferable to D_2? Anyone who opts for the latter will regard a universe of the sort depicted in Figure 6.1 to be shrinking rather than growing, and by their lights, this is the correct description.

The question that we need to consider is whether Broad's preference for D_1 over D_2 is warranted. What the discussion so far has shown is that the preference cannot be justified by appealing to differences in block-size, but this does not mean that it cannot be justified at all. Let us call the direction of time provided by D_1 the *block-arrow*. The *world-arrow* is the apparent direction of time, the direction provided by the physical and psychological asymmetries that we examined in Chapter 4 (e.g. the tendency for entropy to increase, the apparent absence of backward causation and trace-evidence of the future, our apparent "experience" of temporal passage). Our ordinary understanding of the difference between "earlier" and "later" is rooted in these content-asymmetries. Consequently, if it could be shown that the world-arrow and the block-arrow point in the same direction, either necessarily or contingently, Broad's preference for D_1 would be fully justified: the claim that times at which the universe is larger are also *later* would have a clear rationale.

There are grounds for thinking that the block-arrow and the world-arrow do point in the same direction. We have trace-evidence of the past but not the future, and Broad's hypothesis provides an explanation of why this is so: the times we call "future" are not real as of the times we call "present", whereas the times we call "past" are. However, for now-familiar reasons, it does not follow from this that the world-arrow and the block-arrow *are* in fact pointing in the same direction, let alone that they must do so.

Suppose that we lived in a content-symmetrical Gold universe, of the sort discussed in §4.6. If we further suppose that our universe is a growing block, we naturally assume that world-time and block-time coincide, and hence that the block was smaller in the (world) past. Let us suppose that we are right about this. Now contemplate the fate of the yet-to-exist inhabitants of the other half of our universe, whose world-time runs in the opposite direction to our own. These unfortunates will also naturally assume that block-time and (their) world-time are running in the same direction, but they will be mistaken. Their world will be constructed in reverse: the block is expanding *into* their past. Events they will regard (quite properly, by the criteria of their world-time) to be in the future will have occurred at earlier block-times, and events they regard as having occurred in the past will in fact be unreal (they have yet to occur in block-time). Since there is absolutely nothing in the experience of these people to suggest that this is the way things are, for all we know *we* might be in a precisely similar condition. Perhaps it is *we* who have futures set in stone and unreal pasts.

As this example clearly illustrates, the apparent direction of time is grounded in the content-asymmetries that underpin the world-arrow, but there is no logical guarantee that the world-arrow and the block-arrow point in the same direction, and if the laws of nature in our universe are time-reversible, there is no nomological guarantee either. And this opens the way for the perturbing possibility that the large-scale structure of our universe might be very different from how it appears: we might have unreal pasts but real futures!

The version of the growing block theory developed by Tooley rules out this possibility. Tooley believes that there are powerful positive reasons for believing the growing block theory to be true; reasons that go beyond "intuitive appeal". His central argument (in rough outline) runs thus:

1. Events in our world are causally related.
2. The causal relation is inherently asymmetrical. Effects depend on their causes in a way that causes do not depend on their effects.
3. This asymmetry is only possible if a cause's effects are not real as of the time of their cause.
4. Causes occur before their effects: "X is earlier than Y" means (roughly) that some event simultaneous with X causes some event simultaneous with Y.
5. Our universe must, therefore, be a growing block.[5]

Tooley can thus deny the equivalence between the growing and shrinking descriptions of a dynamic block universe. Descriptions such as 1 and 2 (p. 75), which mention only the fact that the sum total of reality is different as of different times, are not complete, since they make no mention of the pervasive causal asymmetry, which runs right through and along the times and events that comprise our universe. It is this causal asymmetry that provides time with its direction, and ensures that the universe really is growing rather than shrinking.

I cannot attempt a proper evaluation of Tooley's complex arguments here, so I will restrict myself to a couple of brief remarks. Even assuming that Tooley's claims about the causal asymmetry are correct, is there any guarantee that the directions of block-time and world-time will coincide? It might seem so, since the directions of block-time (as defined by increasing size) and causation are locked together. But in fact, the issue is far from decided. If the physical laws governing our world are time-symmetrical (as they may well be), then given the appropriate initial conditions (e.g. those obtaining at the central region of a symmetrical Gold universe) these laws could generate a time-reversed version of our world. Or rather, they could generate a universe whose world-time runs in the opposite direction to causal-time in Tooley's sense. Tooley's causal relation may provide time with a direction, but if all physical processes are time-reversible, this "causal-arrow" is of an invisible (or noumenal) variety, since it is wholly independent of observable patterns of change and variation, and hence of *causation* as we actually employ the notion. If our universe is of the Gold type, our time-reversed counterparts will take causation to run in the opposite direction from us: from our point of view, what they call "causes" occur later than their effects. According to Tooley, the *real* direction of causation may be as we describe it, or as they describe it, but one of us is wrong. Tooley's view of the causal relation may not be incoherent, but it is certainly suspect.

In a more speculative vein, proponents of growing block models should consider this question: is there any reason to rule out the possibility of a block that is expanding in both directions at once? If new slices of reality can be created, why suppose that they can only be created at one end of a dynamic block? Is there any a priori reason to suppose that a universe cannot grow outward in two directions from a common core? In a *doubly* dynamic world of this sort there is no reason to suppose that causation, even as conceived by Tooley, runs in only one direction. But if such a universe is possible, we can consider it from the other temporal perspective, according to which the universe is shrinking from both ends. The continual annihilations of the shrinking universe are, we can suppose, governed by a principle of this form: let us say that adjacent events are D-related; unless an event is D-related to other events at adjacent times in both temporal directions it undergoes spontaneous annihilation. Since, by hypothesis, the "outermost" events in either direction are *not* D-related on one side, they undergo annihilation, thus exposing the next innermost events, which in turn undergo annihilation, and so on. Tooley envisages a universe that is *positively unstable*, a universe that spontaneously increases by a process of continual creation. I am proposing a universe that is *negatively unstable*, a universe that spontaneously *decreases* by a process of continual annihilation. What reason is there for favouring one vision over the other?

6.6 The eternal past

Quite independent of the considerations just advanced, there is an objection to growing block models of a very obvious nature. If we assume that the world- and block-arrows point in the same direction, to a substantial degree the growing block model agrees with the natural view of time: the future is unreal, the present is distinctive, the past is fixed. But someone might argue thus:

> When an event becomes past it is *gone*, it has ceased to exist. Napoleon doesn't just not exist at the present time, Napoleon has ceased to exist *completely*. But if the growing block model is true, everything that existed in the past is just as real as what exists now. This is plainly false.

The problem is the status of the past. Most of us believe that there is a definite and unalterable fact of the matter as to what happened at every point in the past. Many of us also believe that the past has utterly ceased to exist. The growing block theorist denies this.

The permanence of the past might also provoke a concern of this sort:

> From our perspective, there are thousands of years of human history; these past times are just as real as our time. The people living at these past times believe that their time is present, but we know they are wrong; they are definitely in the past. But, by all appearances, there is nothing to distinguish our time from past times: there is no distinguishing property of *presentness* that past events lack. So given that lots of people in the past have experiences just like ours and are *wrong* about living in the present, isn't it possible that *we* are wrong when we believe we are living in the present? Indeed, wouldn't it be an astonishing coincidence if the present were located *now*? Isn't it more likely that we are somewhere in the middle of the growing block? Worse still, might it not be that our block has finished growing? In which case even now our world is indistinguishable from a static B-world. Either way, we are eternally embedded in a static past.

This objection is flawed, but it also contains more than a grain of truth. If our universe does have a finite future, and is also a growing block, it will not spend longer (an eternity) in its completed state, since time will cease to pass with the creation of the final events. Moreover, if we live in a growing block world, we can be certain that we exist at a time that is present: since every event is present at the time it occurs, for any event E at t, as of t, E is present. However, this doesn't fully satisfy the intuitive sense that the present is privileged. For although E at t is present as of t, for any time t^* that occurs later than t, as of t^*, t is not present but in the past; t and its contents are still real, but are no longer present. Although every momentary event enjoys a momentary privilege of being present, every moment remains in existence while being buried ever more profoundly in a static and eternal past. Although it is true to say that as of *this* time the future is unreal, it is also true that the events that happen at this time will remain in existence, just as they are, if and when the future becomes real.

Anyone who finds the prospect of being permanently (or timelessly) entombed in the past distasteful (or even terrifying) will be attracted to the *presentist* conception of time, according to which both past and future are unreal.

6.7 The varieties of presentism

Presentism is the doctrine that nothing exists that is not present. Presentism amounts to a dynamic view of time if this claim is combined with another: reality consists of a *succession* of presents.

Although there are signs of renewed interest in the doctrine, it has not featured prominently in recent philosophical discussions.[6] The self-professed presentist John Bigelow puts this down to the impact on the modern mind of the four-dimensional view:

> Presentism was assumed by everyone everywhere, until a new conception of time began to trickle out of the high Newtonianism of the nineteenth century. The Christian's Holy Bible says that there is no new thing under the sun but this is not true . . . The so-called fourdimensionalist theory of time was something genuinely new when it gradually came into being last century.
>
> (1996: 35)

Once one becomes aware of the static block view, and the scientific considerations (especially Einstein's special theory of relativity) that can be mustered in its favour, denying reality to everything but the present can easily come to seem absurd. Bigelow offers the following observation by way of support for his claim:

> the exponential explosion of timetravel stories in the popular media, beginning late in the nineteenth century, is an indication that a very new conception of time is brewing in the Zeitgeist. The utter absence of any timetravel stories whatsoever prior to the nineteenth century is a profoundly puzzling fact . . . I suggest that this is at least partly explained by the utter universality of presentism prior to the nineteenth century and by the utter absence of any rivals to presentism.
>
> (1996: 35–6)

If you believe that the past is wholly unreal, you would only push a button that would send you backwards in time if you were of a suicidal frame of mind. It may well be that we are no longer *astonished* by the very idea of backward time travel in the way that our predecessors once were. Even if many of us also have doubts about whether it really is possible to travel in this way, the fact that time travel is a commonplace in contemporary science fiction may have produced a change in attitudes towards the past, even if only to the extent of making presentism seem less obviously true than it once did.

In any event, in what follows we will be focusing on the metaphysical issues surrounding presentism. We shall see that the doctrine comes in different forms, some a good deal more believable than others. But before proceeding it is worth mentioning two of the issues that any presentist must confront. The first is familiar: how long is the present? As noted in connection with the growing block model, the least problematic answer is that the present is strictly momentary, or of finite but very brief duration. The second is new. A prime motivation for presentism is the desire to stay close to common sense, but common sense tells us that, even if the past isn't real, there are definite *facts* about what happened in the past. How can this be if the past is unreal? If the past doesn't exist, how can there be true statements about it? The problem is especially acute if we subscribe to the *truthmaker principle*: any contingent statements about the world that are true are true by virtue of something in the world that *makes* them true – their truthmaker. If the past doesn't exist, then the truthmakers for statements about the past don't exist, so it seems that all statements about the past must be false, or at least that none can be true. Needless to say, this is radically counterintuitive.

Faced with this problem, the presentist could simply bite the bullet and accept that there are no true statements about what happened in the past, but this is very hard to accept. Another option is to hold that statements about the past do have truthmakers, but these truthmakers all exist within the present. If I say "Dinosaurs once walked on the Earth", what makes this sentence true is not the existence of dinosaurs in the past, but rather the presently available evidence for thinking that there used to be dinosaurs, such as fossils. This *reductionism* about the past has two unpalatable consequences. First, the past becomes inconstant. If we suppose (as most presentists do) that there is more than one present, if different presents include different traces of the past – as they surely would – these presents have different pasts. Secondly, since facts about the past are constituted by what exists *now*, the truth about the past is restricted by whatever relevant evidence now exists. If we take "evidence" here to mean traces that would enable human investigators to reconstruct accurately a past occurrence, the consequences of this restriction are dramatic: the bulk of the past would vanish. The claim that only a minuscule percentage of dinosaurs left fossilized remains would be false, since the *only* dinosaurs that existed would be those that left fossilized remains!

But we need not read the restriction in this way. We might reasonably hold that the past includes everything that a being with god-like powers of observation and calculation could infer about it, given a complete description of the present state of the world down to the sub-atomic level. The truths about the past are not confined to what *we* can discover or infer. From this standpoint, if our universe is deterministic, the totality of facts about the past available to the realist and the reductionist may well coincide. But if, as currently seems more likely, our universe is not deterministic, even a being with god-like powers might be unable to infer very much about what happened even a short while ago. The same will apply (at least if we downgrade our imaginary being's powers somewhat) if chaos theorists are right in their contention that the occurrence of macroscopic phenomena can depend on minuscule variations in initial conditions; the so-called "butterfly effect".

The idea that the past is dependent on what exists in the present may seem absurd and objectionable, but some presentists have looked on the bright side:

> Facts whose effects have disappeared altogether, and which even an omniscient mind could not infer from those now occurring, belong to the realm of possibility. One cannot say about them that they took place, but only that they were *possible*. It is well that it should be so. There are hard moments of suffering and still harder ones of guilt in everyone's life. We should be glad to be able to erase them not only from our memory but also from existence. We may believe that when all the effects of these fateful moments are exhausted, even should that happen only *after* our death, then their causes too will be effaced from the world of actuality and pass into the realm of possibility. Time calms our cares and brings forgiveness.
>
> (Lukasiewicz 1967, quoted in Tooley 1987: 235–6)

This last thought is consoling, but is it one that we can believe? We shall be encountering other presentist responses to the problem of the past in due course.

The claim that nothing exists that is not present is a slippery one, for it can be interpreted in different ways; presentism comes in different forms. Not all of these presentisms constitute dynamic models of time, but by taking a brief look at these non-dynamic forms we not only improve our appreciation of what the presentist standpoint involves, we can also see more clearly the problems that face dynamic forms of presentism.

6.8 Solipsistic presentism

The most radical version of presentism is the doctrine that nothing exists that is not present and that only one present ever exists; *this* one. This "solipsism of the present moment", as it is sometimes called, is invulnerable to decisive refutation but impossible to take seriously.

If only *this* present ever exists we are left with the problem of explaining why it *seems* that we have a past. Why do we have memories of a recent past? Why are there so many *signs* of a past if the past doesn't exist, and never has? There is a general principle of reasoning, "argument from best explanation", according to which it is reasonable to believe the best explanation of a phenomenon. Our world contains all the appearances of having had a long past; isn't the best explanation of these appearances that *there was a past*?

More generally, if only this thin reality-slice exists, why is it *complex and organized*? The big bang theory of the origin of the universe, coupled with the theory of natural selection, explains why the world is at it is: how the current complexity developed from initial simplicity. There is still a lot that is mysterious – we do not know why the big bang occurred, why the initial energy existed, why the laws of nature are as they are – but there is a good deal that is explained. By contrast, the doctrine that the universe consists of a single brief slice of highly organized reality, which simply popped into being from nothing, explains nothing and leaves everything mysterious.

These points provide good reasons for rejecting solipsistic presentism. Bearing them in mind, let's move on.

6.9 Many-worlds presentism

The presentist says that nothing exists that is not present, but the presentist is not obliged to say that only one present exists. One option for "pluralist" presentists is to say that, although only what is present exists, there exists a succession of presents, with each present departing from existence to make way for its successor. I will examine this view in §6.10 and §6.11. A second option is to say that reality as a whole includes many static presents that are *not* temporally related to each other, and so do not succeed each other in any way. Since each of these presents is a very brief self-contained world in its own right, I call this doctrine "many-worlds presentism". (It should be noted that this version of presentism is quite distinct from the "many-worlds" interpretation of quantum theory, although there may be a connection, as we shall see shortly.) For the rest of this section I will call the sum total of these many worlds "REALITY".

Since, on this view, REALITY consists of a timeless ensemble of momentary worlds, it makes no sense to suppose that new worlds can come into being in time; REALITY encompasses all the worlds that there will ever be, and so cannot grow. This kind of many-worlds view is thus static rather than dynamic (in the manner of the growing–shrinking block), but it is also distinct from the position of the static block theorist. The latter also holds that there are many presents, at least to the extent that every event in the block is present at the time when it occurs. But the block theorist also holds that all these presents are temporally related to one another, which is denied by the many-worlds presentist. Furthermore, if we suppose that things belong to the same "world" if they are spatially and temporally related, then anyone in an extended static block who says "My world consists of only what is happening *now*" is saying something false. This claim is true if uttered in one of the short-lived worlds posited by the many-worlds presentist.

Despite these differences, there are similarities between the two positions.[7] Block theorists hold that past, present and future are equally real. So, too, does the many-world presentist, after a fashion. If the solipsist presentist wants to admit truths about past events, reductionism is the only option. The pluralist presentist is not so restricted; since other presents exist, there are potential truthmakers for statements about other times. Depending on the precise contents of REALITY, there may even be truthmakers for all those statements about the past (or future) that common sense tells us either are true or could be true.

The easiest way to appreciate how this might be so is as follows. Start by imagining that we do have a past, and a future, and that all coexist within a static block. Now imagine this block carved into many brief slices, each corresponding to the condition of the world at a particular time, and transport these slices into timeless REALITY; to capture the fact that there are no spatial or temporal relations among the slices, imagine that they are randomly shuffled like a deck of cards. Next, consider all the ways that the slices *could* be brought together and stacked; each of these imaginary "decks" corresponds to a different possible history. We know (or believe) that, in our world, for the most part change is gradual and orderly: we don't find whole cities vanishing or suddenly appearing, furniture stays where it is put, cats don't change into frogs. Consequently, any ordering of the slices that fails to capture this continuity will not be an accurate representation of our history. Indeed, it is perfectly plausible to suppose that if we start from our present, and find those slices that resemble it most closely, and place them in the order that minimizes overall differences between immediate neighbours, we would find ourselves with a unique ordering of slices, an ordering that would correspond with what we believe about our past and (although this is less certain) anticipate about our future. If we extend this "best match" method, and try to order *all* the slices, we might well be able to reconstruct the block we started with in its entirety.

Reductionism is thus avoidable. A statement about the past (or future) is true if the following conditions are met:

- REALITY contains one or more slices that the statement accurately describes.
- The time at which the events are said to occur corresponds to the location of the slice(s) in the ordering generated by the "best match" method.

Of course, while it is *possible* that REALITY includes slices that contain the events we believe to have occurred, there is no guarantee that it does; the past may be more impoverished than we believe. But it may also be far richer; REALITY may contain many more presents than are required to build our history (and future). It may contain enough presents to build billions of world-stories, some similar to our own, some very different. Since the best-match procedure might well situate your current present in more than one world-story, you might well have several different pasts, and several different futures. Indeed, if REALITY is rich enough, all your dreams will come true. But this boon comes at a price: so, too, will your worst nightmares.

The ability of many-worlds presentism to accommodate (albeit in its own distinctive way) our common-sense beliefs about the past means that it is far easier to accept than solipsistic presentism. But in another respect it is drastically counterintuitive. I for one find it hard to believe that I am living a life that is distributed across many different worlds, and that I spend only the briefest of times in any one world. The presentist could, of course, respond by saying that all parts of my life coexist in REALITY, and it is just that these parts are not related (spatially, temporally, causally) in the ways usually assumed.

It may seem that the many-worlds view is vulnerable to the objection levelled at the solipsist. If it is mystery-mongering to suppose that REALITY consists of just one complex reality-slice containing all the appearances of having a long and complex past, isn't it even more problematic to suppose that REALITY consists of *many* such slices, all independent of one another? But the presentist could reasonably respond thus: "Is the existence of these many unconnected 'presents' any more mysterious than the existence of the many connected 'presents' that constitute the static block?" This point aside, pluralist presentists have a further card up their sleeve: developments in contemporary physics.

In *The End of Time* (1999), Barbour tries to persuade us that time does not exist. What he really means is that time is not like we commonly think, for the positive conception of reality he defends (so far as I understand it) is very close to that of many-worlds presentism. What is relevant to the objection that we are considering is that Barbour provides an explanation for why REALITY (he calls it *Platonia*) should contain complex world-slices of the sort that we are acquainted with. Given any physical system, the basic quantum laws – enshrined in Schrödinger's wave equation – tell us how that system will evolve over time. In fact, the equation delivers many different *possible* ways that the system might evolve, and provides no way of distinguishing which of these possible evolutions will be actual. Adherents of the "many worlds" interpretation of quantum theory take the view that it is a mistake to suppose that just one of these solutions to the equation becomes actual; they maintain that they are all equally real, and that as a consequence reality has a branching structure, and is vastly richer than we normally assume. Barbour himself adopts this many-worlds interpretation – applied to a Schrödinger equation specifying all the (physically) possible ways the universe could have developed from an initial big bang – and then takes a further step.

We know that standard quantum theory cannot be complete since it is not integrated with Einstein's general theory of relativity, which is currently our best theory of gravity. There have been many attempts to bring relativity and quantum theory together, but most attempts have failed. One of the few noteworthy (although limited) successes is the *Wheeler-DeWitt* equation.

> . . . what is this equation, and what does it tell us about the nature of time?
> The most direct and naïve interpretation is that it is a stationary Schrödinger equation for one fixed value (zero) of the energy of the universe. This, if true, is remarkable, for the Wheeler-DeWitt equation must, by its nature, be the fundamental equation of the universe . . . [it] is telling us, in its

most direct interpretation, that the universe in its entirety is like some huge molecule in a stationary state. (Barbour 1999: 247)

The universe, thus represented, is *still*: it does not evolve for it does not change over time (since time is absent), and so change itself is rendered illusory (or at best emergent). This "timeless" universe is nonetheless richly structured, for it comprises an infinite number of momentary three-dimensional "worlds", one for each point of the phase space generated by the Wheeler-DeWitt equation.[8] Barbour argues that we have the impression that we live in an evolving universe because there are momentary fragments of this timeless whole (those we inhabit) that contain apparent traces of earlier and later occurrences. But in reality there is no "earlier" and "later", since these three-dimensional world-fragments are temporally unrelated to one another.

As Barbour concedes, the Wheeler-DeWitt equation is surrounded by controversy. Nonetheless, the "canonical quantum gravity programme", which delivers this equation, is one of the leading contemporary approaches to quantum gravity.[9] Barbour is certainly not alone in believing that this programme poses a threat to time, for as we saw in §3.8, it is widely accepted that it does; Kuchař writes of the "quantum calamity, known as the *problem of frozen time*" (1999: 178).

What *is* controversial is Barbour's response to this "time problem"; other theorists are optimistic that they will find time "hidden" or "buried" in the Wheeler-DeWitt equation.[10] But it is clear that from the perspective of physics (if not common sense), his view is by no means absurd. If something akin to the Wheeler-DeWitt equation provides the most accurate and fundamental representation of our world, then perhaps, as Barbour urges, our world does consist of a vast ensemble of temporally unrelated "moments", each of which consists of a three-dimensional configuration of physical bodies.

Needless to say, there is much that is speculative and controversial here, but the conclusion is nonetheless surprising: many-worlds presentism is a doctrine that at least merits serious consideration.

6.10 Dynamic presentism

The version of presentism that has greatest intuitive appeal is a combination of the following claims:

- nothing exists that is not present;
- at any time only one present exists;
- each present is followed by another;
- successive presents are causally related.

This doctrine, *dynamic presentism*, does not suffer from the problems that afflicted solipsistic presentism; if the current present was causally influenced by the immediately preceding present, which in turn was causally influenced by the present that immediately preceded it, and so on back, then we have a good explanation of why it seems that our world has a past, and how our world came to be. Since each present is related, temporally and causally, to past and future presents, the dynamic presentist's world is not the scattered unconnected collection posited by the many-worlds theorist.

However, this version of presentism seems deeply problematic. Causation is a two-place relation, and the form of causation needed to connect successive presents relates events (or objects) existing at different times. According to Bigelow:

> a two-place relation can only be manifested when it holds between two things, and in order for this to be so there must be two things which stand in the relation. And in saying "there must be two things which stand in the relation", one is really asserting that "there must exist" two things – one is committed to the existence of these things. The principle of the existence entailment of a relation is an a priori truth.   (1996: 39)

The claim that a two-place relational property can only hold between two things that exist – and hence that relations are "existence-entailing" does seem plausible, especially if the claim is restricted to what are sometimes called *concrete relations*, the sort of relations that are non-intentional and that can only hold between spatiotemporal (rather than abstract) things, relations such as temporal priority and causation. If relations of this sort are construed as genuine connections holding between the items they relate, and this is a natural way to construe them, it is hard to see how they can be instantiated unless both of the objects they connect exist. In which case, it is absurd to suppose that something that exists can be directly causally related to something that does not exist. But this is precisely what the dynamic presentist is committed to. If the causal relation is existence-entailing, then so, too, is temporal priority if the latter notion is analysed in causal terms. Even if we reject such an analysis, a strong reason for supposing that temporal priority is existence-entailing remains. If E_1 is earlier than E_2, both E_1 and E_2 must belong to the same world. But if, as the dynamic presentist maintains, E_1 is non-existent as of the time when E_2 occurs, how *can* both these events belong to the same world? Surely non-existent events are not parts of any (actual) world; this is precisely why they are non-existent. Ordinarily we draw a distinction between those possible times that *were* actual, and belong to our past, and those that are *merely* possible and that do not belong to our past (or future). If only one time is real, and this time lacks a genuine connection to any other time, it is hard to see how this distinction can be sustained. In which case, the notion that we have a unique past and future must be abandoned. Presentists, it seems, lack the resources to bind different times into a single universe.

If we accept that causality and temporal priority are existence-entailing, dynamic presentism is vulnerable to what Bigelow calls the "argument from relations" (1996: 37):

1. If dynamic presentism is true, then events in earlier presents are causally related to events in later presents.
2. Dynamic presentists also claim that only one present is real.
3. Causation is existence-entailing.
4. It follows that if one present is causally related to another, both must be real.
5. Hence 2 is false, and so is dynamic presentism.

Since an analogous argument could be stated in terms of temporal priority, it seems clear that the presentist is confronted with a disastrous dilemma. Either there are no genuine relations between successive presents, in which case dynamic presentism collapses into many-worlds presentism (if not solipsism), or there *are* genuine relations between different presents, but since these relations are existence-entailing, dynamic presentism collapses into the static block view.

6.11 Compound presentism

By way of a response to the preceding argument, the dynamic presentist could try to persuade us that relations such as causation and temporal priority are not existence-entailing after all. While it would be interesting to see what could be done along these lines, in what follows I will briefly explore an alternative route. It may well be dynamism can be salvaged, albeit at the cost of a small (but significant) compromise.

Dynamic presentism is vulnerable to the "argument from relations" because, in effect, it is indistinguishable from solipsism of the present moment. So why not opt for a compromise, and allow that more than one present can coexist? William James once characterized our immediate experience of time in this way:

> Objects fade out of consciousness slowly. If the present thought is of A B C D E F G, the next one will be of B C D E F G H, and the one after that of C D E F G H I – the lingerings of the past dropping successively away, and the incomings of the new, are the germs of memory and expectation, the retrospective and the prospective sense of time. They give that continuity to consciousness without which it could not be called a stream. (1952: 397)

Irrespective of the merits of this as a description of our consciousness of time, might it not be that time itself has a structure of this sort? In James's example, seven distinct elements coexist, but such extravagance is not needed; the required overlap requires only two elements. An approximate description of a world of this sort might run thus:

> The sum total of reality consists of at least two coexisting very brief reality-slices (each spatially three-dimensional). Suppose A and B are two such, and that A exists at t_1 and B at t_2. One of these slices, A, is annihilated and a new slice of reality, C, comes into existence, and with it a new time, t_3. Slice B is annihilated and D is created, along with t_4; and so it goes on.

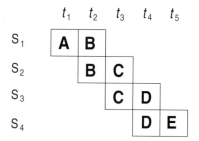

Figure 6.5 Compound presentism. A doubly dynamic reality: the sum total of reality is not limited to a single present, and absolute becoming is matched by absolute annihilation.

In Figure 6.5, S_1–S_4 refer to distinct sum totals of reality, and t_1–t_5 to distinct times. Not all of these times coexist (temporal dynamism cannot be fully captured in a static diagram). Two questions need to be kept distinct:

1. What events occur *at* t_2?
2. What events are real *as of* t_2?

The answer to 1 is "the events in B", whereas the answer to 2 is "the events in A, B and C". The latter answer may seem odd. How can the events in A, B and C be real as of t_2 if there is no sum total of reality that includes each of A, B and C? But of course, a "sum total of reality" consists of a collection of times and events that (timelessly) coexist, and A and C do not coexist in this fashion, despite the fact that there is a sum total comprising A–B, and another comprising B–C.

There is a second confusion to avoid. Looking at Figure 6.5, is easy to fall into thinking that two things happen at t_3: C comes into being just as A is annihilated. This is wrong: t_3 and C come into being together (thus avoiding overdetermination), and this instance of "becoming" does not occur simultaneously with the annihilation of A. The latter "occurrence" does not happen *at* any time at all, least of all t_1 (which is annihilated along with A, again avoiding overdetermination). However, although the annihilation of A does not happen at t_3, there is a sense in which it then "becomes real", since the sum total of reality *as of* t_3 does not include A, whereas the sum total as of t_2 does.

I will call this position "compound presentism". Since neither A nor B (nor any other reality-slice save for the first and last, if there are such) exists all by itself, neither is a "present" in the strict sense of the dynamic (or solipsist) presentist. Nonetheless, the model we are considering (in this guise at least) is much closer to that of the dynamic presentist than the static block, so the label is not without its merits. To avoid confusion, I shall refer to sum totals such as [A–B] and [C–D] as "extended presents".

Compound presentism has one significant advantage over its more isolationist kin: it is immune to the "argument from relations". If two reality-slices A and B coexist, there is no problem in supposing that genuine relations hold between them. But the model is open to another objection, and a potentially fatal one. It is natural to suppose that for any three items, X, Y and Z, if X coexists with Y, and Y coexists with Z, then X coexists with Z; that is, that the relationship of coexistence is transitive. If coexistence *is* transitive, compound presentism is incoherent; if B coexists with both A and C, then A and C must coexist with one another, and the same would apply to A and D, and also to A and E. We are back to the static block.

This problem is insuperable, but only if we assume that coexistence is transitive in all contexts. The compound presentist can reply:

Coexistence is certainly *symmetrical*, but it needn't be transitive; to suppose otherwise is simply to deny the dynamic nature of time, which involves precisely the coming-into-being and departing-from-being of times and events. Co-existence is transitive in *space* (at a given time), but to impose transitivity on time amounts to an unjustifiable spatialization of the latter.

This reply should be treated with respect. The only obvious way that coexistence *could* fail to be transitive is if events can cease to exist and come into being. The

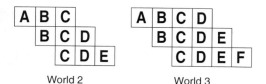

World 2   World 3

Figure 6.6 **Extended presents of different durations.** The duration of the extended present provides the compound presentist with a measure of temporal dynamism that does not require an external meta-time.

failure of transitivity is *inevitable* if the passage of time involves annihilation and creation, a doctrine long ago advocated by Heraclitus: "This world neither any god nor man made, but it always was and is and will be, an ever-living fire, kindling in measures and being extinguished in measures" (quoted in Barnes 1982: 61).

Thus far we have been considering a case in which the extended present has two very brief parts, but there are many other possibilities. Two are depicted in Figure 6.6.

These worlds are indistinguishable in all respects save one: the extent of their extended presents. The events *within* these extended presents are of precisely the same durations in both worlds, but the periods of time relative to which a given moment of time is real are different.

Since the longer a world's extended present is, the closer that world is to a static block, there is a sense in which worlds with shorter extended presents are "more dynamic" than worlds with longer extended presents. One might even go so far as to say that the length of the extended present provides a measure of the rate of time's "passage". But since this "rate" is determined solely by relationships of coexistence within a world there is no need to posit an external "meta-time" to render it intelligible.

The status of the past in compound presentism is ambiguous, since "the past" comes in two forms: the *real* and the *unreal*. In our original case, if we assume that t_1 occurs earlier than t_2, which occurs earlier than t_3, then A belongs to the real past of B, but the unreal past of C. More generally, the real past of a time t consists of those earlier times and events that coexist with t in a single extended present, whereas the unreal past of t consists of those earlier times that do not coexist with t, but are linked to t by what we can call "indirect coexistence". Two times are linked by indirect coexistence if they are at either end of a succession of times, each of which is related to its neighbours by direct (genuine) coexistence (i.e. they are linked by a succession of partially overlapping extended presents).

The status of the real past is unproblematic; it is precisely the same as in the static block model.[11] Of course, there is the additional complication that the extent of the real past can be different in different worlds; moreover, in worlds where the extent of the extended present is inconstant, different times within the same world will have real pasts of different lengths. But these complications are irrelevant to the ontological status of the real past: it is just as real as the (extended) present.

But what of the unreal past? How can statements about the unreal past be true, given that the relevant truthmakers no longer exist? Can reductionism be avoided? It may be that it can. With reference to our original case, consider some event E, which occurs in A at t_1, and let P be some (true) proposition about E. Since t_2 coexists with t_1, P is unproblematically true at t_2. But what of t_3? As of t_3 both t_1 and the events that occurred at it are no longer real, so P lacks a truthmaker. However, t_2 falls within the real past of t_3, and this makes a difference. Suppose that we say the following:

What is true (or factual) as of a time *t* is also true (or factual) as of all other times that coexist with *t*.

Call this the "factual inheritance principle". If we subscribe to the factual inheritance principle then since P is true as of t_2, and t_2 coexists with t_3 in a single extended present, then P must also be true as of t_3, even though the relevant truthmakers are no longer real as of this time. The factual inheritance principle allows truths (or facts) about earlier times to *accumulate* at later times, and thus be transmitted *over* time. Everything that is true (or factual) as of t_3 is also true (or factual) as of t_4, and the same applies to all later times. The reality of the past is thus secured.

This argument, if successful, shows that there can be facts about times and events that are no longer real. It doesn't follow that all these facts are known or knowable. Indeed, this is one of the main advantages of this form of presentism: reductionism and anti-realism about the past are avoidable. The key assumption is the factual inheritance principle; it is thanks to this principle that we are entitled to maintain that facts about the immediate past do not cease to obtain when the past they concern departs from reality. Is this principle true? It is hard to see how to argue for it, but there is no denying that it is plausible, for it corresponds with common sense, although, as we shall now see, it is not as innocuous as it may appear.

Compound presentism seems to offer dynamists everything they could want but, as with the growing block model, on closer examination it turns out to be a good deal more hostile to common sense than first appearances suggest. I have taken the direction of time for granted. I assumed that A occurs *earlier* than B, that C comes into existence *later* than A, and so forth. But this assumption is entirely unjustified. There is nothing in the simple universe described above that supports the assumption. Suppose we follow Broad and say:

> If event X *precedes* event Y, then there is a time at which the sum total of existence includes X but not Y, and a time at which the sum total of existence includes both X and Y.

Reading Figure 6.5 from left to right and top to bottom, this yields the result that A occurs earlier than B. But reading from right to left, bottom to top, Broad's definition yields the result that E occurs earlier than D, since there is a time at which the sum total of existence includes E but not D, and a time at which the sum total of existence includes both E and D. If there is a temporal direction in this universe, it is not due to the pattern of annihilation and creation which constitutes the dynamic of time itself. This pattern is perfectly symmetrical, as is clear from the even simpler case shown in Figure 6.7. This universe consists of four times, whose contents are: a triangle; a triangle and a square; a square and a circle; and a circle. Nothing in this pattern of occurrences justifies thinking that these different states of affairs unfold in one temporal direction rather than the other.

If compound presentism is true, the sum total of reality is different as of different times, and the sum totals at adjacent times partially overlap, but this pattern of overlap is symmetrical in both temporal directions. If you see two adjacent lights flashing on and off quickly [on, off], [off, on], [on, off], [off, on]

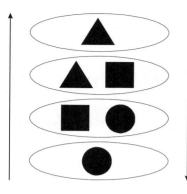

Figure 6.7 Dynamism without direction. In a compound presentist world, the sum total of reality is different as of different times, but in cases where the pattern of becoming and annihilation is symmetrical, the resulting dynamism lacks an inherent direction. Nothing in the pattern shown here favours the hypothesis that this (very simple) universe is developing in the up-direction rather than the down-direction, and vice-versa.

. . ., after a short while you have no sense of the bulb on the right being on *before* the bulb on the left; there is just an impression of rapid *alternation*. The dynamic model of reality offered by the compound presentist is akin to this. There is an alternation of times, but no *passage* of time, at least if passage is taken to be a process that unfolds in a particular direction. Consequently, any apparent direction within time must be a product of the asymmetries that exist among the contents of time, contents that are themselves sustained by a perfectly symmetrical process of creation and annihilation.

The symmetry inherent in the compound presentist's model may not seem disturbing in itself, but it has an implication that is. The argument sketched out above for realism about the past can be run in the other direction, to reach the conclusion that the factual domain of any time *t* includes everything that occurs *later* than *t*: the factual inheritance principle transmits factuality in both directions. If statements about the past have determinate truth-values, so, too, do statements about the future. It is a surprising irony that this most dynamic model of time may have precisely the same implications for future truth as the static block.

To this interesting result we can add another, the principal conclusion of this chapter: there may well be alternatives to the static block model of time.

Many-worlds presentism is the least dynamic of these alternatives. Instead of a single connected block universe we are presented with a fragmented block universe, consisting of unconnected worldlets. The many-worlds presentist could be construed as offering a dynamic conception of individuals, since as a continuant persists through time it (in effect) jumps from world to world. But since the worlds in question constitute a timeless ensemble, the jumps in question are of a singularly undynamic nature. It would be more accurate to say that persisting individuals are *scattered* entities, scattered across spatiotemporally distinct worlds. However, two genuinely non-static conceptions of time have emerged: the lopsided dynamic block; and compound presentism, with its shimmering alternations of times. If these models are metaphysically intelligible, to discover the nature of time in our universe we will have to look beyond metaphysics. Perhaps there are empirical considerations that will decide the issue (we will be looking at one such in Chapter 7).

There are two further conclusions to note. First, a dynamic model of time does not require transitory tensed properties; no such properties are to be found in any of the dynamic models that have emerged as legitimate contenders. Secondly, the questions "Is time dynamic?" and "Does time have a direction or arrow?" have to be kept distinct. At first view, time in a growing block universe seems to possess an inherent and directed dynamic of its own, a dynamic generated by the creation of times and events, but on closer scrutiny this inherent directedness evaporates. The same applies in the context of compound presentism: the dynamism generated by the creation and annihilation of times and events does not "run" in one way rather than another. Consequently, if time has a direction, in any significant sense of the term, this is due to the contents within time and their modes of connection, not time itself.

7 Time and consciousness

7.1 The micro-phenomenology of time

In this chapter I redeem a pledge made in §3.2 and take a closer look at our experience of time. I will not be concerned with what we believe and feel about time and its passing at different stages of our lives; my focus will be on the character of our experience from moment to moment within our streams of consciousness. My approach will be phenomenological: the task is the tightly circumscribed one of trying to formulate a *description* of our short-term experience of time that is clear, accurate and intelligible. Other interesting problems, such as how our brains manage to perform the impressive feat of integrating the inputs from our different senses to produce a real-time representation of our usually changing surroundings (the typical time-lag between initial stimulation and experience is under half a second) will be ignored.

The micro-phenomenological topic is of interest in its own right – arriving at a clear account of the temporal features of conscious experience is a non-trivial task – but it may also have wider implications, for it may well be that our experience of time reveals something (if not everything) about the nature of time itself. Of particular interest here is the B-theory. Many have voiced the suspicion that time as manifest in consciousness possesses dynamic features that cannot be explained with the resources available to the static block theorist:

> Should we simply shrug the human experience of time aside as a matter solely for psychologists? Does the time of an altered state of consciousness have no relevance at all to the time of Newton or Einstein? Does our impression of the flow of time, or the division of time into past, present and future, tell us nothing at all about how time *is* as opposed to how it merely appears to us muddle-headed humans?
>
> As a physicist, I am well aware how much intuition can lead us astray . . . Yet as a human being, I find it impossible to relinquish the sensation of a flowing time and a moving present moment. It is something so basic to my experience of the world that I am repelled by the claim that it is only an illusion or misperception. It seems to me that there is an aspect of time of great significance that we have so far overlooked in our description of the physical universe.[1]   (Davies 1995: 275, original emphasis)

Before we are in a position to judge whether this is so or not we need a clear idea of what these features are.

Let us start by gathering some basic phenomenological data. Our typical experience is a combination of persistence and change. You are, let us suppose, lying in a deckchair gazing at the sky, where you can see a few white clouds moving very slowly against the unvarying blue backdrop. You see the occasional bird overhead. Sometimes they circle around for a while, and often they fly right on by. When you do see a bird in flight you are directly aware of its motion: the flapping of wings; the sometimes fast, sometimes slow changes of position. You are as fully and immediately aware of these movements and changes as you are of the unchanging blue of the sky. The same holds quite generally for other modes of consciousness. We *see* change, but we also *hear* it (a rapid sequence of notes played on a piano), *feel* it (slowly move your finger across the palm of your hand) and *imagine* it (recreate in your imagination the final shot in a tennis tournament – a ball goes out, a player leaps with joy, the crowd roars). In addition, there can be changes in what we smell and taste, and in our emotional state (a surge of anger). And then there is the inner soliloquy, the conversation we have with ourselves in thought, conducted sometimes in words, sometimes in images, sometimes in neither, which rarely ceases during our waking hours.

All this is obvious: consciousness is alive with change and variation. More elusive is the subtle but distinctive sort of dynamism that is characteristic of *un*changing sensations. Return to the deckchair scenario. For some moments now you have been staring at an empty region of blue sky and nothing has changed. Your inner monologue has (if only briefly) ground to a halt, you have seen no movement, your visual field is filled with an unvarying expanse of blue. But is your consciousness entirely still or frozen? Have you come to a complete stop? No. Throughout this period you remain conscious, and conscious of the blue presence *continuing on*; you have a (dim, background, passive) awareness of the blue constantly being renewed from moment to moment. This passive awareness of continuation and renewal is perhaps more vivid in the case of auditory experience. Imagine hearing a sustained but unwavering note played on a cello: you hear a continuous and continuing flow of sound. This feature – call it "immanent phenomenal flow" – is possessed by all forms of experience (think of the burning sensation on the tongue caused by biting on a chilli pepper), and is a dynamic feature of experience that is independent of changes of the ordinary qualitative sort (the chilli-induced burning is felt as *continuing on* even when its intensity and qualitative character remains constant). Phenomenal flow is especially important in the case of easily overlooked but ever-present bodily experiences: the feeling of one's limbs being disposed one way rather than another, the feeling of being vertical rather than horizontal or upside down, the feelings of pressure on one's skin caused by clothes, seats, shoes and so on. When we are not moving about, our bodily experience is largely unvarying in quality, and so goes largely unnoticed, but it remains a constant – and constantly *flowing* – presence within our consciousness.

This short survey of the phenomenological data suggests that any adequate account of the temporal features of experience must accommodate two facts:

- our experience of change is as direct and immediate as our experience of colour or shape;
- our experiences possess the feature of phenomenal flow.

In what follows I will consider a variety of attempts to account for these features – or rather the first, since the second is not always recognized – and, by locating their deficiencies, move towards a clearer understanding of what they involve.

7.2 Memory based accounts

There is no denying the crucial importance of memory to the macro-phenomenology of time but, as we saw in §3.2, some philosophers (e.g. Mellor) go further and argue that memory is also entirely responsible for the character of phenomenal temporality over the short term. As soon as it is seriously considered this claim starts to look very implausible.

Suppose you hear a fast sequence of notes played on a piano: C–D–E. You hear each note as possessing a short but discernible duration, but you also *hear* the succession: C-being-followed-by-D, and D-being-followed-by-E. How could the memory theorist account for this? One option would be to say "When you hear D, you remember hearing C; when you hear E, you remember hearing D", but this won't do. Hearing D while remembering having heard C is compatible with C having occurred *at any time*, perhaps hours or years ago. To solve this problem, the memory theorist might forward this proposal:

> When I hear D, I remember hearing C, while believing that I have *just heard* C; when I hear E, I remember hearing C and D, and believe that C occurred before D, and that D occurred just before E.

But this won't do either. The problem is the invocation of *belief*. It is true that we can form accurate beliefs about the order and duration of events without using a clock. However, what sort of belief is the memory theorist relying on here? There are beliefs that are merely latent (e.g. your belief that Berlin is the German capital, which you possess even when asleep), and there are beliefs whose content is conscious (e.g. you hear someone ask "Is Bonn still the German capital?" and you think "No, it's Berlin!"). Since we are trying to explain our *consciousness* of change, the relevant beliefs must be conscious. But this is implausible. The conscious beliefs (or thoughts) just aren't to be found; not always, not in every experience of change. If they were, then given that we are continually experiencing change in bodily sensations, auditory and visual perceptions, thoughts and mental images, our consciousness would be flooded, horribly *clogged*, with thousands of boring beliefs about what happened before what. But it isn't.

Faced with this objection the memory theorist might try a different tack, and say:

> First I hear C. I then hear D, the experience of which is automatically accom-panied by a short-term memory-image corresponding to my hearing C. I then hear E, and as I do I have a short-term memory of C-being-followed-by-D.

This is better, since the account is only using *experience* in trying to account for our experience of change, rather than thought or belief. But the account is nonetheless problematic, for two reasons.

First of all, the memory theorist is positing a short-term memory of an *experience* of succession: "C-being-followed-by-D". The sort of experience that the memory-account is meant to explain away is in fact being *presupposed*: we cannot remember what we have not already experienced. If the memory theorist is prepared to admit that we are directly aware of succession when we remember and imagine, why not admit that we are directly aware of succession in ordinary experience?

The second difficulty runs deeper. We can distinguish two claims, one weak and one strong. The weak claim is that experiential memory plays a central and indispensable role in temporal awareness. The strong claim is that temporal awareness is *wholly* the product of experiential memory. The various proposals that we have considered thus far are all versions of the weaker claim. Why? Because of the example we have been working with: the sequence of tones C–D–E. Each of these *individual tones* has (it was stipulated) a short but noticeable duration – a duration that is directly experienced. If phenomenal temporality is *wholly* the product of memory, there can be no direct experience (or memory) of duration or change whatsoever. This means that our experience of even a single brief tone must be explained in terms of involuntary short-term memories. But memories of *what*? The answer must be a succession of strictly durationless experiences. My experience of the tone C consists of a large (infinite?) number of momentary experiences, each (except the first) being accompanied by a large number of nested involuntary short-term memories of other momentary experiences. And what holds of the single tone C also holds of the experience of C-being-followed-by-D.

This proposal suffers from a very severe plausibility problem. On the one hand it is hard to believe that we are not immediately aware of some duration in experience. Is a strictly durationless auditory experience even possible? On the other hand we are being asked to believe that our experience of duration depends on vast numbers of nested momentary memory-images (for it should not be forgotten that the short-term memories must themselves be durationless). This, too, is very hard to believe. On hearing the succession of tones C–D–E, are we aware of vast numbers of constantly changing momentary memories? I think not. The weak claim has some plausibility; the strong claim has none whatsoever.

7.3 The pulse theory

Since memory based accounts cannot explain our experience of change or phenomenal flow we must look elsewhere. One option runs thus:

> Our streams of consciousness consist of short pulses of experience, each of some finite duration (perhaps half a second). These pulses correspond to what is often called "the specious present" – *specious* because the present of experience is not the durationless present of mathematical physics. During a single pulse change is directly experienced: we are conscious of all parts of the pulse or, to put it another way, all parts of the pulse are *co-conscious*. There are no direct experiential relations between successive pulses: these are linked only by memory.[2]

So, for example, if I hear C–D–E–F, my experience could consist of two pulses, P_1 and P_2 thus:

$$P_1 \qquad\qquad P_2$$
$$[C–D] \qquad [E–F]$$

It might seem that if P_2 comes straight after P_1, and during P_2 I remember hearing P_1, then my experience would seem continuous. I would experience C-followed-by-D-followed-by-E-followed-by-F. But closer consideration suggests otherwise.

If the pulse theory were correct we should be able to discern *two* distinct forms of transition within our consciousness: change *within* pulses and change *between* pulses (e.g. C–D, D–F). After all, while C and D are experienced together – they are fully co-conscious – D and E are not. Surely this should be a noticeable difference? But in fact we notice no such thing; every transition between any two parts of the scale is experienced, and experienced in just the same way.

In fact, there is no reason to think that it would even *seem* to us that P_2 is experienced as following on from P_1. Given that there are no experiential links between P_1 and P_2, the fact that P_2 occurs *just after* P_1 is irrelevant to the experiential character of the supposed perceived succession; the only link between P_2 and P_1 is the memory that P_2 contains of P_1. Hence, even if P_2 happened a hundred years after P_1, the pulse theorist is committed to saying that these two pulses would constitute an experience of succession, which seems plainly absurd.

The same point can be made differently. Suppose that you and I are simultaneously hearing the same sequence of notes. Our streams of consciousness have this form:

You: [C–D]   [E–F]

Me: [C–D]   [E–F]

Since there are no direct links between the first and second pulses, there are two other streams of consciousness that, according to the pulse theorist, are just as real:

S_1 [C–D]   [E–F]

S_2 [C–D]   [E–F]

If the pulse theory is true, our streams of consciousness are intermingled. Or rather, there is no *purely experiential* fact of the matter as to which experiences belong to which streams. Again, this seems absurd.[3]

The pulse theory may be flawed, but, considering that, it has nonetheless served a useful purpose since a general feature of our experience has come to light. Each phase of a typical stream of consciousness is *phenomenally bonded* to its immediate predecessor and successor: there is a direct experiential link between adjacent co-streamal phases. An adequate account of temporal experience must accommodate phenomenal bonding.

7.4 Awareness and overlap

An alternative approach takes as its starting point an important distinction: it is one thing to have a succession of experiences, another to have an experience *of* succession. If you were to hear C and then D and then E, you would have undergone a succession of experiences; if you were to hear C-being-followed-by-D and D-being-followed-by-E you would have had an experience of succession. The distinction is plainly valid. Moreover, it is clear that we have experience of succession, rather than mere successions of experiences. What account can we give of this?

One solution runs thus. When we are *conscious of* a sound, we are *aware of* it: consciousness consists of an awareness of phenomenal contents, items such as sounds, colours, bodily sensations, thoughts, and so on. Hence there is a distinction between *phenomenal contents* and *acts of awareness*. Clearly, if we are to be aware of succession, then a single act of awareness must embrace more than one momentary content: it must embrace a succession of contents. One way for this to occur is illustrated in Figure 7.1.

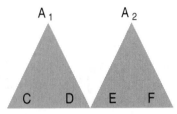

Figure 7.1 Multiple contents falling under single acts of awareness.

We are to suppose that the experience of C–D–E–F involves two (momentary) acts of awareness; one of these acts takes in C–D, the other E–F. However, there is an obvious difficulty here. Since there is no *experienced* connection between D and E, the account is simply the pulse theory in another guise, and we have already seen why this theory is inadequate. But there is an obvious remedy for this difficulty. What if the scope (and so the contents) of the two acts *overlapped* in the manner indicated in Figure 7.2?

A model of this sort was advocated by Broad (1923). I have shown just three acts of awareness (apprehending four auditory contents), but this was merely to present a simple picture. In principle there could be many further acts between A_1 and A_2 (and so likewise between any two acts separated by a noticeable temporal interval) and Broad suggested that this was the case.

By virtue of the fact that successive acts of awareness apprehend numerically the same contents, this *awareness-overlap* model is no longer open to the objection that it is just the pulse theory in another guise. But it may still be vulnerable to the objection that it provides an unrealistically fragmented account of our streams of consciousness. While successive acts provide awareness of the same contents, these

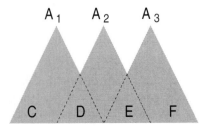

Figure 7.2 The "awareness-overlap" model of our experience of time. Phenomenal bonding is secured by distinct acts of awareness apprehending (numerically) the same items of content.

acts are themselves wholly distinct episodes of experiencing. There is unity (or rather, partial identity) at the level of content, but there is no unity or continuity whatsoever at the level of awareness. However, there is no need to press this objection, for the model suffers from a more obvious and devastating problem. Consider note D in Figure 7.2. This content falls within the region of overlap, and so is apprehended as a whole by both A_1 and A_2. As a consequence, the sound will be heard twice over, when in actual fact it is heard just once. This "repeated contents" problem is even worse if there are additional acts occurring between A_1 and A_2, for the same sound will be heard over and over again! Since the same will apply for any brief contents that are experienced as wholes by different acts of awareness, the awareness-overlap model is clearly inadequate.

7.5 The two-dimensional model

In later years Broad adopted a very different account of phenomenal temporality, an account similar to one advocated by Husserl at one stage.[4] Again, the underlying idea is quite simple. The repeated contents problem arises because one and the same content is apprehended by successive acts of awareness. To solve the problem we deny that numerically the same content is apprehended by successive acts. Instead, we say that each momentary act consists of an awareness of a different momentary content together with *representations* of the preceding contents. These representations are not full-blooded *re-presentations* of the original contents. If they were, then when hearing tone E together with representations of the preceding C and E we would hear three simultaneous tones – a chord – rather than a succession. This doesn't occur, the argument runs, because past contents are not presented as they initially occurred, but under a distinctive temporal mode of presentation. Thus Broad introduces the notion of *presentedness*. When a content is initially experienced it possesses this quality to the highest degree, as it is re-presented with subsequent contents it possesses less and less of this quality, until in the end it drops out of awareness altogether.

To get an idea of how this works in practice it will be useful to look at an artificially simple example. A subject hears a succession of brief sounds, which we label C–D–E–F–G–H–I–J–K. These are apprehended by nine different acts of awareness. Let us further suppose that there are six degrees of presentedness, Pr_1, Pr_2, . . ., Pr_6 (ordered by decreasing intensity). In Figure 7.3, presentedness is indicated by font size (the larger the font, the higher the presentedness).

The hearing of C, let us suppose, was preceded by a period of silence. A_1 is a direct awareness of C, which is experienced as possessing maximum presentedness, Pr_1. A_2 is a direct awareness of D, along with an awareness of C possessing presentedness of level Pr_2 – as a consequence, C is apprehended as being "just past". A_3 presents E with maximum presentedness, D with Pr_2 and C with Pr_3 – as a consequence, D is experienced under the mode "just past", and C "a little more past". And so it continues. C features in A_3–A_6 with ever diminishing presentedness, and by the time A_7 occurs this content has vanished from the subject's immediate awareness (or specious present), and from this point on is accessible only through memory. A_6–A_9 are typical specious presents: the subject's

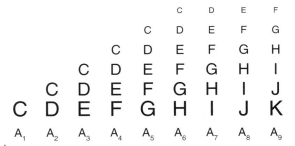

Figure 7.3   The two-dimensional (Broad-Husserl) model of temporal experience. The same experiential contents (in this case brief tones) are apprehended as possessing different degrees of "presentedness" in successive acts of awareness.

span of direct awareness is filled with contents of varying degrees of presentedness. As can be seen, each content enjoys its brief moment of maximum presentedness, and then sinks gradually into the past. This example is unrealistically simple. In reality there would be further acts occurring, for example, between A_1 and A_2, and additional degrees of presentedness (so as to enable contents to slide smoothly into the past).

Broad and Husserl fill out the details in different ways (e.g. Husserl suggests that each momentary awareness includes "retentions" of previous contents along with "protentions" of anticipated future contents), but the basics remain the same. The dynamic character of experience is explained by positing what is, in effect, a *two-dimensional* time: there is the earlier–later ordering of acts of awareness (A_1–A_9), and the earlier–later ordering of contents presented to these acts (e.g. in A_7 this runs from D (Pr_6) to I (Pr_1)). This is by now a familiar manoeuvre – recall the meta-time required to sustain a moving present – but since the posited additional dimension is located within consciousness rather than the world itself, it is not open to the objection that we have no reason to believe that such a thing exists. Indeed, the claim will be that we *do* have reason to believe that this extra dimension exists, since this is what is required to explain what we find in our experience. Nor is the model vulnerable to the objection that it generates an infinite regress. A regress arises if we posit a present that moves along the time line; the model we are considering posits no such thing. According to the two-dimensionalist, the specious present consists of a momentary awareness with a complex content that *seems* temporally extended. The appearance of future-directed motion is a consequence of the ways the complex contents of different acts are interrelated; for example, tone C is represented in the contents of successive acts of awareness, but seems increasingly more past in each. This slippage of contents into the phenomenal past creates the impression that our immediate experience is advancing into the future. Since nothing moves along the time-line, and the impression of passage is not a product of contents being created or annihilated, the model is entirely compatible with the B-theorist's static conception of time.

As well as providing an explanation of how we can have an immediate experience of change, the model provides an account of *phenomenal flow*, the impression we have that a qualitatively unvarying sensation is continuously refreshed and renewed. Figure 7.4 illustrates the experiential structures that result from hearing

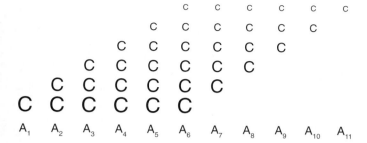

Figure 7.4 A two-dimensional account of phenomenal flow.

an extended C-tone: in each of the successive momentary awarenesses A_1–A_6 a "new" C-content is presented, and these are retained, albeit in modified form, in subsequent awarenesses. The subject's consciousness soon becomes flooded with nothing but the sound of this tone, which is experienced as enduring in the present whilst simultaneously sinking into the past.

The two-dimensional model is ingenious and possesses explanatory potential, but it is also seriously problematic. The difficulties fall under three headings: phenomenological inaccuracy; problematic presentedness; and consciousness fragmented.

Phenomenological inaccuracy

If the two-dimensional model were true, we could expect that:

- experiences would never end abruptly, they would always *linger* on for a short while as they are represented in successive awarenesses as possessing diminishing presentedness;
- at each moment we would be aware of our current experience together with a constantly shifting complex of representations of recent experiences, and so our consciousness would be *choked* with the residues of recent experiences.

But this is wrong on both counts. As far as I can tell, my consciousness is not clogged up with fading residues of prior experiences, and experiences do not always linger; they sometimes *do* end abruptly. It could be objected, "Ah, but the residues are very brief. That's why you don't notice them. Likewise, for the shifting constellations of representations." This reply is unconvincing. If we cannot discern the posited representations, there is no reason to suppose that they have any phenomenological reality, and hence no reason to suppose that they contribute to our *actual* experience of change and persistence.

Problematic presentedness

For the two-dimensional model to be viable we have to be able to make sense of the property that creates the extra dimension: *presentedness*. This is (presumably) an intrinsic property, the intensity of which can vary, and which is responsible for an experience's seeming to be actual, or just past, or slightly more past, and so on. As we have already seen in connection with those accounts of world-time (rather than phenomenal-time) that posit transitory intrinsics, it is far from obvious that such a property exists, or could exist; if anything the problem is more severe when the bearers of the supposed properties are components of experience. What quality, added to a pain sensation, would make that sensation seem to be *in the past* rather than in the present? If I am now experiencing some sensation, won't the experience seem to be occurring *now* irrespective of any peculiar qualities it might possess? Someone might say, "Presentedness is not an additional property that experiences have. It is akin to Hume's 'force and vivacity'. It is simply a measure of the intensity of an experience; as experiences lose presentedness they become progressively fainter, until they fade away altogether." But this won't work. Do some parts of the image in Figure 7.5 appear "more past" than others?

Figure 7.5 A problem for the notion of presentedness.

Consciousness fragmented

Perhaps the most serious objection to the two-dimensional model is the simplest: the account is profoundly *atomistic*. Each momentary awareness constitutes a distinct episode of experiencing in its own right, and so, from a purely experiential view, each awareness is entirely isolated from its immediate neighbours. The awareness-overlap account atomizes awareness, but does not reduce our streams of consciousness to entirely distinct moments of experiencing, since successive acts are held to apprehend numerically the same contents. But according to the two-dimensionalist, successive acts apprehend numerically distinct contents. The model thus suffers from the same problem as the pulse account: it fails utterly to accommodate the reality of phenomenal binding. Since it is obvious that our experience is *not* confined to isolated momentary capsules, we must look elsewhere for a realistic account of time-consciousness.

7.6 The overlap theory

Recall the repeated contents objection to the awareness-overlap model: since different acts of awareness apprehend the same contents, we would be aware of every content many times over – but we are not. The two-dimensional theorists try to avoid this problem by keeping the "momentary act – temporally extended content" schema while multiplying contents: successive momentary acts are not aware of exactly the same contents, but only similar contents (and these contents have an apparent, rather than real, temporal extension). We have seen that this approach is flawed. But as Foster has noted on several occasions (1982, 1985, 1991), there is a far simpler solution to the repeated contents problem.

Suppose that I hear the succession C–D–E. Further suppose that I am directly aware of C–D and also D–E, but I am not directly aware of the entire sequence of notes (although we are directly aware of change, the span of our direct awareness is brief). There are two ways to account for this:

1. I have two completely separate acts of awareness: A_1 with content C–D and A_2 with content D–E.

But this entails that I hear D twice, which I don't. To avoid this problem we can say the following:

2. I have an awareness A_1 with content C–D and an awareness A_2 with content D–E, but these acts of awareness are not distinct: A_1 and A_2 partially overlap, and the part of A_1 that apprehends D is numerically identical to the part of A_2 that apprehends D.

The problem of repeating contents is now solved. Since A_1 and A_2 overlap, there is only one experience of D. We have two temporally extended experiences that share a common part – C–D, D–E – but that remain distinct because they do not completely overlap.

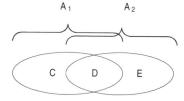

Figure 7.6 The key to phenomenal continuity: temporally extended, partially overlapping experiences.

Of course, if A_1 and A_2 are to partially overlap, these acts of awareness cannot be momentary, they must have some temporal extension. We must, therefore, abandon the doctrine that an awareness of change (or succession) consists of a temporal spread of content being presented to a single momentary awareness. Instead, we say that an awareness of change or succession is itself temporally extended. Since any two acts will overlap to precisely the same degree as their common contents, it is natural to suppose that acts have precisely the same duration as their contents, in the manner indicated above.

The idea that the succession C–D–E involves only two episodes of awareness is an unrealistic simplification. It is more plausible to suppose that there will be a sequence of "experiencings" whose contents differ in just noticeable ways, and so that overlap almost completely; see Figure 7.7, which represents the experience of watching a light rapidly growing dimmer. In this case, very short contents (such as the phase indicated by an arrow) will figure in a larger number of different experiences – but again, since the parts of these experiences that apprehend this content are numerically identical, the content is only experienced once.

Thus far I have continued to talk as though there is a distinction between phenomenal contents and the acts of awareness that apprehend them. This distinction

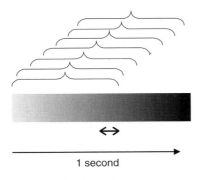

Figure 7.7 Observing a light fade. Many overlapping experiences are involved in even a brief phase of a typical stream of consciousness.

1 second

is unavoidable if (in the manner of the awareness-overlap account) we hold that at any given moment we are aware of a temporal spread of content. But since we are now supposing that contents and awareness run concurrently, the distinction is doing no work, and unless there is some other reason for maintaining it (and I cannot see that there is), we are free to dispense with its services. In place of a two-level view of experience we can adopt a one-level or *simple* view, according to which phenomenal contents (such as sounds, colours, pains) are intrinsically conscious items: they do not need to be apprehended by a separate awareness to be experienced, the contents themselves are experiences in their own right. Instead of supposing that different contents (such as a sound and a flash of light) are experienced together by virtue of falling under a single awareness, we say that these contents are "co-conscious"; that is, they are joined by the basic relationship of "being experienced together". This *simple* overlap theory – which makes no distinction between awareness and contents – entails that co-consciousness not only connects simultaneous experiences, it also ranges a short way *over* time.

An interesting result now emerges. Consider again the brief stream of consciousness consisting of C–D–E. Let us suppose once more that the subject experiences C-being-followed-by-D, and D-being-followed-by-E. In saying this I am assuming that the subject is directly aware of successions involving only two tones: by the time E occurs, C has dropped out of the subject's specious present. We thus see that C is co-conscious with D, and D is co-conscious with E, but C and E are *not* co-conscious. Since the specious present is of limited duration (probably less than a second), the relationship of co-consciousness *over time* (diachronic co-consciousness) cannot be transitive. If it were transitive, then every part of a continuous stream of consciousness would be co-conscious with every other part. Since this is clearly not the case, transitivity is restricted to the brief confines of the specious present. This may seem a surprising result, but on reflection it seems both natural and inevitable.

The overlap theory accommodates our direct experience of change, and respects the phenomenal bonding constraint: experiences occurring at different times (such as C and D) are linked by precisely the same relationship as experiences occurring at the same time: direct co-consciousness. But there remains a problem. Co-consciousness may not be transitive, but it is symmetrical: if C is co-conscious with D, then (necessarily) D is co-conscious with C. If it is co-consciousness that binds together the successive phases of our streams of consciousness, and this relationship is symmetrical, why does our consciousness *flow* in a particular direction? Why do we hear "C-being-followed-by-D" rather than "D-being-followed-by-C"? Why does a persisting tone seem to *continue on* in a particular direction?

There seems to be only one solution: the contents that are symmetrically joined by co-consciousness must themselves possess an inherent directional dynamism. The C-tone is not a static auditory quality, but a flowing quality, likewise for D and E. This *immanent flow* is an essential ingredient of any auditory content, just as essential as timbre, pitch or volume. It is because these contents possess this inherent and directed dynamic character that we hear C as enduring for a brief while and then seamlessly running on into D, which in turn runs on into E. The same applies to bodily sensations, such as pains or tickles, as well as to olfactory and gustatory contents. It applies to the visual case also. Suppose that I see a bird fly overhead. It crosses my field of vision in a second or so. I do not merely see the bird occupying a succession of different locations at different times. I see the bird *moving*. This perceived movement is itself a dynamic and directed intrinsic feature of my visual experience, and just as essential to it as colour, size or shape.

[margin note: How does time move?]

At the outset of this chapter I suggested that an adequate account of our short-term experience of time had to be able to accommodate both our direct awareness of change, and our direct awareness of endurance, or phenomenal passage, a feature of experience that is independent of qualitative variation. If phenomenal contents quite generally possess the characteristic of immanent flow, we have a complete account of phenomenal temporality:

- We are directly aware of change because co-consciousness extends over time.
- The specious present is of short duration because co-consciousness extends only a short way over time and is not transitive (at least in the diachronic case).
- Consciousness flows in a particular direction because phenomenal contents possess inherently dynamic and directed contents.

It might be objected that to posit immanent flow as a characteristic of experience over and above their other phenomenal features reeks of expediency: isn't this precisely what is required to save the overlap model from phenomenological inadequacy? It is, but since the feature in question unquestionably exists, *not* to recognize it would guarantee phenomenological inadequacy. Moreover, it is wrong to think that immanent flow is a separate and additional ingredient that different kinds of experience possess. I use the term merely to indicate that different kinds of experience all have something in common: they all have a directed and dynamic temporal character. As soon as one's attention is drawn to this fact, it is obvious that this feature is built into experience in all its forms.[5]

Putting all this together, we now have sufficient resources to state in a general way the conditions under which experiences are *co-streamal*, that is, belong to the same stream of consciousness: experiences that are directly co-conscious are co-streamal; so, too, are experiences that are *indirectly* co-conscious, that is, form part of an overlapping chain of directly co-conscious experiences. Since experiences possess the temporally asymmetrical feature of immanent flow, we can take a further step and provide co-streamal experiences with a temporal ordering. In the case of an experienced succession such as C–D–E, not only does each individual tone possess immanent flow, but these flows combine to constitute a continuous flow that runs the length of the succession: C is experienced as *flowing into* D, and D as *flowing into* E. Generalizing, let us say that for any two total states of consciousness E_x and E_y that are linked by direct diachronic co-consciousness, E_x is *phenomenally prior* to E_y if and only if some part of E_x flows into some part of E_y but not vice-versa. If E_x and E_y are not directly co-conscious, but are co-streamal, then E_x occurs earlier than E_y if and only if E_x and E_y occur at opposite ends of a chain of total states of consciousness,

$$E_x-E_n-E_{n+1}-E_{n+2}-\ldots-E_y$$

the adjacent members of which are related by phenomenal priority. We can further say that experiences belonging to total states of consciousness related by phenomenal priority belong to the same *phenomenal time*.

7.7 The phenomenal arrow

The arrow of phenomenal time has considerable autonomy in relation to the others that we have encountered. If you were to fly a spaceship into the time-reversed half of a Gold universe and visit an inhabited Earth-like planet, you would perceive all manner of bizarre goings on, but you would continue to feel your experience flowing on in the usual manner, *forwards*, even though the world around you seems to be running in reverse. Even if Gold universes are not physically possible, they are certainly clearly and easily *imaginable*, and this fact alone establishes the autonomy of phenomenal flow with respect to the ordinary patterning of events.[6]

For an equally bewildering scenario we do not need to posit such widespread reversals of the familiar content-asymmetries. It suffices to consider a case where just one is reversed: memory. You wake up one day and are puzzled to discover that whereas you can venture only the haziest speculations as to what you were doing yesterday, or at any points in your past, you find yourself quite certain about exactly what you will be doing for the rest of the day, and for much of the remainder of your life. You still have what seem to be memories about your own experiences but, as you soon realize, the events you now "remember" have yet to occur. As the days and weeks pass, you learn that these future-oriented "memories" are just as reliable as their past-oriented counterparts: what you "remember" happening does happen. Despite this difference, the character of your experience remains essentially the same: the direction of immanent flow remains future-directed; you find your phenomenal present sliding forwards towards future events that you know will occur.[7] Since phenomenal flow is an intrinsic feature of experience, its direction is independent of that of memory. It also seems independent of the knowledge-asymmetry, since it is hard to see how possessing detailed and accurate records (as well as personal "memories") of future events could impinge on the direction of phenomenal flow.

Newton-Smith (1980: 207) considers a variant of this scenario: you wake up and find that you can "remember" past and future experiences equally well. A person in this predicament, Newton-Smith suggests, would find time to be entirely lacking in direction. He is mistaken. Provided this person's experiences continue to possess immanent flow, time would still seem to have a direction; the same direction it always had.

A world such as our own, a world inhabited by conscious beings, is a world *some* of whose contents – experiences – possess an inherent temporal direction, a feature that may well be unique to experience. There is nothing to prevent our defining the earlier–later relation for the world as a whole in terms of phenomenal priority, in the manner sketched above. Indeed, given the overriding importance of immanent flow in our experience of temporal passage, this is in many ways a natural move to make. Of course, if we do make this move, we should not be overconfident about the universality of its appeal; if we live in a Gold universe there may well be others who have a different sense of where time is going.

It has often been said that in thinking about time we must guard against our tendency to think in spatial terms (a tendency that may well be rooted in the dominant role that vision plays in our cognition), and no doubt there is some truth in this, but I suspect that there is a further and very different reason why thinking clearly about time can be difficult. Immanent flow is such a pervasive feature of our consciousness that it is hard to think of *anything* that does not possess this feature, *time included*, for not only do our thoughts possess it as we think them, but so do any mental images that we call up. To take just one example, in trying to think about compound presentism we naturally think in terms of a series, $[P_1-P_2]$, $[P_2-P_3]$, . . .,which in the medium of our conscious thought becomes "P_1 and P_2 exist, *then* P_2 and P_3 exist". And already the damage is done: an unthinking projection of the immanent flow of consciousness on to the world has occurred, for this alternation in the contents of reality is being apprehended as possessing an inherent directionality of its own. Even if we realize that it is possible to view this sequence from another perspective we are liable to think to ourselves "Ah, so it could go $[P_3-P_2]$ *then* $[P_2-P_1]$" and again the immanent directedness of consciousness is transferred where it doesn't belong, and we will find ourselves thinking "Well, it must surely go one way or the other!" The thought that it *doesn't* go one way or the other, that all that is really going on is that the sum total of reality is different as of different times, and that these times do not "run" in any particular direction, does not come naturally at all.

7.8 Further consequences

If the overlap account of phenomenal time is correct, at least in the essentials, we may not be able to draw any conclusions about time in general, but given certain minimal assumptions there are conclusions that we can draw about the time in which we live.

Two versions of presentism are incompatible with the fundamental structures of our streams of consciousness that we have uncovered. Presentists invariably assume that the present is of short duration; either strictly momentary, or so brief as to be incapable of including change. Let us assume that this is the case. According to the solipsistic presentist, the sum total of reality consists of just one present. If this were so, it would be impossible to directly experience change; since we do directly experience change, we know that reality does not consist of a single (minimal) present. According to the many-worlds presentist, reality consists of a vast number of momentary presents, all of which are entirely isolated from each other. The fact that we directly experience successions, such as C–D, shows that we do not inhabit a reality of this kind. Suppose that C occurs in a present P_1 and D in present P_2. If C is co-conscious with D it is clearly wrong to say that P_1 and P_2 are related at most by similarity. Diachronic co-consciousness is a perfectly real mode of connection; it seems quite absurd to suppose that experiences related in this way could belong to entirely separate and unrelated worlds. More relevantly, it seems impossible to deny that experiences related by diachronic co-consciousness are *temporally* related, and so exist within the same time. This will be so irrespective of the precise relationship between phenomenal time and ordinary (world) time.

Of course, this is assuming that the overlap theory of phenomenal temporality is essentially correct. The two-dimensionalist theory is perfectly compatible with many-worlds presentism, since it denies that there are any genuine phenomenal connections between successive phases of a stream of consciousness, but as we have seen, there are good reasons for rejecting this theory. I am also assuming that there are temporal relations (of the ordinary, non-phenomenal variety) between experiences and other events. Consequently, if experiences E_1 and E_2 are co-conscious, and hence temporally related, so, too, are all the events (experiential and non-experiential) that are simultaneous with E_1 and E_2. But this is hardly a controversial assumption (ignoring relativistic complications).

What of the other models of time that we have considered and found metaphysically intelligible? For better or worse, the overlap theory seems entirely neutral with respect to these.

Consider the static block. The overlap theory tells us that co-consciousness extends over short periods to create temporally extended "phenomenal presents", that successive phenomenal presents overlap, and that the contents of these phenomenal presents have an inherent dynamic character. That our experience is structured in this way guarantees that we will have the impression that our consciousness is confined to a forward-moving present even if, in reality, past and future experiences are just as real as present ones. From the four-dimensional perspective, a stream of consciousness is akin to a glowing filament embedded in a long glass block; the filament is aglow throughout its entire length, but this is not discernible to the subject of these experience, who is only ever aware of those tiny stretches of experience joined by co-consciousness.

As for the dynamic block model, the overlap theory is quite compatible with streams of consciousness growing – along with the rest of reality – in an incremental fashion. It also seems quite compatible with streams of consciousness gradually shrinking. This may seem surprising: wouldn't a person notice if their experiences started being deleted? There is no reason to think that they would. From the standpoint of any given present we have no direct awareness of any future experiences that we might have, and so could hardly be affected by their disappearance; likewise, the annihilation of an experience isn't itself something that is experienced, it is simply the ceasing-to-be of something that *was* an experience. This said, if it could be shown that immanent phenomenal flow depended on new experiences being created and added on to a pre-existing stream of consciousness, then we could rule out both the static block model, and the version of the dynamic block model that fails to recognize a distinction between growth (via becoming) and shrinkage (via annihilation). The *growing* block model in its original form would be vindicated on empirical grounds. But if the current analysis is correct, and phenomenal flow is an inherent feature of experiences themselves and entirely independent of experiences coming-to-be or ceasing-to-be, this option is not available. Whether this analysis of phenomenal flow *is* correct is another issue, and one to which those attracted to the growing block might want to devote attention. It is certainly tempting to think that the inherent directedness we find within our streams of consciousness is linked to new phases of experience coming-into-being. But since it also seems that this same directedness is an intrinsic feature of our experience, it is hard to see how this could in fact be the case.

Dynamic presentism of the compound variety is not only compatible with the overlap model, but in some ways it is a precise analogue of it. Whereas the overlap theorist holds that a stream of consciousness consists of extended phenomenal presents that partially overlap, the compound presentist holds that *the world* consists of extended presents that partially overlap. And whereas to make sense of this the overlap theorist is obliged to deny the transitivity of co-consciousness, the compound presentist is obliged to deny the transitivity of co-existence. But the similarities end here. The time of the compound presentist consists of an alternation of moments and events that lacks an inherent direction: we could as easily regard the sequence $[P_1–P_2]$, $[P_2–P_3]$ as running forwards or backwards. By virtue of the dynamic patterning immanent in experience, a typical stream of consciousness *does* have a definite direction.

So by appealing to empirical (rather than metaphysical) considerations we are in a position to rule out two of our five contenders. It is often assumed that the only sort of empirical evidence that can further our understanding of time is of the scientific variety. As should by now be plain, this is a mistake. Scientific theories attempt to explain and predict the observed behaviour of material things, often by reference to invisible forces and entities; there is no denying that such theories can have important implications for how we think of matter, space and time. But we can also learn something about time by examining *experience itself*, rather than what experience reveals about the non-experiential parts of our world.

You may be inclined to think "Fine, but what has this foray into the phenomenological really delivered? The models of time that we have eliminated were never serious contenders anyway, since of the five they are by far the least plausible." If so, there are two points to bear in mind. There is no denying that solipsistic and many-worlds presentism are highly implausible, but might this not be because they both conflict so dramatically with what we find in our immediate experience? In which case, by working our way towards a clearer picture of how experiences are interrelated in streams of consciousness, we have put the intuitive case against these doctrines on a firmer foundation. Secondly, as we saw in §6.8, the time-problem in quantum gravity theory has led some scientists to take many-worlds presentism very seriously indeed; it could even be that the *only* empirical data that can refute this doctrine is of the phenomenological variety.

As for the question of what science *can* tell us about (our) time, we will be considering this again in chapters to come.

8 Time travel

8.1 Questions of possibility and paradox

Space offers something that time does not: unconstrained freedom of movement. Up, down, north, south, east, west, and all points in between, not only are we able to move in all spatial directions, but we are free to move back and forth as we wish, and at different speeds. Time is miserly in comparison. As it is by nature one dimensional, it would be churlish to condemn it for offering us only two directions, but the additional constraints it places on us can seem unnecessarily severe. Although we might be said to "travel" into the future simply by persisting, we are condemned to the monotonous rate of a second per second, with no option of skipping over times that we might prefer to avoid; and as for the past, it is utterly out of bounds. There is a sense in which we are *captives* of time, but not of space; there is no temporal counterpart of the car.

This most basic of differences between the dimensions naturally gives rise to some questions. Is travel through time impossible because of the nature of time? If so, why? One answer is implicit in Bigelow's observation (see §6.7) concerning the rarity of time travel stories prior to the late nineteenth century. Since it is hard to see how one could travel to the past or future unless these times *exist*, presentists will be inclined to regard time travel as impossible (just as growing block theorists will dismiss leaps into a non-existent future), but it would be moving far too quickly to assume that our inability to move as we wish through time entails the falsity of the static block model. It is certainly true that anyone who believes that times have the same ontological status as places has some explaining to do; if the difference between "now" and "then" is akin to that of "here" and "there", there is clearly a *prima facie* problem since we can move more or less as we like between here and there, but not between now and then. But in fact there are several responses to the problem available to the static theorist.

- Time travel involves insuperable paradoxes of a broadly logical sort, and so is metaphysically impossible.
- Time travel is not metaphysically impossible, but is forbidden in our world by contingent physical laws.
- Time travel is neither metaphysically impossible nor physically (or nomologically) impossible: it either does or could occur in our universe.

We will spend most of this chapter exploring these claims. Until further notice I will be assuming the static block model; we will be in a better position to assess where dynamic models stand *vis-à-vis* the possibility or otherwise of time travel when we have a clearer idea of just what the obstacles are.

The time machines that we are familiar with from science fiction fall into several categories. Some give access to the past but not the future, others do the opposite and others do both. Some can travel any temporal distance and others are limited; for example, the "wormhole" time machines being seriously explored by contemporary physicists permit backward travel, but only as far as the earlier mouth of the hole.[1] Some machines cater for round-trips. Others cater only for the outward leg, but travelling to a distant time by this route is a risky business, for unless you have a second machine available for the return journey you will be stranded. Some machines work by *time-slides* and others by *time-jumps*. If you embark on a time-slide between 1900 and 2000 you will pass through all intervening times, but at a faster-than-normal rate: your trip only takes (say) a week instead of a century. A time-jump, on the other hand, would take you directly from 1900 to 2000, and the twentieth century would pass in your absence.

Interestingly, the possibility of one form of time travel is not seriously disputed, and that is *time-slides into the future*. Such trips seem neither physically nor logically problematic. On the logical side, the problem of self-defeating loops ("What if you killed your grandfather before he sired your mother?") that plague backward travel do not arise, since your presence in the future cannot, in itself, causally undermine what has already happened. On the physical side, there are several ways to achieve the desired effect that seem compatible with known natural law. Putting someone in a deep-freeze that effectively halts all biological processes and then successfully thawing them out at a later date is perhaps the most obvious way to slide someone forwards into the future. When the subject awakes no time has passed *for them* but fifty or a thousand years might have passed for the rest of the world. Relativity theory provides several methods for futureward slides. Anyone who embarks on a lengthy spaceship voyage of (say) ten years' duration at a sizeable fraction of the speed of light would find on their return that hundreds or thousands of years had passed on Earth. Fast travel is not the only way to slow down time. Spending a while orbiting a very strong gravitational field (such as produced in the vicinity of a black hole) has equally dramatic consequences. (We will be exploring the theoretical background of such effects in Chapters 16–19.)

The fact that slides into the future are comparatively easy increases the pressure on the four-dimensionalist. Why, if there is no ontological difference between past, present and future, are slides into the past not equally possible? And if they are possible, why are they comparatively more difficult? To make any progress with these questions we will have to untangle a number of physical and metaphysical issues.

8.2 Misconceptions and multidimensions

One objection to time travel, in either direction, can be speedily dealt with. I step into the time machine, and two minutes later I am three thousand years in the past. Isn't this a direct contradiction? Before I step into the machine it is true to say that two minutes from now (in the future) I shall be thousands of years in the past. How can one and the same event – my stepping out of the machine – occur later than now and earlier than now?

Rather than revealing time travel to be impossible, this objection merely reveals what time travel, should it occur, necessarily *involves*. In any time travel story we have to distinguish two frames of temporal reference. One frame is based on the time travellers themselves; time as measured by the watches they wear, or by their mental and biological processes. This is commonly called "personal" or "proper" time. There is one such time for each individual time traveller. The other frame is that of Earth (or our starting point, or the universe as a whole); call this "normal" or "external" time. In ordinary life our personal time coincides with external time, but if we embark on time travel the two systems of time reckoning will diverge. Unless we have independent reasons for supposing that this divergence cannot occur, we cannot use it to argue that time travel cannot occur (see Dainton (1992) for more on the ways personal and external times can diverge). In fact, we have good reason to believe that it *can* occur. Relativity theory predicts that less time will pass for the returning space traveller than will pass on Earth (a prediction that has been empirically confirmed); more generally, objects moving relative to one another all have their own (diverging) proper times. We can apply precisely the same distinction to the time travel case: when I step out of my time machine three thousand years in the past, I am stepping out three thousand years in the past as measured by external time, and a couple of minutes in the future as measured by my personal time. Admittedly, backward time travel generates deviations between personal and external time that are more radical than those permitted by special relativity – we shall see in Chapter 16 that special relativity limits the frame-dependence of temporal ordering (to "space-like" separated events). But the underlying principle is the same: those in different states of motion have their own temporal frames of reference.

Another common misconception, a more serious one, is embodied in arguments such as the following:

> But couldn't we *travel into the past* and witness for ourselves the Egyptians building pyramids? . . . There is a problem, however. Those centuries have passed; the events have happened; they're over. And they all happened *without you* – you weren't even born until the twentieth century. The ancient civilizations came and went, and you weren't anywhere in that picture. How then can you say that you could now go *back* there and *be* among the people then living? It all happened without you, and now you want to say that you could go back and participate in it, that is, that it did *not* happen without you. It did happen without you and it didn't – surely, that's a contradiction.
>
> The past is what has happened; not even God could make what has happened *not have happened*. Let's say the pyramids were built by 100,000 laborers. Now if you "go into the past" and join these laborers, there would be 100,001 laborers. If it was 100,000, it wasn't 100,001; there's a contradiction in saying it was both. (Hospers 1997: 121)

There are two key assumptions made here: the past cannot be changed; and backward time travel involves changing the past. Are these assumptions true? It depends on the nature of time. There are possible worlds in which the past cannot be changed, but in such worlds time travel does not involve changing the past; there are also worlds where the past *can* be changed, and time travel is a way of bringing this about.

The idea that the past could be changed is highly counterintuitive. Not only does it entail two conflicting but true accounts of what occurred at a given time, but it is quite likely that changes in the past produced by a time traveller's arrival will have wider ramifications, and require the wholesale replacement of the subsequent timeline. Consider a scenario familiar from the *Terminator* films. A nuclear war in the near future is provoked by out-of-control Strategic Defense Computers; humanity is reduced to a few ragged bands of survivors; the struggle commences and human rebels discover the secret of time travel and travel back to prevent the construction of the computers. They succeed, and history is immediately changed: the nuclear war does not take place; the billions that perished *don't* perish, but happily live out their lives; one entire stretch of human history simply vanishes and another stretch is immediately created to take its place. Is this scenario absurd? I am inclined to think that it is.

But don't forget, we are currently assuming the block view to be true, and in this context it is simply a mistake to suppose that backward time travel *would* bring about changes in the past of the problematic sort just envisaged. If the universe consists of a single block of times and events, what occurs at a given time is fixed and unchangeable. This does not mean that you cannot go back to 3000BC and assist with the building of the pyramids; what it means is that if you do travel back and assist, *that* you do so is as true now as it was then. It is true at *all* times, including those that occur before you set off. You may have been born in 1975, but unbeknownst to you and your parents this was not your first appearance in the world's affairs: you first entered history as a thirty year old, assisting with the building of the pyramids several thousand years previously. In short, everything that you will ever do as a time traveller is *built into the past* before your first journey. Prior to your departure you do not remember being in Egypt all those years ago, but this is because the time you spend there lies in your (personal) future. Of course, if you left any traces of your visit – perhaps you carved your initials on a sarcophagus – these may well be discovered prior to your departure. It is one thing to *affect* the past – to contribute to what occurred at the times in question – quite another to *change* it. You certainly affected the past – your initials are a testament to that, and there are Egyptian slaves who were glad of your help – but you did not change it. The building of the pyramids only occurred once, and you were there at the time.

So the argument fails. But the mistake on which it is founded is easily made. In imagining how things were in ancient Egypt we envisage ourselves *being present* there; since we naturally assume that times later than the present are unreal, it seems absurd to think someone from the twentieth century could put in an appearance; the future from which they would supposedly emerge is nonexistent. Consequently, if someone from the twentieth century *were* to travel back, it seems that they will be emerging into a history that originally unfolded in their absence. All this shows, however, is that backward time travel is problematic in the context of certain dynamic models of time. Since we are currently working within the confines of the static block view, this is an irrelevant result.

In fact, as I suggested above, there are models of time in which the past can be changed, and some of these models are dynamic. Presentists (and others, for that matter) who take an anti-realist view of the past reject the idea of a static unalterable history. If facts about what occurred at earlier times are constituted by the evidence (traces, records, etc.) that exists in the present, then if the evidence that exists at t_2 is different from that which exists at t_1, the latter time will have a different past from the former. However, since time travel and presentism are uneasy bedfellows, this loophole offers little assistance to the would-be time traveller, and none whatsoever to those who wish to visit the past *in order to change it* in some particular (known) respect. More congenial to those in the latter category is the two-dimensional model depicted in Figure 8.1.

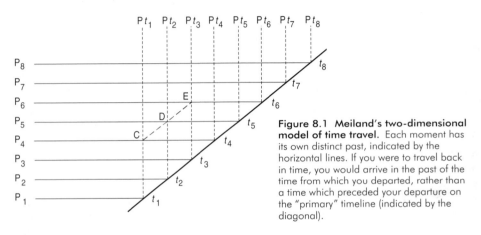

Figure 8.1 Meiland's two-dimensional model of time travel. Each moment has its own distinct past, indicated by the horizontal lines. If you were to travel back in time, you would arrive in the past of the time from which you departed, rather than a time which preceded your departure on the "primary" timeline (indicated by the diagonal).

In Figure 8.1, t_1–t_8 are present moments, each a year apart. The line P_1–t_1 represents the past when t_1 is present, and similarly P_2–t_2 represents the past with respect to t_2. The point where the line Pt_1–t_1 intersects with P_2–t_2 represents the location of t_1 in the past of t_2. As can be seen, each present has *a past of its own*, and these pasts are potentially different. Suppose that at t_4 a newly born child, Sam, is sent back three years. Where does Sam arrive? At t_1? Yes and no. Sam does not arrive at t_1 (i.e. at the terminus of the line P_1–t_1) – there were no time travellers present at t_1; Sam arrives three years earlier *in his own past*, at point C. He remains there for two years: a year later he is at D, three years earlier than t_5, and yet another year later he is at E, three years earlier than t_6. Sam is not present in the past of t_2, nor that of t_3; he is, however, present at the corresponding times in the pasts of t_5 and t_6.

This two-dimensional model of time is due to Meiland (1974), who explores its ramifications in some detail – but I leave these to the interested reader to discover for themselves. It can be interpreted in two ways: as either a static two-dimensional configuration, or a dynamic one. In the latter case, as of t_3 no later times are real, and the sum total of reality includes all those presents that occurred prior to t_3 and their associated pasts, but no subsequent presents (or pasts); reality grows not just by an accretion of presents, but by presents and pasts – entire timelines come into being with each new moment.

Our world could be like this, but in the absence of any positive reason for thinking it is, the hypothesis of a one-dimensional time is rightly favoured on grounds of ontological economy. The model is of interest, however, for it provides an example of how backward time travel could involve changing the past without generating paradoxical consequences. In a world where each present has its own past, there is no necessity for the past of one present to be the same as the past of another. Moreover, in such model it is not easy for a time traveller to do something in the past that threatens his own existence. Suppose baby Tom's presence at D results in the couple who discover him deciding not to have a child of their own; instead they adopt the mysterious baby they found in their garden a year earlier. This baby is, of course, none other than Tom himself, the baby they have just decided not to have. However, their decision does not threaten the existence of Tom, who was conceived and born on the P_4–t_4 timeline; their decision is made on the P_5–t_5 timeline. Contrary to what Meiland suggests (1974: 173), however, there may in fact be ways in which a traveller could undermine their own existence in this model. What if Tom's time machine were to land on his own mother at point C, killing her three years before his own birth?

There are models of reality in which even this sort of eventuality is unproblematic: multiverses. In these realities there are many parallel dimensions, each containing a distinct spatiotemporal system, some of which resemble one another. If a way of moving between dimensions were to be discovered, a traveller could arrive at what *seems* to be a period in their own world's past.

In Figure 8.2 each column depicts a portion of the history of a different universe. Tom, our would-be time traveller, wants to prevent the outbreak of a devastating war, and so travels back into what he believes is his own past. He sets off from the A-universe and arrives at the B-universe, at a time that strongly resembles an earlier time in the A-universe. Indeed, until the time of Tom's arrival, the A- and B-histories are exactly the same, but diverge thereafter due to Tom's arrival: no (apparent) time traveller arrives at the corresponding time in the A-universe. Alas, Tom fails to stop the war, and so decides to make a further attempt, by travelling still further back in time. He arrives at the indicated point in the C-universe, which, until this time, exactly resembles the A- and B-universes. Tom succeeds in preventing the war, and lives in the C-universe for the remainder of his days. This success comes at a cost. Tom's activities do not just prevent the outbreak of war but they result in his parents never meeting, and so Tom (or his C-counterpart) is not born. But since *our* Tom originated in a different universe there is no threat of paradox, and no threat (this) Tom's existence. Close variants of this model of backward time travel can be realized in a number of ways. Instead

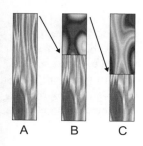

Figure 8.2 Interdimensional travel. A trip which takes you to another universe might seem like a voyage into your own past, provided you arrive in the other universe at a time when it resembles your own universe at a time prior to your departure. A, B and C are distinct universes, the arrows indicate your interdimensional travels. B and C closely resemble A until the time of your arrival – thereafter what happens is very different.

A   B   C

of travellers arriving in parallel universes they might land in different limbs of a *branching* universe. It may well be that their arrival produces the limb in question. The version of presentism that takes reality to consist of innumerable momentary reality-slices that are related only by resemblance (cf. §6.9) can also be viewed as a many-worlds model, although the "worlds" in question are of a very distinctive kind: highly fragmented and overlapping.

There are writers who believe that time travel, if it is possible at all, must take this form. Deutsch and Lockwood (1994) argue that backward time travel within a single universe leads to irresolvable paradoxes, and so can only occur *between* universes; as firm adherents to the many-worlds interpretation of quantum mechanics, they are confident that such universes exist (Lockwood 1989, Deutsch 1997). Not everyone shares their confidence – there are less extravagant interpretations of quantum mechanics that posit only one world – but we can agree with them on one point. If we do live in a multiverse that permits movement between worlds, then the sort of time travel they envisage may be possible – depending on the availability of suitably similar worlds – and if so, it would not be threatened by paradoxes that afflict time travel within a single world. However, whether the envisaged dimension hopping really amounts to *time* travel is open to question. It certainly doesn't involve visiting one's *own* past.[2] Irrespective of the terminological proprieties, it is time travel within a world that has greatest philosophical interest, precisely because of its potentiality for paradox – which leads us to our next topic.

8.3 Self-defeating loops

The most serious threat to backward intra-world time travel is familiar from science fiction: the self-defeating loop. Someone who travels into their own past could do something to ensure that they themselves will not exist at the time of their departure, such as kill a parent, grandparent, or even their own younger self. Successfully carrying out this action makes it impossible for the action itself to occur, which is a paradoxical state of affairs if ever there was one.

There are ways of telling the tale that can tempt us into thinking that the impossible is possible. The 50-year-old time traveller accidentally kills his 10-year-old self, and then as the realization dawns, he sits down, resigned to his inevitable fate: he knows the past four decades of his life are steadily being wiped from history, his twenties and thirties are probably already gone, he has only a few moments left before annihilation. But while we may be able to visualize such a process – at least seeing a film that tells a story along these lines – it could never occur. Someone who dies aged 10 does not live to see fifty. Someone who succeeds in travelling back in time *could not possibly* kill their younger self, or do anything else that would make it impossible for them to travel back. And yet they surely *could*. If you were to travel back, what could prevent your buying a gun and blowing out the brains of your earlier self, whether on purpose or only accidentally? Hence the real paradox seems to be this: there are actions that a time traveller both can and cannot perform.

Bearing in mind what has already been said about the fixed character of a block universe, a four-dimensionalist wishing to defend time travel may elect to meet the challenge head-on:

You exist now, and have existed for the past thirty years. Even if you enter your time machine tomorrow, and go back in time and try to kill yourself as an infant, we can be certain that you will not succeed simply because you *didn't*. Since in actual fact you survived your childhood, it is an unalterable fact *as of now* that any attempts made on your life during this period fail. So there is no paradox. Since it is logically impossible for you to commit auto-infanticide, it is not true that you both can and cannot kill your earlier self.

While endorsing this general line, Lewis (1976) takes a further step and argues that the paradox-mongers are guilty of equivocation. To say that someone both can and can't do X is to make two different claims, and these claims are only contradictory if "can" has the same meaning in both. Lewis suggests that it doesn't. It is not *logically* possible for a time traveller to kill his earlier self, since he didn't, so in this sense he can't do it. But it is not *practically* impossible – he has the strength and know-how, he is at the right time and place – so there is a perfectly legitimate sense in which he *can* perform the act in question, it is just that, for whatever reason, he doesn't (or didn't).

> To say that something can happen means that its happening is compossible with certain facts. *Which* facts? That is determined, but sometimes not determined well enough, by context. An ape can't speak a human language – say Finnish – but I can. Facts about the anatomy and operation of the ape's larynx and nervous system are not compossible with his speaking Finnish. The corresponding facts about my larynx and nervous system are compossible with my speaking Finnish. But don't take me along to Helsinki as your interpreter: I can't speak Finnish. My speaking Finnish is compossible with the facts considered so far, but not with further facts about my lack of training. What I can do, relative to one set of facts, I cannot do, relative to another, more inclusive set. Whenever the context leaves it open which facts are to count as relevant, it is possible to equivocate about whether I can speak Finnish. (1976: 143, original emphasis)

Likewise for whether you can kill your earlier self. In an ordinary case of attempted assassination, in judging whether the would-be killer is able to pull off the job, we take the range of relevant facts to include the power of his weapon, his distance from the target, his marksmanship, resolve, concentration – we would not dream of including facts about the *future* in our assessment. By this delineation of the relevant facts, you are perfectly able to kill your younger self. You don't succeed, of course, but there is nothing paradoxical about that: we often fail to do things that we are capable of doing. Luck has a part to play in any success, and without luck all the preparation in the world will come to naught. However, if we consider the same question from the perspective of a more inclusive set of facts, a set that includes facts about the future – which, unusually, in this case are known – then it is clear that you can't succeed. Your younger self lives to do a great many things, not the least of which is take a trip into the past.

8.4 Global consistency constraints

It is hard to fault the logic of this reply, but it is also hard to believe that there is nothing more to be said. Consider an example that avoids the psychological and moral complications of self-murder.

You are a committed block theorist and, although you have no doubts about the unalterable character of the past, you decide to put it to the test using a time machine that has recently fallen into your hands. There is a tree in your back garden on which you carved your initials (and the date) some twenty years ago. You decide to travel back fifteen years and try to cut it down. Since the tree evidently survives, you have no doubt that you will fail, but you want to find out what happens when you make the attempt. So you load a chainsaw, an axe, some dynamite and various other destructive implements into the time machine, enter the appropriate coordinates, and moments later find yourself in your back garden, as it was fifteen years ago. Picking up the tools you walk up to the tree, verify that it is the right one – yes, the date and initials are clearly to be seen – switch on the chainsaw and set to work.

What happens next? Do the laws of logic prevent the saw from cutting into the tree? It doesn't seem possible; only *material* forces and obstacles could do this. So something else must happen. Some irrational fear stays your hand at the last moment. Or the saw breaks down, and when you pick up the axe you pull a muscle and find yourself unable to wield it. No matter what you try, no matter how often you try, all your attempts fail. Now, assuming that there is no conspiracy here – no one has found out about your plans and has gone to extravagant lengths to thwart your attempts – your long succession of failures is purely coincidental. There is no unifying explanation; each failure has its own independent cause, and the cause is always local – the fact that the tree exists tomorrow (and for many years to come) is not what caused your chain saw to fail (it was a faulty electrical connection) or made your dynamite damp (that was a passing dog).

Hence the problem. Isn't this series of coincidences *very peculiar*, to say the least? Is it reasonable to think such a sequence *could* occur, not only in this instance, but in every instance in which a time traveller attempts to undo the future? Nahin writes "he *must* fail because he *did* fail. To demand an accounting for the specific why of failure before accepting the failure is as misguided as a stranded motorist refusing to believe his car won't start until he knows why" (1993: 196). But there is bad luck and there is bad luck. Confronted with many immensely improbable sequences of coincidences (one for each sustained attempt to create a self-defeating loop), the idea that *bad luck* is the only factor at work seems ridiculous. Unless some more plausible explanation is forthcoming, the rational response would surely be to dismiss the possibility of backward time travel.

Coincidences do happen, and from time to time very big coincidences happen, or at least they could; so this objection does not demonstrate that time travel is *logically* impossible. But it does merit a response. Can we at least render it intelligible why block universes that contain lots of time travellers intent on creating self-defeating loops must also be universes with lots of highly improbable coincidences? Perhaps.

Imagine that I am a God-like being who has decided to design and then create a logically consistent universe with laws of nature similar to those that obtain in our universe, laws that cannot be broken (so miracles will never occur). On a perverse whim, I have also decided to include within it time travellers of the troublesome sort. Since the universe will be of the block-variety I will have to create it *as a whole*: the beginning, middle and end will come into being together. Now, given that I will be introducing troublesome time travellers, all of whom will make many attempts to bring about self-defeating loops, to ensure that their attempts fail I have to incorporate a good many coincidences into my design. There is no other option, since I am constrained by both logical consistency and the "no miracles" decision. As a consequence, the existence of these coincidences is not coincidental: they *must* be there if the universe is to exist at all, given that the universe is going to contain troublesome time travellers. So we see that, in this context at least, coincidences are unsurprising; we have an explanation of why they have to exist.

But what if our universe was not designed in this way, but does include troublesome time travellers and the associated coincidences? Do we still have an explanation for why these coincidences exist? Well, assuming that our universe is a static block, even if it never "came into being", it nonetheless exists (timelessly) as a coherent whole, containing a globally consistent spread of events. At the weakest level, "consistency" here simply means that the laws of logic are obeyed, but in the case of universes like our own, where there are universe-wide laws of nature, the consistency constraint is stronger: everything that happens is in accord with the laws of nature. In saying that the consistency is "global" I mean that the different parts of the universe all have to fit smoothly together, rather like the pieces of a well-made mosaic or jigsaw puzzle. A troublesome time traveller can be thought as a particularly complicated and awkward piece (or collection of pieces) of the mosaic, with parts oddly spread through time. The pieces that surround the parts of the traveller's life when he is trying to create self-defeating loops will not be unusual in themselves – they will be events that are lawfully related to their surroundings – but they will also constitute an improbable series of coincidences. Or rather, they will when viewed from a local perspective; from a global perspective, given the need to accommodate all stages of the traveller's life – including those that occur *later* than the events in question – it will be obvious that events of that sort *had* to occur. A universe that conforms to our laws, is a globally coherent whole and contains awkward time travellers will *necessarily* include the problematic coincidences. But since, given these constraints, they *had* to occur, the coincidences are not in fact problematic: we have a reasonable explanation for why and how they exist.[3]

If this strikes us as odd it is because we are unused to thinking of the universe as a vast spatiotemporal mosaic, but if the universe *is* a vast spatiotemporal mosaic then, given the reality of the future, the future determines the past as much as the past determines the future. The constraints that later events place on earlier ones are not always causal (although it can be in the special case of backward time travellers). It is more typically a matter of coordination: the future events exist in the same universe as the earlier events, in a coherent, smooth-fitting, law-abiding whole. So *given* that the future is in a certain way, there are only so many ways the past *can* be.

8.5 Bilking

Backward time travel (as usually conceived) involves *backward causation*: for example, the time traveller's arrival in the past is the causal consequence of an event in the future – the pressing of the button that activates the machine.[4] But backward causation comes in other, less dramatic guises; for it involves nothing more (or less) than an earlier event being causally dependent on some later event. The earlier event could be the flashing of a light bulb and the later event the pressing of a switch; there is no need for objects to be translated through time. Can we make sense of backward causation? The so-called "bilking" argument suggests not (cf. Flew 1954, 1956, 1957), and is worth exploring for the additional light that it casts on the issues that we have just been considering.

In Dummett's classic discussion (1964), a tribal chief performs a ritual dance every day during the six-day period in which the young men of the village are away hunting, even though it takes a good two days to reach – and so return from – the hunting grounds. Asked why, he says "So as to ensure the young men perform bravely on the hunt." The chief, it seems, is convinced that his dancing significantly affects the chances of the young men acting bravely. The question is put "But why do you continue to dance for the final two days, when you know the hunt is over and they are on their way back? Surely dancing on the first four days would be enough?" The chief's response is that experience has shown otherwise. Another chief in a neighbouring village tried to cut corners in this way and the hunters behaved ignobly; the same happened on the two occasions when he himself fell ill and was unable to dance for the final two days. On the other hand, whenever the dance is performed on all six days, the hunters usually turn out to have acted bravely.

Finding these claims suspect we decide to subject them to the rigours of the experimental method, and we set about the task of designing a series of tests that will reveal whether the chief's dancing does exert a backward influence on the behaviour of the hunting party. The tests we come up with are both simple and effective; so much so that it seems that we needn't bother actually carrying them out, for, no matter what the result, the backward causation hypothesis is refuted.

Consider the following scenario. We repeatedly send someone out with the hunt equipped with a radio, which allows us to know whether the hunters have behaved bravely before they return to the village. On occasions when the hunters have behaved in a cowardly way, we pass the news on to the chief, and persuade him to dance anyway – in a spirit of cooperation he agrees. Now consider the possible results. If the chief dances successfully, and does so on numerous occasions when the men have not been brave, his claim that his dancing produces bravery is undermined. At the very least, his dancing is not sufficient for the bravery. But suppose instead that he tries to dance, but repeatedly fails for no apparent reason (his limbs keep locking up). This result does not refute the hypothesis that bravery and dancing are causally correlated, but it serves instead to cast doubt on the *direction* of the causal influence, for it now seems plausible that it wasn't his dancing that was causing the bravery, but rather the young men's bravery that was *making it possible for him to dance*. The chief's dancing was not, as he thought, an action that it was in his power to perform or not perform as he saw fit; the hunter's acting bravely was a necessary causal precondition for his dancing. To strengthen our case we could carry out another experiment. We could try to *prevent* the chief from dancing on occasions when (thanks to our advance knowledge) we know that the hunters have acted bravely. If we succeed on repeated occasions, the hypothesis that dancing causes bravery is undermined; the dancing has been shown to be irrelevant to how the hunters act. But suppose that, despite our best efforts, the chief finds himself compelled to dance and we just can't stop him. The hypothesis that the dancing and the bravery are causally correlated is not refuted, but we now have reason to believe that the earlier acts of bravery are causally *sufficient* for the dancing.

The general policy adopted in this case is known as "bilking". Confronted with an alleged case of backward causation, we repeatedly observe the presence (or absence) of the alleged earlier effect (E), and then try to prevent (or produce) the alleged later cause (L). It seems that, no matter what the result, the hypothesis that backward causation occurs is falsified. Either there is no correlation between the two at all – L occurs without E, and E without L – or else the bilking attempts fail, in ways that suggest that the later event is in fact causally dependent on the earlier event.

Is the bilking argument decisive? Not at all, but the reasons why not are less than entirely straightforward.

As Horwich argues (1987: 93–9), in cases where the bilking policy is successfully implemented (either the chief dances when the young men didn't act bravely, or he doesn't dance when they were brave), a response is available to the chief. He could reasonably insist that although his dancing increases the probability of bravery, and there is lots of prior evidence for this, it is not infallible – it is not nomologically sufficient for bravery in *all* conditions – and that since the conditions in which his dancing *is* effective are very delicate, no doubt the establishment of the new channels of communication disrupted them. However, while this reply does show that the backward causation hypothesis is not automatically refuted by the successful implementation of the bilking procedure, it does leave us with some puzzling *coincidences*. There are many ways that the bilking policy could be implemented (using radios, carrier pigeons, different stories fed to the chief, etc.), yet it turns out that every implementation changes the background conditions in such a way that the chief's dancing proves either ineffective or unnecessary, as the circumstances require. Coincidences such as these are not impossible, but they are definitely suspicious.

But suppose that the bilking policy doesn't succeed, and we cannot produce L when E doesn't occur, and when E does occur, we cannot prevent L from occurring. There are two ways for this to come about.

First, it could be that the types of event in question are nomologically connected. The proponent of the bilking argument construes this connection in a particular way: the bravery causes the chief to dance; the absence of bravery results in his losing his ability to dance. But why assume that the no-bravery–chief-can't-dance and bravery–chief-compelled-to-dance combinations establish the nomological dependency of the later event on the earlier event? If E is necessary for L, then L is sufficient for E; if E is sufficient for L, then L is necessary for E. Regarded in a neutral light, these results are just as compatible with the hypothesis that the bravery depends on the dancing as they are with the hypothesis that the dancing depends on the bravery. To maintain that in such cases forwards causation is more likely than backwards, on the grounds that causation *normally* acts forwards rather than backwards is clearly begging the question. However, as Horwich points out, the unorthodox interpretation is in one respect problematic. The earlier effects – the acts of bravery – all have their own *independent causal antecedents*, such as the young men's upbringing and the growing food shortage, which are in themselves sufficient to ensure that they would act bravely on this occasion irrespective of what occurs subsequently. So, assuming the efficacy of the dancing, we have a case of causal overdetermination. We are also presented with more puzzling coincidences. The chief's dancing may make the men act bravely, but it isn't responsible for the circumstances leading up to the hunt, which are *also* responsible for their bravery, and yet the two are correlated: whenever the chief dances, conditions prior to the hunt are such that the men will act bravely anyway.

The second reason why bilking attempts might fail is familiar from our earlier discussion of self-defeating loops – strangely *convenient* coincidences – such as when the young men don't behave as bravely as they might, the chief twists his ankle and can't dance, or the message from the hunting grounds is garbled in transmission and we end up believing that the young men *were* brave and so don't ask the chief to dance; there are many possibilities. The same applies to the cases where the young men act bravely: through some mishap or other, the chief always ends up performing his dance.

Horwich suggest that these results, taken together, show that the bilking argument fails: it turns out that there aren't any simple, easy-to-perform experiments that can decisively refute a backward causation hypothesis. It may well be that particular instances of backward causation are surrounded by various unexplained coincidences, but in the light of our discussion of self-defeating loops, this is to be expected. If backward causation does obtain in a given case, anyone who attempts to implement a bilking policy will be attempting to create a self-defeating loop and, since such loops cannot exist, their attempts will invariably fail, even though there will usually be no single unified explanation for these failures. The cases where it is hardest to believe this could come about are those in which a person is compelled to act in one way rather than another. A time traveller from the future tells you that on a certain date you will order such and such a dish in a particular restaurant: he has seen you do it. To thwart fate you decide to bilk. No matter what the cost, you are determined that you will be doing something else on the date in question. Perhaps you succeed; if so, your informant was not a genuine time traveller, or else was lying. But if he wasn't, if he really did see you ordering that dish in that restaurant on that day, then, no matter what you do, events will conspire to ensure that you end up doing precisely this. As for *why* events so conspire (or seem to), there is only one explanation: global consistency constraints.

8.6 Quantum retroaction

Anyone who remains unconvinced by the defensive strategies just outlined should not conclude that backward time travel (or causation) is an absolute impossibility:

there may well be universes where such travel is possible, but only in such a way that it is impossible to generate self-defeating loops. For example, the physical laws that make backward travel possible are such that objects always arrive billions of light-years away from where they set off – a distance that makes it impossible for the travellers to causally interact with their earlier selves. Time travel in this guise undeniably has less dramatic potential, but it is also far less problematic. And the same applies to instances of backward causation, where the earlier effect is so distant from the later cause that bilking is not an option.[5]

However, there is a further possibility that we have not yet considered, a way in which backward causation (and travel) could occur *locally* without the threat of self-defeating causal chains arising. The loophole was discovered by Dummett in the course of his exploration of the dancing chief case, but I will focus on Price's recent discussions of it, which broaden both the scope and the interest of the basic idea.

Unwilling to believe that the highly improbable coincidences required to protect backward causation from successful bilking could occur in our universe, Price believes that it is reasonable to conclude that physical law does not permit instances of backward causation that *could* be bilked.[6] However, from this it only follows that backward causation is impossible if bilking is always possible, and Price sees no compelling reason to believe that it is. To enact a bilking policy you must first discover whether the alleged earlier effect E has in fact occurred (only then can you decide to prevent or encourage the subsequent cause L). But suppose E is such that either (i) it is nomologically *impossible* to discover whether it has occurred before L occurs, or (ii) it is possible to discover whether E has occurred, but only by carrying out procedures that are themselves causally relevant to whether E occurs or not. In such circumstances the bilking strategy cannot be implemented successfully.

For most ordinary past events neither (i) nor (ii) obtains. It is possible, at least in principle, to discover whether the event in question obtained, and by means that do not influence the event in the relevant respects; that is, the event would have happened (or not) irrespective of our attempts at detection. It is, for example, plausible to suppose that the behaviour of the hunting party would not be affected by a concealed observer whom they did not realize was present. But there are conceivable circumstances in which this is not so. The case that follows combines (apparent) backward causation and inaccessibility of the relevant sort.

Imagine a gambling machine that drops a ball into one of two closed boxes, X and Y. Players bet on which box contains the ball, and choose which box is opened. The machine is so constructed that the boxes cannot be opened simultaneously, and the only (physically possible) way of finding out which box contains the ball is by opening the box; X-rays and so on either don't work or cause the device to self-destruct. As it happens, it is always the chosen box that contains the ball, and gamblers soon realize that they can make a lot of money by opening the box on which they have placed a bet. As Price notes, "It would seem natural for players to speak of choosing to open box X, *in order to ensure* that the ball *had* dropped into it" (1984: 310, original emphasis), thus invoking backward causation.

We are not obliged to adopt this view as there are two alternative explanations. The first is the *discontinuity option*: if the ball has dropped into box Y, it moves instantaneously to box X when the latter is opened. The second is the *indeterminacy option*: until box X is opened the ball is in an indeterminate condition – it is located in neither box – and opening X instantly causes the ball to acquire a determinate location. If we are prepared to countenance instantaneous travel, or objects with indeterminate properties, we might adopt either of these accounts. Or, alternatively, we could accept the reality of backward causation: the ball falls into box X because this is the box that the gambler *will* open. It would be wrong to reject this account on the grounds that backward causation could lead to paradox or (nomologically) impossible coincidences, for in this case they can't. Since there is no way of finding out where the ball has fallen until the box is opened, bilking is not an option.

This example is of more than curiosity value, for in key respects it is analogous to situations that are commonplace in quantum physics. According to quantum orthodoxy, the state of a quantum system is indeterminate in certain respects until a measurement is conducted; the act of measurement results in a "collapse" (of the wave-function) yielding determinate properties. It is very odd to think that a particle can fail to have a definite location or momentum (and just as odd to think that it can have a combination of different locations and momenta) but, in fact, experimental findings make this a plausible view to hold.

Price cites an example due to Putnam (1979). Suppose that S is a system containing a large number of atoms, R and T are two incompatible properties and A and B are two different states about which the following are true.

1. When S is in state A, 100 per cent of the atoms have property R.
2. When S is in state B, 100 per cent of the atoms have property T.

Quantum mechanics predicts that there is a state C of the system – a special kind of combination of A and B – of which the following are true.

3. When S is in state C, 60 per cent of the atoms have property R.
4. When S is in state C, 60 per cent of the atoms have property T.

And these predictions are confirmed by experiment. Needless to say, this is a very perplexing result. If ten marbles are placed in a box, and you know that they are red or green, and if you discover that six are red, then you would reasonably conclude that four must be green; how could you count out six red ones and also six green ones? We are saved from an outright contradiction because it turns out that it is impossible to test both 3 and 4 for a given system S. If we test 3 the system is disturbed in such a way that 4 cannot be tested; likewise *mutatis mutandis* if we first test for 4.

Putnam suggested that we can only make sense of such a situation by rejecting "The Principle of No Disturbance" (ND), according to which performing a measurement on a given system does not disturb what is being measured, so the value of the relevant quantity is very nearly the same an instant before the measurement is made as when the measurement is actually made. This is an understandable reaction. Since it is impossible for 60 per cent of the atoms in S to have *both* R and T prior to the measurement – the properties are incompatible – it seems reasonable to suppose they have neither until the measurement is conducted, in which case, the act of measurement must itself be causing the system to enter a determinate state. However, Price argues, in concluding that we *must* reject ND, Putnam is ignoring an alternative explanation, one that relies on backward causation. The procedure for measuring R is different from the procedure for measuring T, so it could be that if we decide to put 3 to the test, the fact that an R-measurement *will* be carried out causes the system to enter the predicted state – 60 per cent of the atoms having R – right at the start. The atoms thus have determinate properties *before* they are measured, but only because they are causally influenced by a future event. And the same would apply if a T-measurement were to be conducted. The situation is a probabilistic analogue of the gambling machine: the ball falls into X because X will be opened; if the gambler had bet on and opened Y, the ball would have fallen into Y instead.

Backward causation can be deployed in other quantum mechanical contexts. So-called "EPR-correlated" particles (which we shall be encountering again) are typically pairs of particles that are produced by a single source event, which then speed away from one another while remaining intimately related: if at some later time a measurement is conducted on one of the particles, the other is instantaneously affected in a particular way. Or at least, so runs the orthodox interpretation. This result is particularly problematic since instantaneous interactions seem to be ruled out by Einstein's special theory of relativity.[7] But there is no need to postulate instantaneous action-at-a-distance if we introduce backward causation into the picture, and stipulate that the future measurement retroactively determines the states of *both* particles at the moment of their creation.[8]

This radical proposal is highly controversial, but so, too, is every other current interpretation of quantum theory, and it does have the merit of replacing several bizarre phenomena (indeterminacy, non-local influence) with just one: backward causation. There is also the striking fact that, since the rules governing quantum systems render bilking impossible, the principal objections to backward causation can find no grip. Is this merely a coincidence?

8.7 The inexplicable

Backward time travel opens up the possibility of loops of yet another kind; not self-defeating (or at least not obviously or directly), but nonetheless peculiar and intriguing in their own right. Since they also give rise to new issues, they are worthy of a brief mention.

- *The Marriage Decision* Mary is torn between two suitors. She can't decide whether to marry Tom or Jack. So she travels to the future and finds out that she is happily married to Tom. She then travels back, and marries Tom for this reason.
- *The Discovery* You are looking for a lost sock one day, in the depths of a drawer, and come across a book that you haven't seen before. It seems to contain details for a time machine. You dismiss it as a prank, but years later, now an eminent scientist, you realize that the design in the book is a serious piece of physics. You build the machine, and find that it works. After years of travelling back and forth and many adventures, you realize that you still don't know how the book with the design for a time machine got into your drawer. So you travel back to your home on the day you originally found the book to

see who plants it. The minutes pass, no one comes, and you remember that your original self will return in about five minutes. Suddenly you realize what has to happen. With a feeling of unease, you place the book in the drawer (happily you had it on you) and quietly withdraw from the scene.

• *The Big Loop* Many years from now, a transgalactic civilization has discovered time travel. A deep-thinking temporal engineer wonders what would happen if a time machine were sent back to the singularity from which the big bang emerged. His calculations yield an interesting result: the singularity would be destabilized, *producing* an explosion resembling the big bang. Needless to say, a time machine was quickly sent on its way.

There are many similar tales.[9] Are they coherent? Although there are no glaring inconsistencies, closer scrutiny reveals that some loops are more dubious than others.

A useful starting point is a distinction between loops in which objects come full circle, and those that involve only information. Some object-loops are certainly impossible. In *The Discovery* a young man finds a book in his drawer, reads it, carries it about for a number of years, and then replaces it in the drawer for his younger self to discover. Unless the book is made of a miraculous material that is entirely immune to the ravages of ordinary wear and tear, the book that is replaced will be in a different physical condition from the book that is discovered; since they are supposed to be one and the same, this is impossible, so the envisaged scenario could not take place. This may seem odd: what is to *prevent* your handing the book over to your younger self? But this is just the problem of the self-defeating loop in another guise: since you can't do the impossible you *didn't* hand the book over, and since you didn't, you can't. *The Big Loop* also involves an object – the time machine itself – but since the machine does not coincide with itself in two different conditions, the paradox of the book does not arise. The loop here is of the causal rather than object-involving variety.

Now consider a variant of *The Discovery*. One day, in your youth, you receive a mysterious email. It seems to provide the details of how to build a time machine. Years later you send the mysterious email to your earlier self. Here there is no object paradox, but there is still a puzzle. Where did the information about how to build a time machine come from? It just seems to *be there*, present throughout a closed loop. Opinions differ as to how problematic this is. Lewis sees no difficulty:

> But where did the information come from in the first place? Why did the whole affair happen? *There is simply no answer.* The parts of the loop are explicable, the whole is not. Strange! But not impossible, and not too different from the inexplicabilities we are already inured to. Almost everyone agrees that God, or the Big Bang, or the entire infinite past of the Universe, or the decay of a tritium atom, is uncaused and inexplicable. Then if these are possible, why not also the inexplicable causal loops that arise in time travel?
>
> (1976: 140, original emphasis)

In the light of *The Big Loop* the big bang might *not* be entirely inexplicable, but this quibble aside, Lewis seems to have a point. Deutsch and Lockwood (1994) disagree, arguing that informational loops violate a metaphysical constraint that they attribute to Popper: *knowledge comes into existence only by evolutionary, rational processes*; solutions to problems do not spring fully formed into the universe. But is this "no free lunch" constraint on information really metaphysical? According to quantum theory it is *nomologically* possible (although extremely improbable) for complex objects to appear from nowhere, random distillations from the quantum foam; classical thermodynamics carries a similar implication. An edition of the complete works of Shakespeare is an object like any other. Does the fact that it carries information mean that it is an object of a special kind, and distinguished as such by physical law? It seems unlikely. And there are other ways in which Popper's constraint could be violated. Set enough monkeys randomly tapping on typewriters long enough, and sooner or later a book containing any information you like would be produced. If information can spring into being in these ways, why not in the time travel cases too?

8.8 Voyaging in dynamic time

Time travel may be possible in a static block universe, but what of dynamic worlds? On the face of it the prospects are dim. The seriousness of the "no destination" problem – as Grey (1999: 56) aptly calls it – is most obvious in the case of presentism. Even if we follow the dynamic (or compound) presentists, and suppose that one present succeeds another, it is hard to see how travel to another time period is possible, since at the moment of departure there are no times other than the actual present where one might arrive. The situation is rather different for the growing block model. Having constructed (what I believe to be) a working time machine here in the year 2001, I might think "Since the future is not real as of now, I can't travel there. Fine. But the past is real as of now, so there is nothing to prevent me from paying it a visit. I'll start with the sixteenth century!", which sounds reasonable enough. However, the situation looks very different from the perspective of the selected time in the sixteenth century: as of *this* time the future is not yet real, so nobody could arrive from it; since nobody *could* arrive from it they didn't; and so my time machine can't possibly materialize in the sixteenth century. Or at least, it could only arrive by changing the past, which – for growing block theorists – is simply not possible.

Is the "no destination" problem insuperable? It certainly seems so if we confine our attention to time jumps *within* a dynamic world, but if we lift this restriction, and consider more fanciful scenarios, matters are no longer so clear-cut. Suppose it were possible to build a machine that could *leave* this universe (and so our time) and re-enter it at some different location. Were you to depart from the twenty-first century, spend some (personal or proper) time in this atemporal "limbo" and then re-enter our universe in the sixteenth century, it is not clear that the original objection still stands: for you are no longer arriving directly from a non-existent time. If it is conceivable that someone originating outside this universe could materialize at some location in the sixteenth century (for the sake of the argument, grant that it is), why couldn't *you* do likewise? Why should it matter that you spent some time in this universe before leaving it?

It is probably fruitless to spend time trying to work out whether scenarios as far-fetched as this really make sense, especially when there is a slightly less far-fetched possibility to consider. Even if *jumps* through dynamic time are impossible, the same may not apply to *slides*.

Recall the Gold universes introduced in §4.6. I noted in §6.5 that, if we lived in a finite and symmetrical universe, we might one day encounter – or at least find signals originating from – inhabitants from the "other side", who, from our perspective, are living their lives backwards. If we assume that our universe is a dynamic block, we naturally suppose that it is growing in the direction of our future, and so the lives of the time-reversed beings are being built up, layer by layer as it were, in the "wrong" order. But of course, from *their* perspective, it is *our* lives that are being miraculously put together in the wrong order. Unless there is some underlying invisible asymmetry in such a world, there will be no fact of the matter as to which of these views is correct. Suppose that a spaceship from the other side penetrates our space, and manages to shield itself from dangerous interactions with our matter (which might well cause explosions of the matter/anti-matter sort). Those on board could observe life on Earth from afar, and by the judicious use of suspended animation, selectively "visit" any times in our history that they wished to see. They would, in effect, be travelling into our past. Now for the speculative leap. Aware of the existence of these "watchers", we decide to make contact of a physical sort; we devise a way of changing our matter into its time-reversed counterpart (perhaps by generating a field that modifies the appropriate quantum signatures – a ploy not unknown in science fiction). Being an unusually adventurous historian, you volunteer for the process, which in due course is carried out successfully, and join the time-reversed ship. With your new companions, you embark on a voyage back through Earth's history – a history to which you have *always* been a witness.

I am not claiming that a scenario of this sort is a genuine physical possibility. But it does illustrate that backward time travel, of a sort, is at least conceivable in worlds where time is dynamic rather than static. The same method could be employed in presentist worlds of the dynamic variety.

8.9 Real time

Once again our results are inconclusive. Had we found an incurable logical contradiction in the very idea of backward time travel, we would be in a position to rule that any physical theory that permits such travel must be false on a priori grounds, but since no such contradiction was found, we are not in this position. From a purely conceptual point of view, forward time travel seems entirely unproblematic, backward travel likewise, provided it is restricted in ways that prevent the establishment of self-defeating loops; even unrestricted backward travel can occur in universes where coincidences are improbably common. And this is just for a one-dimensional static block time. In multiverses, and worlds where time is two-dimensional, the past is changeable (after a fashion) and backward travel is not especially problematic.

Does any of this help us in pinning down what *our* time is like? The logical possibility of time travel saves the block theorist from potential embarrassment, but from the mere fact that time travel is possible we cannot conclude that we live in a block world: we have seen that certain forms of time travel are available in dynamic universes, and it could be that time travel in our universe is nomologically impossible. So there are two questions we need to consider. What forms of time travel are nomologically possible in our universe? What models of time are these modes of time travel compatible with?

In addressing the first of these questions we have no option but to take current physics as our guide. We know that the relevant theories are not fully adequate, but we have no clear idea as to what will replace them, and the implications of potential successor theories for time travel are even more obscure. Moreover, unless our current theories are completely misguided, they at least provide *some* indication of what can and cannot occur in our universe. Although many physicists doubt that our laws of nature do permit time travel and/or backward causation, not all share this view, and there is a fair amount of consensus on how the laws *might* allow these phenomena to occur. There are two main routes.[10]

We have already seen how some theorists appeal to backward causation to explain the bizarre behaviour of some quantum systems. Since the causal influence is transmitted via quantum fields, we are dealing here with perturbations (whether wave-like or particle-like) that travel backwards through time. The Wheeler–Feynman absorber theory of electromagnetic radiation is another example of the same sort of phenomenon: the "advanced" wave that this theory postulates travels backwards in time; likewise tachyons, hypothetical particles condemned always to travel faster than light, and backwards in time. Although there is no evidence that such particles exist, it is at least arguable that their existence is compatible with current theories. For the sake of the argument, let us suppose that there are waves or particles that travel backwards through time. Can we conclude from this that we live in a block universe? No. If particles could jump *instantaneously* from one time to another, it is hard to see how we could resist the conclusion that the times in question coexist. If such a particle were to emerge now, having jumped from a future time, we would have to accept that this future time is as real as the present. But since the phenomena in question travel *continuously* through time, their existence at one time does not entail the existence of any later time. The growing block theorist will regard the history of an alleged backward travelling particle as being built up, slice by slice, in just the same way as an ordinary particle. The physical properties of such a particle might be odd, but in itself this does not mean that the future times to which they point are actual.

There is a second way in which backward time travel might come about, one that is potentially more damaging from the point of view of dynamism. According to Einstein's general theory of relativity (which we shall be looking at in more detail in Chapters 18 and 19), the shape of the spacetime fabric is affected by the presence of mass (or energy). There are certain spacetime structures – created by very special distributions of mass – that permit backward time travel (the structure of both space *and* time is mass-dependent). Some of these structures arise naturally in universes very different from our own, for example spacetimes entirely filled with rotating matter-fields. But there may be ways to create the conditions required for backward time travel in universes such as our own. One of these hypothetical mechanisms is potentially lethal to the dynamist: *wormhole time machines*.

A wormhole is a tunnel that makes shortcuts through space possible. Think of two ways of getting round a mountain: the long winding road that skirts the mountain, and the tunnel that cuts directly though it. The mountain tunnel is a hole through a three dimensional object; it is hard to visualize a tunnel through three dimensional *space*: where would the tunnel go? Tunnels between different regions of space certainly *are* possible (at least logically and mathematically), but to visualize them we would need to imagine a three-dimensional space curved back on itself, which is something we can't do. So focus on a two-dimensional space; think of a piece of paper folded over so that the two opposite ends nearly touch one another. A two-dimensional being confined to the paper wouldn't realize that one side of his world is very close to the other, and this proximity would be of no use, since it only exists in the third dimension, to which our being has no access (indeed, we can suppose that the envisaged two-dimensional space is not embedded in a three-dimensional space). But given the shape this two-dimensional space now has, we could easily create a shortcut by extending the space, for example by bridging the gap with a small flap or tube of paper. Figure 8.3 illustrates a wormhole bridge between Earth and Epsilon Eridani, with one spatial dimension suppressed; the "long way round" is a distance of 10.5 light-years (or roughly 60 000 billion miles) but the wormhole route is only 4 miles.

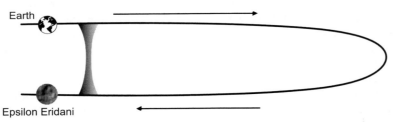

Figure 8.3 A wormhole shortcut between two distant planets.

Wormholes of this sort can cut down travel times, but they are not time machines. To turn a wormhole into a time machine, a temporal discrepancy between the two mouths has to be created, so that passing through the wormhole amounts to taking a shortcut between *times*. We know from Einstein's special theory of relativity that this is easily done: it would suffice to take one of the wormhole mouths on a high-speed round-trip (cf. Thorne 1994: Ch. 14). If a temporal wormhole a hundred metres long connected London in 1960 with London in 2010, the fifty-year interval could be crossed in a matter of minutes, or even seconds. By establishing a network of such wormholes, it would be possible to travel back and forth between distant times, and times in between. Figure 8.4 is somewhat misleading, in that the wormholes appear no shorter than the spacetime regions that they connect; this is an unavoidable artifact of trying to represent a variably curved four-dimensional spacetime on a two-dimensional surface.

Since a structure of the sort depicted in Figure 8.4 could only exist in a block universe, should we ever find ourselves able to take this sort of shortcut we can be certain that the times in question all exist. This form of time travel thus opens the way to an *empirical* way of determining whether the past and future are as real as

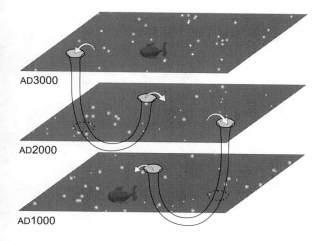

Figure 8.4 A network of wormhole time machines.

the present. Unfortunately, since the jury is still out on whether the laws of physics will permit such wormholes to exist, and even if they do there is no guarantee that we will ever be able to create them ourselves (or stumble across them on our travels), it seems unlikely that we will be carrying out the experiment in the near future. If we want answers to the great conundrums posed by time, we have no option but to employ more traditional methods of inquiry, with all their limitations – for the time being.

9 Conceptions of void

As I mentioned at the outset, there are various metaphysical issues connected with space: structure at the very small and very large scales, dimensionality, uniqueness, topology and geometry. But, as with time, I am going to concentrate on a single central topic: is the *substantival* view of space correct, or is the *relational* view correct? Is space an object in its own right, in addition to the material bodies that occupy it, or does it consist of a network of relations between material bodies?

This question is quite unlike that of the passage of time. We could make a good deal of progress on the latter without considering what science has to say; purely metaphysical considerations ruled out some models of time and empirical facts concerning experience ruled out others. The existence of space is different; purely metaphysical considerations don't get one very far, and little that is relevant can be gleaned from considering the structure of our experience. Scientific considerations will thus be entering the scene very quickly. In this preliminary survey I will set the stage by distinguishing some very general ways of thinking of space, a discussion that will yield some general constraints on a satisfactory account of it. I will then try to clarify the substantivalist and relationist views.

9.1 Space as void

For many of us these days, a conception of space that has considerable intuitive appeal runs thus: space is an infinite expanse of *featureless emptiness* within which physical bodies are located and move. It is not surprising that we are drawn to this idea, for we are brought up to think of the Earth as a planet revolving around the Sun, which is just one star out of billions, strewn through the vastness of galactic and intergalactic space. The so-called "outer space" that separates the planets, stars and galaxies is, we are told, a hard vacuum, empty save for the odd molecule, perhaps one per cubic metre. An initial characterization of this view of space might run along these lines:

> *The Void Conception:* Space in itself is nothing at all; it has no intrinsic features of its own, it is mere absence. Objects can be separated by different spatial distances – London is closer to Paris than it is to New York – and we know this because of the different amounts of time it takes to travel or transmit signals between them; we cannot directly measure magnitudes of space, since space is itself featureless void.

Although for most of us this naïve picture is later qualified when we learn of the vast quantities of electromagnetic radiation being pumped out by the stars, the underlying idea of space *itself* as empty nothingness, devoid of intrinsic features, not only remains in place, but retains a good deal of appeal. After all, if we removed all particles and fields from a region of space, what would be left? Absolutely nothing.

If we do assume that space is featureless emptiness, several further ideas can seem compelling. Since there is always more nothingness beyond any alleged limit to the universe, space cannot have an outermost limit, and so space surely must be infinite in all directions. If space is infinite, there is surely just *one* space, for if there were two distinct spaces, where could the other one be located? Not *within* our space, clearly, but since our space has no boundary, it is hard to see that this other space could exist outside our space. The question "What structure does space have?" seems empty: if space is just *nothing*, how can it have a structure?

Now, although this view of space may seem quite natural, it was not always so. It is certainly true that the equation of space with pure nothingness resembles some of the (early) Greek atomists' conception of the void within which atoms are located. According to Torretti, the "atomist's void, though boundless, is not an expanse that *contains* everything, but rather the nothingness *outside* all things" (1998, original emphasis).[1] But the void soon fell out of favour. According to the cosmologies of Aristotle and Ptolemy, which dominated scientific thought until the late Middle Ages, the universe is divided into two regions: the heavenly and the sub-lunary. The heavens consist of a nested series of rotating shells, carrying the Moon, planets and stars; the Earth is at the centre of the universe, surrounded by the lunar sphere; the sub-lunary realm is filled throughout with matter, water, fire and air. Since Aristotle believed the notion of a vacuum to be incoherent, he argued that it made no sense to suppose that space continues on past the outermost sphere of the heavens. In fact, Aristotle often talks not of space but of *place*, which he defined as the inner surface of the first *motionless containing body*.[2] Descartes also rejected the void, arguing that interplanetary space was filled with a subtle fluid. Although Newton, the most influential "modern" thinker on these questions until Einstein, rejected Descartes's material plenum in favour of an infinite absolute space, he was also firmly committed to a substantival view of this space. As he said in *De Gravitatione et Aequipondio Fluidorum*: "space is eternal in duration and immutable in nature . . . Although space may be empty of body, nevertheless it is not itself a void; and *something* is there, because spaces are there, although nothing more than that" (Huggett 1999: 113).[3]

In rejecting the void these various thinkers were often motivated by different considerations, which I will not enter into here (although we will be looking at Newton's position in some detail later on). I will restrict myself to pointing out two respects in which the void conception is problematic. The exercise is well worth the trouble, for isolating these deficiencies will provide us with two criteria for a satisfactory account of the nature of space.

9.2 The unseen constrainer

Suppose we could create a "Flatland": a very thin two-dimensional plane inhabited by two-dimensional creatures.[4] Those who like their imaginary cases fleshed out with some harmless science-fictional detail can take it that Flatland is an intense tightly confined electromagnetic field; the occupants of Flatland – both the sentient Flatlanders themselves, and the objects in their world – consist of intricate self-organizing but potentially mobile patterns of perturbation within this electromagnetic membrane. Current forms of so-called "artificial life" only exist as computerized simulations, but it may not always be so.

The Flatlanders are, we can stipulate, ignorant of the fact that their world is suspended a couple of feet above the floor of a laboratory between the field-generating devices. They are perfectly free to move around in their flat planar world, but they cannot leave it: they can move north, south, east and west, but not up or down. Not only are the Flatlanders unable to move in the third dimension, but they are wholly unaware of its existence; indeed, their language does not contain terms for the directions that we call "up" and "down". Their perceptual systems are geared to their two-dimensional world (their visual experience takes the form of a thin coloured line), and their cognitive systems are attuned to their perceptual experience. As a consequence they literally cannot imagine spaces or objects that have more than two dimensions; although they are fully acquainted with objects such as squares and circles, they can no more visualize cubes or spheres than we can envisage "hypercubes" or "hyperspheres", the four-dimensional counterparts of three-dimensional cubes and spheres.

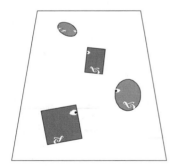 

Figure 9.1 Two conceptions of Flatland. The Flatlanders on the left are free to rove where they like on their plane. The Flatlanders on the right also inhabit a plane, but are further confined to moving back and forth along their world's surface (the thick line). Consequently, these Flatlander's would have to climb over the tree to get past it (likewise for each other). Note also the way their internal organs are clearly visible from the vantage point of the third dimension.

We may not be able to visualize four-dimensional objects but, logically and geometrically speaking, a fourth spatial dimension is perfectly possible, and mathematicians are familiar with the geometrical properties of objects such as hypercubes and hyperspheres.[5] There is thus a very real sense in which we find ourselves in the same predicament as the Flatlanders. If our three-dimensional space existed as a "plane" within a four-dimensional space, someone enclosed in a locked room could escape without opening the door, or burrowing through the walls: they could (in principle) be lifted "upwards" into the fourth dimension and then set down again somewhere else in three-dimensional space, just as we could rescue a Flatlander trapped within a square. Inhabitants of four-dimensional space could peer into our three-dimensional plane and see our internal organs laid out before them, just as easily as we can see the insides of the Flatlanders (see Figure 9.1), a perturbing thought to say the least. And an utterly baffling one. The idea that there could be another spatial axis at right angles to each of the familiar three, and hence another two directions in which to move, is not one that we can make any intuitive sense of; there just doesn't seem enough *room* in space for any other directions. Since we cannot visualize this additional spatial dimension we can only grasp it by analogy. Flatlanders find the idea of a third dimension hard to grasp, but *we* know perfectly well that a third dimension is possible and, indeed, Flatlanders could discover the same for themselves by considering the plight of the inhabitants of one-dimensional "Lineland" and reasoning analogically.

We will be taking a closer look at Flatland later on, but this brief introduction is all that is needed for the simple point I have in mind. If you were to look down on the Flatlanders going about their business, you might find yourself thinking "Ignorance is bliss. How astonished and disturbed they would be if they ever discovered that they were surrounded by an extra spatial dimension, that in addition to the directions they know about – north, south, east and west – there is also an *up* direction and a *down* direction!" The same applies to us. But assuming that our universe is the real thing (rather than a simulation run by four-dimensional beings experimenting with artificial life), this is not something that we have to worry about. There clearly are only three spatial directions, at least at the macroscopic scale.[6] We do not encounter objects appearing as if from nowhere; objects do not vanish into thin air. There are only three spatial axes (backwards–forwards, left–right and up–down) within which we can move.

But now suppose that the void conception of space were true. If space is simply *nothing* why should our possibilities for movement be so tightly constrained? If space is pure void, not only would we be free to wander in the additional directions made available by a fourth dimension, but there would be nothing to prevent our wandering in the limitless different directions made available by 5-, 6-, 12-, 101-, and *n*-dimensional spaces, which are just as possible, mathematically speaking, as 1-, 2- and 3-dimensional spaces. The big bang would not have created a sphere of energy expanding in just three spatial dimensions: it would have been an explosion into *n*-dimensions, and as energy condensed into matter, the resulting objects would themselves have anything up to *n*-dimensions. Clearly, our universe is nothing like this. Every macroscopic object in the known universe not only seems three-dimensional, but its possibilities for movement are similarly restricted. But if space is simply *nothingness*, this would not be the case, or at least, there is no obvious reason why it should be the case.

Of course, the fact of our confinement to three dimensions is not at all evident to us; we naturally suppose that the familiar three dimensions are the only possible dimensions, and so we naturally suppose that nothingness itself can have only three dimensions. But once our eyes are opened to the logical possibility of four-, five- and higher dimensional spaces, the fact of our confinement has to be faced, and any satisfactory theory of space has to explain these unseen constraints in some way.

9.3 Connection in question

You are in a dark room. You know that you are not alone, and you want to locate the other person without making any noise. One way to do it would be to wander about with arms outstretched until physical contact is achieved, but you don't want to run the risk of stumbling over pieces of furniture. Happily, you are equipped with a torch. You shine the torch in different directions, and soon locate your companion. Facts of this sort – that objects can move and collide with other objects, that objects are connected by paths through space along which light can travel – are so familiar that we find them utterly unpuzzling; it is just the way things are. However, this sort of fact *would* be puzzling if the void conception were true. For one thing, as we have just seen, there would be many more possible directions in which objects could be located than we usually suppose, so it might take you a long time to point your torch in the right direction. But the problem runs deeper: if space is mere absence, the very possibility of spatial connection becomes problematic, and with it the notion of spatial distance, since the only way we have of measuring expanses of void is in terms of the time taken to travel between spatially separated objects. To bring this point clearly to the fore, I will again make use of a thought-experiment.

We find it natural to suppose that we live in a single all-embracing spatial frame-work: at any given time, any object, of any kind, no matter where it is located, is some definite spatial distance away from every other object. The idea that space is *necessarily* unified in this way can be undermined by describing an imaginary scenario in which it would be reasonable to believe that one was living in a two-space universe; that is a universe consisting of two collections of spatial locations, S_1 and S_2, such that every member of S_1 lies at some distance from every other member of S_1, and likewise for the members of S_2, yet no member of S_1 lies at any distance from any member of S_2, and vice versa. Probably the best-known scenario of this type is Quinton's Lakeside village story (1962). What follows is based on a variant devised by Newton-Smith (1980: Ch. 4).[7]

The world "Pleasantville" is the sole planet orbiting a solitary star located in otherwise empty space. Pleasantville is a near-perfect sphere. It is always warm and sunny, but otherwise it is much like the present-day Earth. The inhabitants of Pleasantville discover that eating a certain root causes them to vanish, only to reappear some time later. When a root-eater reappears, they invariably bring back tales of another world, "Harshland", a cube-shaped planet orbiting a solitary star located in otherwise empty space, which is always cold and inhospitable. On vanishing from Pleasantville, root-eaters finds themselves on the surface of Harshland, where they amuse themselves as best they can until their return. People who eat the root together are "transported" to Harshland as a group. People who eat the root at different places on Pleasantville end up at different places on Harshland. The various descriptions of the features of Harshland and the events that take place there brought back by different root-eaters invariably agree.

Different interpretations of this scenario are possible. It could be argued that the root induces intermittent existence accompanied by collective delusions, but given the intersubjective agreement on the character of Harshland, the hypothesis that the latter is as real and objective a world as Pleasantville seems more reasonable (and certainly *just as* reasonable). Another option would be to maintain that Harshland occupies some remote region of Pleasantville space, and that the root induces near-instantaneous spatial relocation. But there are no stars to be seen in the night skies of either planet. The Pleasantville folk once travelled far and wide through their universe, but never detected any trace of any star other than their own; they gave up looking when they discovered that light in their universe travels instantaneously, so if there *were* any other stars in their universe, they would be visible. Since the Harshland star is nowhere to be seen, the hypothesis that Pleasantville and Harshland each exist within a separate *space* is certainly a reasonable one.

In response, it could be maintained that the two planets exist in different three-dimensional spaces embedded in a common four-dimensional space (analogous to the way that any number of two-dimensional spaces can be "stacked" like parallel sheets of paper in a three-dimensional space). But, as Newton-Smith points out, unless we are provided with some reason for thinking this to be the case, the hypothesis has nothing to recommend it. If Pleasantville scientists possessed a well confirmed physical theory that explained why it was impossible to travel from one of these three-dimensional spaces to the other, and could provide data on the relative locations of these spaces in the encompassing four-dimensional space, then matters would be different; but, as it happens (we can suppose), no such theory is forthcoming. Yet another option would be to claim that, on disappearing from Pleasantville, the root-eaters remained within the same space, but travelled either forwards or backwards in time, to the remote temporal location of Harshland. But the facts count against this: the scientists of Pleasantville are able to roughly estimate the age of both their planet and their universe. When the age of Harshland is measured by the same procedures, a similar result is found. The idea that Harshland is located within the Pleasantville universe, but at some distant point in the past or future, can thus be ruled out; the time travel hypothesis seems quite *ad hoc*. The one-time/two-space hypothesis is the simplest available.

This story shows that our experience could be such as to warrant the postulation of two distinct spaces, which is an interesting result in its own right, but I introduced the scenario simply to undermine the complacent assumption that *all* objects are *necessarily* spatially related. There could be two objects that are not connected by any spatial path. Bearing this point in mind, let us return to the void conception.

To simplify, let us suppose that extended material objects are *intuitive solids*; that is, their matter completely fills their spatial volume – they are not composed of particles separated by space. Now consider this proposal:

> The idea that the universe could consist of material objects dispersed through a void at different distances from one another is simply incoherent, or at least it is a seriously misleading way of describing a state of affairs that is better characterized in terms of *many worlds*. For consider, each solid object

constitutes a three-dimensional space in its own right. Take any two points in an object O_1. There is a continuous spatial path through the matter in O_1 that connects these points. Likewise for any two points in a distant and distinct object O_2. But is there any spatial path between any point in O_1 and any point in O_2? There is not: since (by hypothesis) these objects are separated from one another by an expanse of void, there is absolutely *nothing* between them, or that links them, so the objects are entirely unconnected. Any two material objects are properly viewed as distinct and independent spatial worlds in their own right. Since there are no spatial paths or distances between two distinct objects, it is impossible for one object ever to collide with another. If an object breaks into two parts, these parts would not drift gradually away from one another, they would each immediately come to constitute isolated worlds in their own right.

Once we accept the possibility of multiple unconnected spaces, this proposal is by no means absurd. Anyone who says that the *only* connection between two spatially separated objects is an expanse of void does seem vulnerable to the objection that a connection of this sort is no connection at all, and hence that there is no reason at all to suppose that the objects in question inhabit the same space; and if they don't inhabit the same space, each object constitutes a space (or universe) of its own.

Once again, we know perfectly well that objects in our world *do* inhabit a common space: objects can collide with one another, light and other signals can pass from one object to another, and so on. What should now be clear is that this is by no means a trivial matter. There could easily be a collection of objects that are neither connected nor connectable in these ways.

So we have a further requirement that a satisfactory account of space must meet. In addition to providing an explanation of the *constraints* on spatially related things, such an account must provide some explanation of the distinctive ways that such things are *connected*. This is something the void conception signally fails to deliver.

9.4 Substantivalism: a closer look

The substantivalist solves the problems of constraints and connections in a direct and intuitively appealing way. Consider:

> *The Fluid World*: an infinite and infinitely divisible three-dimensional liquid plenum, a great ocean. The intelligent inhabitants of this world are jelly-fish; they are mostly liquid themselves, they float and swim about, feeding on jelly-plants and they live in tunnels in jelly-planets. Since they are fully permeated by Fluid, the jelly-fish don't notice or feel the Fluid as they move through it. The presence of the Fluid is completely undetectable.

There is reason to suppose that in this world the Fluid *is* space: it is, after all, that within which other objects are located, but which is not itself located within anything else. No less importantly, the Fluid constrains the movements of physical objects: the various jelly-things move by floating through the Fluid, and they are confined to moving in three dimensions because the Fluid is itself three-dimensional. If in some regions the Fluid is four-dimensional, the jelly-fish will have an extra spatial dimension within which to move; if in other regions the Fluid is two-dimensional, they will (if they can enter such a region) have only two dimensions within which to move. If there are any empty regions to be found in the Fluid, these cavities should not be thought of as *regions* of space, but rather as *holes in space*, and the jelly-fish would only be able move around them, not through them. Similarly, if the Fluid were finite in volume, rather than infinite, then space itself would be finite and have edges beyond which no traveller could pass. As is also clear, there is no connection problem in such a world. Objects are not separated by expanses of nothingness. They are separated by expanses of Fluid, and so there is no difficulty in understanding how they can move closer together or further apart. Should the Fluid undergo fission, and become divided into two unconnected expanses, space itself will have divided into two unconnected regions, and the occupants of each of these regions would be cut-off from one another: there would no longer be any spatial relations, or distances, between the two.

You might be inclined to object: "You are misdescribing this world. The Fluid can't be a space, it's the wrong sort of thing: it is merely an object like any other . . . it's simply too *substantial* to be a space!" This objection is natural, but misguided. The idea that space *could* be substantival or thing-like initially strikes many people as very odd. Isn't saying that space is substance-like akin to saying that something can be nothing, or nothing can be something? But this thought is rooted in the void conception, which as we have already seen is problematic. On the assumption that space *is* substantival, why does the Fluid seem *too* substantial to be a space? Surely, an entity that was entirely featureless, entirely lacking in properties, wouldn't exist at all. In which case, if space is substantival it must have a nature of some kind. If it has a nature of some kind, why can't the Fluid be *a* space? I am not suggesting that our space is akin to this hypothetical Fluid – it may well be that substantival spaces in different worlds have different intrinsic characteristics – but a substantival space must have *some* nature to call its own.[8]

Once this point is fully taken on board, there seems no need whatsoever for a substantival space to be transparent to its inhabitants, or, more generally, for it to be bereft of detectable physical features. Imagine a variant of the jelly-world where moving through empty space is accompanied by a resistance or drag that affects different sorts of object to different degrees; or where motion creates slight ripples in space that result in objects themselves undergoing small but observable movements; or a variant where the Fluid is coloured, and is seen as such by the inhabitants.

There is a general question that any substantivalist must answer that we have not yet considered. When a substantival space contains an object, by virtue of what is the object located within that space? What is the *relationship* between a space and an object within it? It may seem as though nothing interesting can be said here. Isn't it obvious that objects are located *at* places in space, what more is there to say? That there is something more to be said is obvious as soon as the following possibilities are distinguished.

- *Relational substantivalism*: space and material objects are equally basic types of entity; there is a primitive relationship of "spatial locatedness" that holds between objects and places within space.

- *Container substantivalism*: space and material objects are equally basic types of entity; material things are enclosed by or embedded within substantival space, but there is space only outside and between material things.
- *Super-substantivalism*: space is the basic material object; other material objects consist of this space taking on various properties. Objects are thus "adjectival" on space; space is the *only* basic physical entity.

The space of the Fluid World as described above doesn't fall neatly into any of these categories, but is probably closest to the third, since the jelly-things are entirely permeated by Fluid and couldn't exist without it. If we modify the scenario, and suppose that the only occupants of the Fluid are regions of colour, and that these regions consist of the points within the Fluid possessing monadic colour-properties, we have a clearer instance of super-substantivalism. By varying the scenario again, and supposing that the occupants of the Fluid are solid steel spheres, we get an instance of container substantivalism: the Fluid surrounds its occupants, but does not permeate them. If, as in this case, the occupants of substantival space are themselves spatially extended (rather than punctiform), there are really two modes of spatiality in such worlds; there is the space within which material things exist, and there is the space constituted by material things themselves. Illustrations of relational substantivalism are harder to envisage clearly, but easy enough to describe: we simply state that certain material objects bear the location relation to certain points in space. Since these objects are not *in* their space in an intuitively obvious manner, this doctrine might be thought to leave something to be desired, but of course an advocate of this position will maintain that it is our intuitive conception of containment that is flawed.

By way of summing up, the substantivalist's position includes the following claims:

- the physical world does not consist solely of material objects such as stars and atoms, it consists of objects of this sort together with an additional entity: space;
- this space possesses an intrinsic nature of some kind;
- it possesses certain topological or geometrical properties; for example, if our space is Euclidean it is composed of a collection (or manifold) of points that conform to the axioms of Euclidean geometry;
- material objects are related to space in one of the ways mentioned above;
- material objects possess some (and probably *all* in the case of super-substantivalism) of their spatial properties by virtue of being related to space; for example, if two point-like objects are separated by a distance d it is because they occupy points in space that are separated by d;
- the ways material objects can move are constrained by space;
- since objects which are located in the same space are parts of the same physical world, the extent of our space determines the extent of our world.

9.5 Relationism: a closer look

There is no denying that a substantival space constrains and connects, but some find the price excessively high.

The existence of the various substantival spaces that we encountered in §9.4 may have seemed unproblematic, but if so this was in large part due to the way they were introduced into the discussion. I not only stipulated that these entities existed, I also stipulated that they possess material features of a familiar sort and, in doing so, I assumed an all-knowing God-like vantage point on the worlds being described. When it comes to our world we are not able to adopt a perspective of this kind, and there is no obvious evidence that we are immersed in a space of this sort: hence the appeal of the void conception. Even those who believe that our space is substantival space concede that it is not composed of the same sort of stuff as other material things, and is not directly observable in the way other things are. Given these facts, and in light of Occam's razor, it is obviously worth investigating whether we can get by without positing such a thing. This is precisely the task that relationists set themselves.

Different relationists go about the task in different ways. As Earman notes "there are almost as many versions of relationism as there are relationists" (1989: 12). While this is certainly true when it comes to the details, most relationists would agree with the following:

> In claiming that objects inherit their spatial properties from the regions of space that they occupy, the substantivalist is inserting an invisible and redundant intermediary between objects. We cannot observe space itself, but we can observe objects at various distances from one another. The most economical way of making sense of this is simply to say that objects are *directly* related to one another by spatial relations. Instead of appealing to *space–objects* relations, we can appeal to *object–object* relations, where the relations in question are of a spatial sort. These spatial relations should not be thought of as material objects in their own right, but as distinctive *properties*, of a relational sort, that material objects can possess.

On the face of it, the relationist programme has considerable appeal. In effect, the relationist is proposing a modified void conception: space is nothing at all, mere absence, the world consists of nothing but material objects, but these objects have relational properties that perform the same role as the vast entity posited by the substantivalist. Here is how Leibniz characterizes the doctrine:

> I will here show, how Men come to form to themselves the Notion of Space. They consider that many things exist at once, and they observe in them a certain Order of Coexistence, according to which the relation of one thing to another is more or less simple. This Order is their Situation or Distance . . . Those which have such a Relation to those fixed Existents, as Others had to them before, have now the same Place which those others had. And That which comprehends all their Places, is called Space.
>
> (Alexander 1956: 69–70)

But there are clearly questions that need answering. What are these relations? How exactly should we think of them? What is it, exactly, that they relate?

The obvious answer to the last question is "material objects", but there are some potential problems here. Container substantivalism is an awkward doctrine because it posits two modes of spatiality: that of space itself, and that of the solid material things that are immersed in space but not pervaded by it. Relationists who want to avoid this predicament will have to extend the account they offer of the external spatial relations that hold between objects to the interior regions of these objects. This can be done if extended objects are regarded as fusions of point-particles, which themselves are spatially related. Most relationists adopt this view, which conforms with the standard (although no longer universal) account of elementary particles in contemporary physics.

What of the relations themselves? I will have more to say on this topic in the next section, so for now I will restrict myself to the basics.

Generally speaking, relations and relational properties go together, in that, if we say that X bears some relation R to Y, then X and Y can each be said to possess a relational property, either singly, as in "X is R-related to Y", or jointly, as in "X and Y are R-related". There are many different relations (e.g. ". . . is larger than . . .", ". . . is redder than . . .", ". . . is to the right of . . .", ". . . is the sister of . . ."), and different relations can have different properties. Rather than attempt an exhaustive categorization, I will focus on the specific characteristics of spatial relations, conceived as *distances* separating objects. In itself this is a considerable simplification, since – as we shall see in due course – it should not be assumed that any two spatially related items are necessarily at some determinate distance from one another, but this complication need not concern us at this preliminary stage.

The formal properties of a spatial relation are not difficult to specify. Distance is a concrete relation – one that holds between concrete rather than abstract entities – and it is symmetrical, transitive and reflexive (an object is zero distance from itself). Some relations are *internal* and others are *external*: roughly speaking, if X is internally related to Y, then X cannot exist without Y, whereas if X is externally related to Y, it can. Most would agree that spatial relations are external rather than internal. I may be spatially related to you, but if you were to vanish into thin air, I wouldn't necessarily follow, and nor would my intrinsic properties undergo any change (or if they do, it is not because I am no longer spatially related to you, but because of the changes your disappearance makes to gravitational and other fields that affect me).

We can further distinguish between *supervenient* and *non-supervenient* relations. A relation such as "is taller than" is supervenient in the intended sense since the truth of "Jim is taller than Bob" depends entirely on some other property of Jim and Bob, namely their height. To express the point in a picturesque way, in creating a universe in which Jim is taller than Bob all God had to do was to make Jim 1.9m tall, and Bob 1.6m tall; no further property needed to be created to make it true that Jim is taller than Bob. Supervenient properties of this sort are sometimes called "ontological free lunches", for obvious reasons. Given our earlier discussion of the connection problem, it is clear that spatial relations are not properties of this kind.

Suppose O_1 and O_2 are two objects, each of a certain size, mass, material constitutions, and so forth, that happen to be a certain distance apart. Since it seems perfectly conceivable for O_1 and O_2 to exist on their own – as *not* spatially related to one another (or anything else for that matter) – it seems clear that spatial relatedness is not an ontological free lunch. The spatial relation that actually exists between O_1 and O_2 required an additional creative act on God's part, and since this applies quite generally, spatial relations are non-supervenient, As a consequence, relationism is by no means a cost-free doctrine, ontologically speaking: we save on substantival space but pay extra in the coin of spatial relations.

Spatial relations are not only ingredients of reality in their own right; they perform distinctive functions. Bearing in mind "the connection problem" that we encountered in §9.3, they amount to genuine links or bonds between the objects they relate. Not all relational properties amount to genuine connections in this sense: ". . . has the same number of sides as . . ." is a relational property that objects in a different universes can possess. Spatial relations are *potent*, since they constrain or limit the nomologically possible movements of what they relate. Objects in Flatland can only move in two dimensions because all available spatial relations inhabit the same plane; we live in a three- rather than two-dimensional world because the spatial relations in our world have an extra degree of freedom – the laws governing these relations allow them to be oriented with respect to one another in additional directions.

A key difference between relationism and substantivalism lies in the different stances they adopt towards empty space. The substantivalist's account entails the possibility of empty places: space is constituted of points, and since these do not depend on material bodies for their existence they can exist when empty of matter. Since relationists only recognize the existence of material bodies and the spatial relations between material bodies, they are committed to the *non*-existence of unoccupied places and regions. But while the rejection of unoccupied locations amounts to a considerable ontological economy, it is also problematic, for it is the substantivalist's position that conforms with our usual ways of talking and thinking. Suppose we are in empty space; I fire a harpoon at you; I know it will hit you because I have shot it straight at you and there is nothing between us; I know there is a *path through empty space* that the harpoon can follow – if there weren't such a path, how could it reach you? Or to take another example, suppose that we agree to rendezvous next week at the mid-point between Earth and Jupiter. This point is unoccupied. Surely the point exists. How else could we agree to meet there?

Most relationist are prepared to concede the legitimacy of these ways of talking, but they deny that they have the ontological import that the substantivalist grants them. The general relationist line runs something like this:

In practice we all work with more or less precise maps of space. Such a map will typically include references to regions of empty space. But these parts of the map do not refer to anything that exists or that is actual. A map reflects a *representation* of reality that we have created; it embodies our beliefs (theories) about the ways objects *could* be spatially related. Our theories about the ways objects can move in relation to one another tell us that objects could occupy a place mid-way between Jupiter and Earth, so we put this "place" on our map, but there is no such place unless an object is actually at these distances from Earth and Jupiter. As for distances themselves, it may be natural to think of these as *quantities of space*, but this is a mistake. Since empty space isn't an ingredient of reality, there cannot be quantities of it. We can, of course, measure distances between objects, for instance by taking a ruler and moving it around, or sending signals (such as light, sound, carrier pigeons)

back and forth, and measuring the time taken. The success of these procedures can be explained in wholly relational terms: no reference to empty regions of space is required. Not even the substantivalist says that we can
measure our space *directly*, since our space is invisible and intangible. ろ

This position is not obviously incoherent and is certainly worthy of serious
consideration. All that exists are objects and their actual spatial relations; since
objects can change their spatial relations to one another in certain restricted ways,
we can draw up (or imagine) maps (or representations) that reflect these
possibilities, but it remains the case that at any given time only objects and their
actual distance relations exist. All remaining points on the map do not correspond
to anything real.[9]

There is more to be said about these relations, but before moving on to this topic
there is one further point that is worth clarifying. I noted at the outset that
although relationists reject the idea that space is an entity in its own right they do
not claim that all propositions about the sizes of objects or the distances between
them are false. The key relationist claim is that spatial facts of this sort do not
require or involve a spatial substance. The position of the relationist thus
construed can usefully be contrasted with that of the *radical spatial anti-realist*. I
have in mind idealists who deny the reality of the physical world: if reality consists
only of immaterial minds (as Berkeley held), then clearly there are no material
objects and no spatial relations between them. Nonetheless, there are idealists who
maintain that there *are* spatial facts (even though there is no space) just as there are
facts about physical objects (even though there are no such objects – or so they
believe). How is the trick worked? Well, the idealists in question want to leave our
ordinary ways of talking and describing the world intact. They do not advocate
adopting a completely new vocabulary. So if I say that there's a teacup on my table
(true), they would allow that this proposition is true, but maintain that what makes
it true is not a fact about material objects (cups, tables) but rather *patterns in
sensory experience*. They would adopt a similar stance towards propositions about
spatial *relations*.

We clearly need to distinguish two sorts of "relationism". Relationists of both
sorts accept that there are true statements about spatial relations, and hence spatial
facts. The difference is this:

{ • *Realist relationists*: true statements about space are made true by facts about
material bodies and the way they are related.
• *Idealist relationists*: true statements about space are made true by facts about
human minds and/or patterns of sensory experience. }

In what follows, unless I state otherwise I will be concerned with realist relationists
– people who assume that the physical world is real. The distinction between the two
forms of relationism is, however, worth noting at a preliminary stage because of one
central figure in the debate: Leibniz, the founding father of relationism. Leibniz's
various writings on space can be confusing. Although he was himself an idealist, and
some of his pronouncements concerning space are clearly made from an idealist
standpoint, many others are not. When arguing against the Newtonian conception
of (substantival) space, he often seems to be assuming a realist position, at least for

the sake of argument. We will be concerned only with the Leibnizian arguments of the latter sort.

9.6 Two concepts of distance

The relationists' rejection of the reality of unoccupied locations does bring into focus an important aspect of the way they conceive spatial relations: these relations operate, as it were, *across* space (or void) but not *through* it. Imagine building a model of the solar system using plastic balls of different sizes and lengths of stiff wire. Your "planets" are all connected to the "Sun" at the centre of the system by thin wires extending through empty space. Relationists do not (usually) take spatial relations to be entities akin to these wires (not surprisingly, since there is no empirical evidence for the existence of anything resembling them). <u>The envisaged relations are held to connect spatially separated objects directly, without passing (or extending) through the medium of intervening empty space (which, of course, relationists reject).</u> These relations are more akin to action-at-a-distance *forces* than the material connections with which we are familiar in everyday life. However, the force analogy should not be stretched too far; spatial relations do not attract or repel, and standard scientific theories do not posit anything akin to a "spatial force". So far as most relationists are concerned, the connecting relations they posit are among the basic ingredients of reality, ingredients we are obliged to recognize in order to explain the world as we find it in the most economical fashion.

Connections of this sort are certainly not easy to imagine, but does this mean that they cannot exist? Nerlich suggests as much:

> One reason for taking space as a real thing is the strongly intuitive belief that there can be no basic, simple, binary spatial relations . . . Consider the familiar (though not quite basic) relation *x is at a distance from y*. There is a strong and familiar intuition that this can be satisfied by a pair of objects only if they are connected by a path. Equivalently, if one thing is at a distance from another then there is somewhere half way between them. Distances are infinitely divisible, whether the intervening distance is physically occupied or whether the space is empty. (1994a: 19)

If Nerlich is right, relationism simply doesn't make sense. But is he right? We clearly need to take a closer look at the notion in question: spatial distance.

Useful light is shed on this whole issue by a distinction explored in Bricker (1993). We start off with a familiar type of space, a Euclidean plane, which we construe in a substantival way: the space is full of points. Two of these points, P and Q, are 20m apart (see Figure 9.2, A). We now perform some spatial surgery. In the

Figure 9.2 Spatial surgery. Deleting points leaves holes in space – but is the distance between P and Q altered as a result? The answer depends on the conception of distance we adopt.

first case, a circular region of 10m diameter is removed (Figure 9.2, B); in the second case, an infinitely long segment 10m wide is cut out of the space (the points within the region are removed) (Figure 9.2, C).

The question that can now be posed for each post-surgery case is how far apart are P and Q? Someone might answer, "In C they no longer stand in any distance relation; in B they are more than twenty metres apart." Someone else might answer, "In both cases P and Q are still exactly twenty metres apart." Each answer reflects a different conception of distance.

The first answer would be given by anyone who thinks that distance is "shortest path length", where the path is a continuous route through the points in a space. Imagine a Flatland geometer painstakingly measuring off distances between points using a long piece of string and a ruler. He aligns his string along every continuous path between the relevant points, measures its length, and declares the shortest path to be the distance between them (in the case of a tie, the distance is that of the shortest *paths*). The seventeenth-century mathematician Gauss envisaged a procedure of this sort in his work on curved surfaces, which is why Bricker calls this the _Gaussian_ conception of distance. Since in C there is no continuous path between P and Q – the deleted region extends infinitely in either direction – there is no (Gaussian) distance between the points. In B, the shortest path is shown in Figure 9.3, and it clearly adds up to more than 20m. The exact distance is $10\sqrt{3} + 5\pi/3$ m.

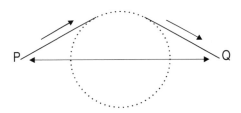

Figure 9.3 The intrinsic distance between P and Q is the length of the straight line (passing through the centre of the hole) connecting them; the Gaussian distance, indicated by the single-headed arrows, is rather longer.

Gaussian paths are themselves parts of space, consisting as they do of collections of points. On this conception, paths are the primary bearers of distance; it is only after lengths have been assigned to the paths that connect them that distances can be assigned to points. This conception of distance lives on: it is the basis of differential geometry, the formalism used in contemporary spacetime theories.

The person who answers "It's still twenty metres" for both B and C clearly isn't working with the Gaussian conception of distance. They are working with what Bricker calls the _intrinsic conception_ of distance: the distance between P and Q (and any other two points) depends solely on the properties of the points in question. For P and Q to remain 20m apart despite the removal of parts of the intervening space entails that the distance between these points depends on nothing but the points themselves and how they are directly related to one another. This direct intrinsic distance is independent of the space within which P and Q are embedded: _all_ the points surrounding P and Q could be removed and they would still be 20m apart. Like its Gaussian counterpart, the intrinsic conception also lives on:

The mathematical embodiment of the intrinsic conception is the abstract structure of a metric space. A metric space consists of a non-empty universe of points together with a family of distance relations (or a single distance function – it matters not) satisfying the axioms of distance . . . The distance relations are taken as primitive, and other features of space – e.g. topological – are defined in terms of the distance relation.   (Bricker 1993: 278)

ᛁThe fact that the intrinsic conception has a home in mathematics does not, in itself, mean that it has a physical application.ᛁA proponent of the Gaussian conception, such as Nerlich evidently is, might argue: "But it makes no sense to suppose that P and Q are still twenty metres apart, especially in C, where the radical spatial surgery left two disconnected spaces. How can P and Q be at a distance from one another if they no longer inhabit the same space?" But while this objection has some intuitive force, it is by no means overwhelming. Someone might reply, "Yes, there is no continuous path linking P and Q, but it remains the case that they are twenty metres apart. This is a fact about P and Q themselves. Why should it be affected by what happens to other parts of space?" Not only does this response also have some intuitive force – I think we can make sense of it at a gut-level, for what that's worth – but it can be reinforced by imaginary tales. For instance, it is easy to envisage a communications system that continues to function despite the spatial deletion (it doesn't work by sending signals along paths through space), and so not only can messages still be sent from P to Q, but the time delay is unchanged – it remains consistent with P and Q being 20m apart – a fact that suggests that the relevant laws of nature are sensitive to intrinsic rather than Gaussian distance.

Bricker starts by assuming that points and distances are real, and then poses the question as to how these points are interwoven to form the fabric of space:

Are there direct ties only between "neighbouring points", so that points at a distance are connected only indirectly through series of such direct ties? Or are there also direct ties between distant points, so that the fabric is rein-forced, as it were, by irreducibly global spatial relations?   (1993: 271)

If the latter, then space retains its overall shape even when sliced into two dis-connected pieces, as in Figure 9.2C. This point-realism amounts to spatial substantivalism, for, as we have seen, the relationist denies the reality of unoccu-pied points. However, although the concept of intrinsic distance *can* be applied to substantival spaces, is there any reason why it cannot be applied to relational spaces? There is no obvious reason why not, and indeed, the notion is just what the relationist requires. From the Gaussian perspective, the individual material bodies scattered through the void are each island universes; since they are not linked by continuous paths through space (unoccupied points do not exist), they are not spatially related, and so at no distance from one another. By introducing intrinsic distances into the picture a unified cosmos is restored. Nerlich might object that the notion of intrinsic distance simply doesn't make sense in a relational context, but it is hard to see why this should be so. If we can make sense of points in a substantival space being linked by intrinsic distances – and there is reason to think that we can – given that these distances are unaffected by the presence or absence of intervening points, we can surely make sense of points (or objects) in a relational space being similarly linked.

9.7 Two conceptions of motion

I suggested earlier that in deciding between the relationist and substantivalist accounts, empirical or scientific issues play a crucial role, and we can now see why. Although relationism and substantivalism amount to very different conceptions of reality, both seem metaphysically coherent, at least when considered on their own terms. It may well be that there are *logically* possible spatial worlds where space is substantival and distances Gaussian, and others where space is relational and distances intrinsic. If so, then in trying to decide whether *our* space is substantival or not we cannot confine ourselves to a purely metaphysical inquiry. We must consider this question: is there some observable feature of our world that requires, or is best explained in terms of, a substantival space?

One crucial scientific issue, which will loom large over the next few chapters, concerns the dynamics of moving bodies (the science of forced motion). Without entering into any details it is important to be clear right at the outset that the different accounts of the nature of space entail very different accounts of what motion itself involves.

- The *substantivalist* says: a body moves if it occupies different locations in substantival space at different times.
- The *relationist* says: a body moves if its distance relations with other material bodies change.

For the substantivalist, at least some motions are *absolute* in this sense: the bodies in question occupy different locations in substantival space at different times. For the relationist, there is no such thing as absolute motion in this sense, since the supposed substantival space does not exist. Consequently, all motion is relative *to other material bodies*.

Although standard, this terminology is potentially confusing. You might think, "The substantivalist is also claiming that motion is *relative*: a body moves when it changes its location relative to substantival space." You would be perfectly correct. For understandable historical reasons, a tradition has developed of calling motion with respect to substantival space "absolute". It is also true that the substantivalist recognizes that bodies move relative to one another (e.g. the distance between bodies A and B increases/decreases). The key difference is that the substantivalist recognizes a sort of motion that the relationist denies: movement relative to substantival space.

The implications of this difference are far-reaching. For now note just one point. Suppose that there are two rocket ships S and T close to one another but millions of light-years from the nearest star; the two ships are moving apart at 10mph. Since no stars are visible, it is impossible to tell how fast either ship is moving relative to other bodies. The captain on S might say, "We are stationary and T is moving away from us at 10mph." The captain on T might say, "*We* are stationary and S is moving away from *us* at 10mph." Who is right? No test will reveal the right answer. Indeed, there are plenty of other possible answers; both ships could be moving apart at 5mph. The different accounts of the nature of space have different implications about what is really happening in situations of this sort.

Substantivalists will say, "There is a fact of the matter about the *real motions* of the two ships, namely how they are moving (if at all) relative to substantial space, even if we don't know what these real motions are." What can relationists say? One thing they can't say is that there is a fact about how fast either ship is moving relative to substantival space. So they must say that the only facts in this situation are the speed at which the ships are moving apart relative to one another, and the speed at which the ships are moving towards/away from other material objects in the universe. If we now suppose that all the other objects in the universe vanish from existence, leaving just our two ships, what can the relationist say? Clearly, there is now just one fact: the ships are moving apart from one another at 10mph. For the substantivalist the presence or absence of other objects is irrelevant, since each ship has an absolute velocity determined by its motion relative to substantival space.

The relationist's conception of motion can seem odd. It is easy to feel that there must be a right answer to the question "How fast is ship S *really* moving?" But then again, it can also seem quite plausible. Who, when travelling in a fast car or plane, hasn't had the feeling "It's hard to believe that I am actually moving!" If we think of things as the relationist does then, when we are in motion relative to the Earth, it is just as true to say that the Earth is moving at several hundred miles per hour and we are stationary as it is to say the reverse, which is just how it seems.

9.8 Matters terminological

In "Time, space and space-time" (1978), Horwich notes that terms such as "absolute", "relative" and "relational" are used in connection with space (and time) to mean very different things. Horwich is certainly right about different writers using the same terms to refer to different doctrines and combinations of doctrines, and this can be confusing. It helps to have some idea of what the various doctrines are. Horwich lists the following "relativist" theses, each of which is denied by a corresponding "absolutist" thesis.

1. Space and time depend for their existence upon the existence of objects and events.
2. Space and time are not substances.
3. The spatial and temporal location of an object or of an event is to be analysed in terms of spatial and temporal relations between those things and other objects and events.
4. All motion is relative motion: an object moves only in virtue of changes in its distance from other objects.
5. Certain features of space and time, such as geometrical properties and congruence relations between intervals, must be defined in terms of the behaviour of objects and events. (This is the thesis of the intrinsic metrical amorphousness of space, maintained by Adolf Grunbaum.)

6. All spatial and temporal facts are to be analysed in terms of non-spatial and non-temporal concepts. (For example, according to the so-called "Causal theory of time", the notion of temporal betweeness may be reduced to various causal relations between events.)

7. Space and time are relative in the sense that certain magnitudes such as duration and distance vary from one frame of reference to another. (This is implied by Einstein's special theory of relativity.)

Many of the claims made in this list have cropped up in earlier chapters, even if the labels used were different. In this chapter we have mostly been concerned with 2, and I have eschewed "absolutism" and "relativism" in favour of "substantivalism" and "relationism" (except when discussing 4), but beware: the latter pair of terms, too, have sometimes been used in connection with most of the other doctrines listed above.

10 Space: the classical debate

10.1 The last of the magicians

Over the next three chapters we will be examining the debates concerning motion and space in the context of classical Newtonian mechanics. Newton himself will be playing the role of arch-substantivalist and Leibniz the arch-relationist, with Galileo and Descartes fitting uneasily between the two. As far as *our* space is concerned, this debate has in some respects been rendered redundant by subsequent developments, since Newton's physics has been replaced by Einstein's, but the earlier debate is well worth studying. Not only does it provide useful preparation – recent developments can only be fully understood against the backdrop of the positions from which they emerged – but it is also of interest in its own right, for in considering the different ways that the substantivalist and relationist can account for a basic physical phenomenon such as motion, we learn a good deal about both the character and explanatory resources of the two competing frameworks, and what we learn may well be relevant outside the narrow confines of the classical Newtonian worldview.

Newton's *Principia* was first published in 1687 (*Philosophiae Naturalis Principia Mathematica*), and in the opinion of many it is the single most important work in the history of science. Within half a century most scientists were Newtonians, and most were to remain so for the next two hundred years.[1] Building on the various developments in the understanding of motion and forces made during the previous centuries, Newton not only developed a comprehensive theory of mechanics (one that is still in use today), but he also devised his theory of universal gravitation and the mathematics of the calculus, which allowed him to predict the movements of the heavenly bodies with great accuracy, explain why planetary orbits were elliptical rather than circular, and much else besides. As Barbour puts it, "So comprehensive was his genius, it appeared to open all doors into nature, to leave nothing really major to discover. Life after Newton seemed a mere walking through the garden into which his genius had directed us" (1989: 629).

Although the importance of the work was immediately recognized, it also gave rise to considerable controversy. In particular, contemporaries such as Huygens and Leibniz were highly critical of the cornerstone of the Newtonian edifice: absolute substantival space. In the famous *Scholium* following definition eight, at the very start of the *Principia*, Newton lays bare his commitments on this score:[2]

> Hitherto I have laid down the definitions of such words as are less known, and explained the sense in which I would have them understood in the

following discourse. I do not define time, space, place and motion, as being well known to all. Only I must observe, that the common people conceive these quantities under no other notions but from the relation they bear to sensible objects. And thence arise certain prejudices, for the removing of which it will be convenient to distinguish them into absolute and relative, true and apparent, mathematical and common.

Absolute, true and mathematical time, of itself, and from its own nature, flows equably without relation to anything external . . .

Absolute space, in its own nature, without relation to anything external, remains always similar and immovable. Relative space is some movable dimension or measure of the absolute spaces; which our senses determine by its position to bodies, and which is commonly taken for immovable space; such is the dimension of a subterraneous, an ariel, or celestial space, deter- mined by its position relative to the Earth. Absolute and relative space are the same in figure and magnitude; but they do not remain always numerically the same. For if the Earth, for instance, moves, a space of our air, which relatively and in respect of the Earth, remains always the same, will at one time be one part of the absolute space into which the air passes; at another time it will be another part of the same, and so, absolutely understood, it will be continually changed.

Absolute motion is the translation of a body from one absolute place into another. Thus in a ship under sail, the relative place of a body is that part of the ship which the body possesses; of that part of the cavity which the body fills, and which therefore moves together with the ship: and relative rest is the continuance of the body in the same part of the ship, or of its cavity. But real, absolute rest, is the continuance of the body in the same part of that immovable space, in which the ship itself, its cavity, and all that it contains is moved.   (Huggett 1999: 118–19)

While conceding that we are perfectly familiar with space, time and motion, Newton points out that for the most part in ordinary life we deal only with relative locations and motions; that is, with how material bodies are spatially related to one another. When on land, we typically assume that the Earth is motionless and so place it at the centre of our frame of reference; we generally take a thing's speed to be how fast it is moving with respect to the Earth's surface. (Imagine the reaction to this line of defence from someone caught speeding: "But your Honour, I was only doing 50 with respect to the Moon!") When on board a ship far from land, we might well take the ship to provide our stationary point of reference, and so find ourselves saying of a lazy crewman, "He hasn't moved an inch all day", even if, relative to the Earth-based frame of reference, he has moved many miles. This choice of object-based reference frame is entirely a matter of convenience: people living on other worlds might well take *their* planet (or star) to be stationary, and their choice would be just as well founded as ours.

Newton does not criticize these practices, but he does warn against the "prejudice" of assuming that all frames of reference are on an equal footing, and hence that all motion is relative. He insists that there is a privileged frame of reference, one that is not arbitrary, one that provides a measure of *real* motion – the frame constituted by absolute space. As Newton makes clear elsewhere, he takes absolute space to be infinite, perfectly uniform and Euclidean in geometrical structure. Although invisible, it is perfectly real, and an entity in its own right, as this extract from the earlier *De Gravitatione* makes clear:

> [H]ence it is not an accident. And much less may it be said to be nothing, since it is rather something, than an accident, and approaches more nearly to the nature of substance. There is no idea of nothing, nor has nothing any properties, but we have an exceptionally clear idea of extension, abstracting the dispositions and properties of a body so that there remains only the uniform and unlimited stretching out of space in length, breadth and depth. And furthermore, many of its properties are associated with this idea . . . In all directions, space can be distinguished into parts whose common limits we usually call surfaces; and these surfaces can be distinguished in all directions into parts whose common limits we usually call lines; and again these lines can be distinguished in all directions into parts which we call points. . . . Furthermore spaces are everywhere contiguous to spaces, and extension is everywhere placed next to extension . . . And hence there are everywhere all kinds of figures, everywhere spheres, cubes, triangles, straight lines, everywhere circular, elliptical, parabolical and all other kinds of figures, and those of all shapes and sizes, even though they are not disclosed to sight. For the material delineation of any figure is not a new production of that figure with respect to space, but only a corporeal representation of it, so that what was formerly insensible in space now appears to the senses to exist. . . . In the same way we see no material shapes in clear water, yet there are many in it which merely introducing some colour into its parts will cause to appear in many ways. However, if the colour were introduced, it would not constitute material shapes but only cause them to be visible.   (Huggett 1999: 111–12)

space as absolute being

It was this doctrine that Leibniz took as his target:

> These gentlemen maintain, therefore, that space is a real absolute being . . . As for my own opinion, I have said more than once, that I hold space to be something merely relative, as time is; that I hold it to be an order of coexistences, as time is an order of successions . . . I have many demonstrations, to confute the fancy of those who take space to be a substance, or at least an absolute being.   (Alexander 1956: 25–6)

The debate between relationists and substantivalists was thus joined.

Newton's *Principia* may have been the defining moment in the development of modern science, but Newton himself was fully a man of his time, and as such was motivated by concerns alien to those that we now tend to associate with the scientist. Far from devoting himself entirely to physics, he was also deeply religious and as a committed (although secret) Unitarian, he laboured long on the project of demonstrating that the doctrine of the trinity was an invention of the early church fathers. Combining work in physics and theology was not uncommon at the time – Leibniz did the same – but Newton's interests were particularly broad, extending as they did far into the arena of the esoteric: he was a devotee of both alchemy and numerology. Not for nothing did Keynes describe him as "the last of the magicians". These broader interests almost certainly exerted an influence on Newton's thinking about space; his work in physics was, in part at least, inspired by a mystical-cum-theological vision of absolute space as a divine attribute. As Barbour aptly says, "Newton is explicitly seeking to demonstrate, through phenomena, the transcendent basis of all motion. But he aims even higher; the boy from Grantham has set his eye on the *anatomy of God*" (1989: 628).[3]

This may well have been Newton's ultimate goal, but there is little in the first edition of the *Principia* to suggest as much: Newton's many arguments – including those concerning absolute space – are entirely scientific in character. Only in one remarkable sentence does Newton explicitly reveal the profound importance that the doctrine of absolute space had for him. The final paragraph of the Scholium terminates thus:

> It is indeed a matter of great difficulty to discover, and effectually to distinguish, the true motions of particular bodies from the apparent; because the parts of immovable space, in which these motions are performed, do by no means come under the observation of our senses. . . . But how are we to obtain the true motions from their causes, effects, and apparent differences, and the converse, shall be explained more at large in the following treatise. For to this end it was that I composed it.   (Huggett 1999: 123–4)

The *Principia* as a whole was intended to show that the distinction between real (absolute) and relative motions is indispensable in an adequate account of the observed behaviour of material bodies. If discerning the anatomy of God was Newton's ultimate goal he set about reaching it in an entirely modern way, with no appeal to wizardry or revelation; he firmly believed the existence of absolute space could be established on empirical grounds alone.[4]

Evaluating the strength of Newton's case is by no means an easy matter, and I will approach the topic from different directions. In this chapter I set the stage by taking a brief look at the relevant doctrines of Galileo and Descartes (the latter was almost certainly Newton's target in the Scholium), and then move on to Leibniz's attacks on substantivalism. In Chapter 11 we encounter the considerations that Newton himself believed were decisive, and survey the main relationalist responses to Newton. In Chapter 12, I consider in some detail the impact on the classical debate of the contemporary concept of "spacetime". While anachronistic, this broadening of scope could also be regarded as perfectly justifiable in an investigation of a broadly *philosophical* sort, since to arrive at the four-dimensional spacetime way of conceiving the material universe requires only a conceptual revision, which Newton or Leibniz could easily have arrived at by thinking alone – no empirical discoveries are needed. Unfortunately, while the four-dimensional perspective sheds useful new light on the whole debate, it also brings problems and difficulties of its own, as we shall see.

10.2 Galileo

The pivotal role that Galileo played in the development of the science of motion is well known and well described elsewhere, as are his troubles with the Inquisition, so I will dwell only on those of his doctrines that are especially relevant to our concerns. In his *Dialogue Concerning the Two Chief World Systems, Ptolemaic and Copernican* (1632) Galileo set out to prove that the Earth moves; specifically, that it rotates on its own axis once per day, and rotates around the Sun once per year. This hypothesis was by no means novel – it had advocates in ancient Greek times – and although interest in it had increased since the publication of Copernicus's *De Revolutionibus* in 1543, there were few who believed the hypothesis could be true, and fewer still prepared to risk ridicule (and the condemnation of the church authorities) by defending it in public. Religious considerations aside, Galileo had to overcome objections from two sources: common sense and Aristotelian physics, which to some extent acted to reinforce one another.

The objection from common sense is simple: the Earth must be stationary because if it were moving there would be all manner of dramatic observable consequences, none of which are actually observed. Take some mud, squeeze it into a firm lump and set it down on a table: the lump remains intact. But if you attach the same lump to a piece of string and whirl it around your head, the lump would quickly fly apart. Wouldn't the same happen to the Earth if it were flying around the Sun? If you put a cup on top of a book and snatch the book away the cup is left behind, stranded in mid-air, just as we would be if the Earth were whirling through space at thousands of miles per hour. We could also expect a powerful wind (move your hand through the air). Birds would not be seen flying in all directions with equal ease; they would have to fly extra hard to catch up with the moving Earth. Objects dropped from towers wouldn't land directly below, at the base, but many miles away. There is no need to continue: the case seems secured.

These powerful intuitions are confirmed and enshrined in Aristotle's physics, where every kind of thing has its own "natural place", its own "natural motion", and things that are removed from their natural place strive to return there. Celestial objects (e.g. the heavenly spheres carrying the planets) are not subject to change or decay, and their natural motion is circular, which explains why they are observed to move around the Earth at constant speeds; they don't need anything to *keep them going* since by their very nature they move in circles. The essential natures of the four corruptible non-heavenly elements – earth, water, air, fire – are defined in terms of their tendencies to move towards their natural places. The natural motion of heavy things (earth, water) is downwards, towards the centre of the universe; that of light things (air, fire) is upwards, away from the centre. The heaviest things are those that sink below those bodies whose natural motion is downwards, whereas the lightest things are those that rise above those bodies whose natural motion is upwards. It immediately follows that if the Earth is already at the centre of the universe it would not move away from it; and since objects when dropped are always observed to move straight down, even at different locations on the Earth's surface, we can be sure that the Earth is already where it belongs. Aristotle did, of course, recognize that material objects can move in directions other than downwards – stones can be thrown upwards, lumps of wood can be set in motion across ponds – but in all such cases there is a force applied, and objects will return to their natural state of rest, in their natural place (or as close to it as they can reach), unless a force is continually applied. Why does an arrow keep moving forwards when it leaves the bow? Shouldn't it fall straight down as soon as it loses contact with the string? Aristotle was aware of the problem and suggested some possible solutions. Either the bow string causes a mobile agitation in the air, which keeps pushing the arrow forwards, or the arrow's movement causes air to loop towards the rear and push it forwards. These Aristotelian arguments may not fully convince, but the underlying principle – that moving objects naturally return to a state of rest in the absence of a continually applied force – does conform to what we observe in daily life.

In arguing that the Earth goes round the Sun, Galileo did not find it necessary to abandon every element of Aristotelian physics and cosmology. Far from rejecting the idea of natural circular motions, he *extended* it. He followed Copernicus in arguing that since the planets (of which the Earth is one) are all spherical bodies, they each possess their own natural circular motions – around their own axis and around the Sun – and moreover, that each planet determines its own "up" and "down" – away from and towards its own centre. Consequently, a ball dropped on Earth naturally moves towards the centre of the Earth, whereas a ball dropped on Mars would naturally move towards the centre of Mars, rather than towards the centre of the universe, which is occupied by the Sun. As for the alleged observable effects of Earthly motion, he deployed numerous brilliantly contrived examples, such as the following:

> For a final indication of the nullity of the experiments brought forth, this seems to me the place to show you a way to test them all very easily. Shut yourself up with some friend in the main cabin below decks on some large ship, and have with you there some flies, butterflies, and other small flying animals. Have a large bowl of water with some fish in it; hang up a bottle that empties, drop by drop into a wide vessel beneath it. With the ship standing still, observe carefully how the little animals fly with equal speed to all sides of the cabin. The fish swim indifferently in all directions; the drops fall into the vessel beneath; and, in throwing something to your friend, you need throw it no more strongly in one direction than another, the distances being equal; jumping with your feet together, you pass equal spaces in every direction. When you observe all these things carefully (though there is no doubt that when the ship is standing still everything must happen in this way), have the ship proceed with any speed that you like, so long as the motion is uniform and not fluctuating this way and that. You will discover not the least change in all the effects named, nor could you tell from any of them whether the ship was moving or standing still. In jumping, you will pass on the floor the same spaces as before, nor will you make larger jumps toward the stern than toward the prow even though the ship is moving quite rapidly, despite the fact that during the time that you are in the air the floor under you will be going in a direction opposite to your jump. . . . The droplets will fall as before into the vessel beneath without dropping toward the stern, although while the drops are in the air the ship runs many spans. The fish in the water will swim towards the front of their bowl with no more effort than toward the back . . . the butterflies and flies will continue their flight indifferently toward every side, nor will it ever happen that they are concentrated toward the stern, as if tired out from keeping up with the course of the ship, from which they have been separated during long intervals

by keeping themselves in the air. And if smoke is made by burning some incense, it will be seen going up in the form of a little cloud, remaining still and moving no more toward one side than the other. The cause of all these correspondences of effects is the fact that the ship's motion is common to all the things contained in it, and to the air also.   (Barbour 1989: 395–6)

There is nothing in the least surprising here for anyone who has spent hours in a commercial airliner and watched normal life unfold around them while moving at many hundreds of miles per hour, but of course no one in Galileo's day had the benefit of such experiences, and the "obvious" had to be painstakingly pointed out and argued for.[5] When it is, the message sings through loud and clear. No observations that we can make within a ship's cabin will enable us to determine whether the ship is stationary, or moving at a constant speed in a straight line (on a smooth sea). If the ship is moving, everything in the cabin – falling drops of water, jumping people, floating smoke, flying insects, particles of air – will have the same forward motion as the ship itself, and so nothing is left behind as the ship moves smoothly on. Since both the ship and its contents share in the same forward movement, in considering the movements of objects within the ship this movement can be subtracted and ignored. What applies to the moving ship applies to the moving Earth.

Galileo's case did not rest solely on analogies of this sort, persuasive though they are. Rather, he used the analogies to illustrate and confirm basic principles of motion for which he provided independent arguments. The ship's cabin example illustrates what we now call the "principle of Galilean invariance", which formulated crudely says that the laws of motion do not distinguish between (and so remain the same within) frames of reference that are moving at uniform speeds in straight lines with respect to one another (we will be returning to this notion). Of Galileo's various discoveries, two are of particular importance: the principle of *inertia* (albeit in a restricted form) and his account of *projectile motion*.

According to Aristotle an arrow will keep moving only as long as it is being pushed; stop pushing and it will fall straight to the ground – as dictated by its natural motion – and the same applies to all material things. Galileo rejected this, and made the startling claim – which we now know to be true – that once an object is set in motion it will continue moving *for ever*, without slowing down and without the need for additional "pushes", unless acted upon by external influences, such as gravity and friction. That it resisted discovery for so long is powerful testimony to the unobvious nature of this principle. Since everything we ever observe going up soon comes down, it is very hard to accept that it is just as natural for a ball to move upwards as downwards. Once the principle *has* been stated and absorbed, the ship's cabin argument seems almost obvious: the objects that share the ship's motion – those that are moving in the same direction and the same speed – will continue to do so unless something impedes them. Newton made the principle of inertia the first of his three laws of motion:

Law I   Every body continues in its state of rest, or in uniform motion in a straight line, unless it is acted upon by an external force.

Law II   The change of motion (acceleration) is proportional to the motive force impressed; and is made in the direction of the force.

Law III To every action there is an equal and opposite reaction; or, the mutual actions of two bodies upon each other are always equal, and directed to contrary parts.

What is striking to the modern (post-Newtonian) mind is the fact is that although Galileo had clearly grasped the essential point – namely that motion is as natural a condition for material objects as rest – he did not state the principle in its Newtonian form. Rather, he endorsed a principle of *circular inertia*: all Earthly objects have a natural tendency to move in circles, just as the Earth does.

Galileo's reasoning ran like this. Imagine a long, smooth plane that first slopes gradually down, then gradually up, and a ball rolling down it, without any impediment from air or surface friction. As the ball rolls down it first gathers speed, but then, as the plane inclines upwards, it gradually slows. Now imagine a variant. The ball has a certain speed, but the plane does not rise, it remains horizontal; why should the ball ever come to a stop, given the lack of impediments? Won't it go on rolling at the same speed for ever? Galileo thus drew the conclusion that a ball rolling along a horizontal plane would continue moving at the same speed for ever, unless impeded by some external force, such as friction. He didn't, however, take the horizontal plane to be a straight line, but rather the spherical surface of the Earth. For his purposes this was all that was required: it is because a ball dropped from a tower possesses a natural tendency for circular motion that it follows the Earth's movement, and falls at the base of the tower. In effect, Galileo's principle of circular inertia is simply a generalization of the Aristotelian principle of "natural" circular motions, and no doubt this made it easier for him and his readers to believe it.[6]

Prior to Galileo it was commonly believed that projectiles such as arrows and cannon balls moved in straight lines. According to "official doctrine", a cannon ball first moves in a straight line in the direction of the barrel, since its natural downward motion is completely overwhelmed by the impetus generated by the explosion. Then, some time later, when the impetus from the explosion is greatly weakened, the natural downwards motion wins the battle and the ball drops vertically downwards like a stone. This account seemed compelling because it was assumed that different kinds of motion – natural and forced – could not coexist. Galileo knew otherwise. Careful experimental work had led him to a correct formulation of the principle of free-fall: the distance traveled by a freely falling object increases as the square of the time. He then relied on his principle of the persistence of horizontal motion, and argued – again correctly – that a cannon ball fired in a horizontal direction (e.g. from a high castle wall) would keep on travelling at precisely the same speed in the horizontal direction, while simultaneously falling in a vertical direction in accord with the law of free-fall. The resulting trajectory, as Galileo demonstrated, is a parabola (Figure 10.1).

This influential piece of analysis played a vital role in the defence of Copernicanism. Galileo argued that a ball dropped from a tower will hit the ground at the tower's base irrespective of whether the Earth is moving or not, and will take the same time to do so. For this to be possible, a ball dropped on a moving Earth must have a complex motion with a vertical component that is exactly the same as that found in the ball falling on a stationary Earth, and for this to be

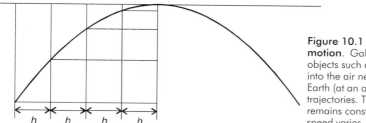

Figure 10.1 Projectile motion. Galileo discovered that objects such as cannonballs, fired into the air near the surface of the Earth (at an angle) follow parabolic trajectories. The horizontal speed remains constant, only the vertical speed varies.

possible the various components of a complex motion must not interfere with one another, which is just the result delivered by Galileo's analysis of projectiles. Precisely the same sort of compounding of independent motions is required in the case of the ship's cabin. The motions of rising smoke particles in stationary and moving cabins are exactly alike in all respects save one: in the latter case there is an additional horizontal component, but since this component is the same for all the particles, there is no observable difference between the two cases.

Perhaps Galileo's contribution is best summed up in his own words: "Motion, in so far as it acts as motion ... and among things which all share equally in any motion, it does not act, and is as if it did not exist". It is a perturbing result, as Galileo himself appreciated: "you are not the first to feel a great repugnance toward recognizing this inoperative quality of motion among the things which share it in common" (Barbour 1989: 391).

10.3 Descartes

Although Newton worked all Galileo's significant discoveries into the synthesis of the *Principia* – indeed, he credited him (not entirely accurately) with having discovered the first and second laws of motion – when he carried out his first important work in the 1660s, the dominant physics, which he was both influenced by and reacted against, was that of Descartes. Since Newton's arguments about absolute and relative space in the Scholium and elsewhere often take Descartes as their target, it will prove helpful to know something of what Descartes himself had to say about these matters.

Descartes was working on *The World* (*Le Monde*) – completed around 1633 and including the essentials of Cartesian physics – at around the same time as Galileo was composing the *Dialogue*, but whereas Galileo was content merely to modify the Aristotelian worldview, Descartes was intent on demolishing it entirely, and largely succeeded. He was inspired by the dream of explaining *everything* in entirely mechanical terms: by universal natural laws governing the movements and interactions of material bodies in space. This mechanical outlook is taken for granted by most contemporary scientists, but it was far from widespread in Descartes's day, and it was a perspective that Newton found appealing.

More specifically, in his construction of the cosmos Descartes allowed himself just two basic ingredients, matter and motion, and three fundamental principles, inertia, the conservation of matter, and the conservation of motion. According to the latter pair, there is a fixed quantity of matter and motion in the universe that can neither diminish nor increase. Descartes here anticipates the contemporary principle of energy conservation. As for inertia, here, too, Descartes subscribed to the principle in its modern (or at least Newtonian) form: objects tend naturally to move in straight lines at constant speeds, and will continue to do so for ever unless acted on by some external force or influence. Whereas previously people had wondered why a stone released from a thrower's arm should continue to move, Descartes turned the question around, and wondered why it should ever stop. He went further and pointed out that the forces needed to move and stop objects are equal: "the action needed to move a boat which is at rest in still water is no greater than that needed to stop it suddenly when it is moving" (*The Principles of Philosophy*, II, 26; Huggett 1999: 96). The lingering Aristotelian influence that led Galileo to subscribe to *circular* inertia has been utterly expelled.[7]

The details of Descartes's scheme, although interesting in their own right, need not detain us long. He took the universe to be a vast (possibly infinite) plenum; that is, there is matter everywhere, and no "empty space". Matter is mobile and comes in portions of different sizes. The smallest particles are very fast-moving and are associated with fire; the medium-sized pieces move less quickly and are spherical (having been ground into this shape by eons of constant motion), the largest pieces move yet more slowly and compose large bodies such as the Earth. Descartes assumed that the space vacated by a moving body would immediately be filled by other matter (much as water immediately rushes in to occupy the places vacated by a moving fish), and since the easiest way for this to happen is for matter to move in continuous circular streams, he argued that after a long period of time matter would tend to congregate in a number of spherical vortices. At the centre of each vortex was a star; our Sun is just one of many. Since Descartes believed that the planets were transported around the Sun by rotating matter-fields, the vortices were huge – at least the size of the solar system. In accordance with the principle of rectilinear inertia, the fast-moving particles at the centre of a vortex are constantly trying to move in a straight line (just like the stone propelled by a sling), but they are prevented from doing so by the surrounding larger, slower-moving pieces of matter, which constrain them to move in circles.

Several key ingredients of Newtonian physics are now in place: the principle of inertia; the conservation principles; and, no less importantly, the fundamental idea that the universe as a whole can be explained solely in mechanical terms. But what of absolute space? This is where things get interesting. In *The World* it seems that Descartes simply assumed that space had a Euclidean structure and existed independently of matter – much in the way that Newton would subsequently do. But *The World* was never published in Descartes's lifetime. Hearing of Galileo's troubles with the Inquisition in 1633, troubles linked to the issue of whether the Earth moves, Descartes preferred to suppress it. Instead, *Principles* was published posthumously in 1644, and contains all the essential ingredients of Cartesian physics that were to be found in *The World*, but with one significant difference: superimposed on this familiar backdrop is an entirely new (and confusing) account of space.

He starts off with a relational–relativist account of both position and motion:

The terms "place" and "space", then, do not signify anything different from the body which is said to be in a place; they merely refer to its size, shape and position relative to other bodies. To determine the position, we have to look at various other bodies which we regard as immobile; and in relation to different bodies we may say that the same thing is both changing and not changing its place at the same time. For example, when a ship is under way, a man sitting on the stern remains in one place relative to the other parts of the ship with respect to which his position is unchanged; but he is constantly changing his place relative to the neighbouring shores . . . Then again, if we believe the Earth moves, and suppose that it advances the same distance from west to east as the ship travels from east to west in the corresponding period of time, we shall again say that the man sitting on the stern is not changing his place, for we are now determining the place by means of certain fixed points in the heavens. Finally, if we suppose that there are no such genuinely fixed points to be found in the universe (a supposition which will be shown below to be probable) we shall conclude that nothing has a permanent place, except as determined by our thought . . .   (*Principles* II, 13; Huggett 1999: 93–4)

There is here no mention of motion with respect to space itself; Descartes is now assuming that all movement consists in bodies changing their distance relations with one another.

The rejection of space as the ultimate frame of reference does not, in itself, entail that all motion is relative. Aristotle did not believe in the existence of space, but nor did he believe that motion is relative: the stationary Earth and constantly revolving heavens (whose distance from the Earth is constant) provided secure points of reference. But in rejecting both space and the closed Aristotelian cosmos, Descartes has cut himself adrift. In a cosmos consisting of nothing but pieces of matter in constant motion, where even the stars are continually jostled from their positions, there are no fixed bearings. The only fixed points are those that *we* choose to regard as fixed.

This position is radical and disturbing, but in the context of the perpetual flux of the Cartesian cosmos it is also quite intelligible (although as Newton was later to point out, it is scarcely consistent with Descartes's principles of motion). But no sooner has he advanced this relativist position than he takes a step back:

If, on the other hand, we consider what should be understood by *motion*, not in common usage but in accordance with the truth of the matter, and if our aim is to assign a determinate nature to it, we may say that *motion is the transfer of one piece of matter, or one body, from the vicinity of the other bodies which are in immediate contact with it, and which are regarded as being at rest, to the vicinity of other bodies.*
(*Principles* II, 25, original emphasis; Huggett 1999: 93–4)

Descartes now seems to be telling us a different story. The relativist position conforms with "common usage" (as Newton also maintained in the Scholium), but there is a truth of the matter: a body *really* moves if it changes its position with respect to a motionless body with which it is in immediate contact. This to some extent agrees with Aristotle, who, it will be recalled, said that we should define an object's "place" in terms of the nearest surrounding motionless body. But as Descartes himself surely realized, relativism has scarcely been vanquished: in the context of the Aristotelian system "motionless" could mean "really motionless", but not so in the Cartesian system. This, of course, explains why we find Descartes defining "true motion" in terms of bodies *regarded* as being at rest; there is no fact of the matter as to which bodies are really moving, only convention. Descartes himself goes on to stress just this point: "transfer is itself a reciprocal process: we cannot understand that a body AB is transferred from the vicinity of a body CD without simultaneously understanding that CD is transferred from the vicinity of AB" (*Principles* II, 29).

Confusing it might be, but this doctrine served Descartes very well *vis-à-vis* the Inquisition. Galileo found himself in trouble for suggesting that the Earth might *really* move. Descartes has immunized himself against any such accusation. Since he held that the material in the vortex that is in direct contact with the Earth is *motionless relative to the Earth*, given his definition of "true motion", it follows that the Earth *does not truly move*, since it is stationary with respect to its immediate surroundings.[8]

It may well be that Descartes did not himself believe the account of motion he develops in the *Principles*; it is hard to tell. But as Barbour says, "What is not in doubt is the most remarkable of the paradoxes we meet in Descartes: that he was simultaneously the effective founder of the two diametrically opposed concepts of motion . . . the absolute and the relative" (1989: 437).

10.4 Leibniz

We now jump forward in time to November 1715, some eighty years after Descartes and thirty years after the publication of Newton's *Principia*, when Leibniz sent a letter to Caroline, Princess of Wales, warning her of the theologically pernicious character of Newton's various doctrines, not least the claim that space is the "sensorium" of God. The eminent scientist and theologian Samuel Clarke responded on Newton's behalf, and the result was the *Leibniz–Clarke Correspondence* (Alexander 1956), which was published in 1717, shortly after Leibniz's death. At various points Clarke seems to have consulted Newton when drawing up his replies.[9] As the *Correspondence* developed, the focus fell increasingly on the existence of absolute space, which, of course, Leibniz resolutely denied. The arguments that Leibniz deployed against absolute space are, as we shall see, a rather mixed bag, but he did succeed in bringing to light a crucial difficulty in Newton's position. Whether the difficulty is surmountable will be the topic of Chapters 11 and 12.

Leibniz pointed out two implications of Newton's physics. Each involves a way the actual world could be differently related to absolute substantival space; each posits a sort of *shift* that cannot be detected.

- *Static shift*: Newton claims that the material universe has a particular location within absolute space. If the material universe were situated somewhere else in absolute space, then, provided all the distances and relative motions between bodies remained the same, everything would appear just as it does. For example, if all the bodies in the universe were located ten trillion miles to the south of where they actually are, there would be no empirically discernible consequences.

• *Kinematic shift*: According to Newton, everything in the material universe has a definite but undetectable state of motion relative to absolute space. If everything in the entire universe had a different state of motion relative to absolute space, provided the distances and relative motions between bodies remain the same, there would be no observable differences. For example, if all the bodies in the universe were boosted to a velocity of a thousand mph in a north-easterly direction, there would be no empirically discernible consequences after the boost.

These shifts generate what we can call "Leibniz alternatives", possible situations with the following characteristics:

• all the material bodies in the universe are differently related to absolute space;
• all the spatiotemporal relations between bodies are the same;
• the situations in question are exactly the same in all observable respects.

The kinematic shift is illustrated in Figure 10.2 for a simple universe containing just three objects in absolute space. The scene on the left illustrates how things actually are; the three objects are moving away from one another in the directions indicated by arrows. On the right are the same three objects, with the same relative motions with respect to one another, but in addition, the whole system has an additional velocity component (with respect to absolute space) in a north-easterly direction. Since absolute velocity is undetectable, so is this boost.

Before proceeding it is important to note that in claiming that absolute velocity is without empirical consequences Leibniz was not guilty of imputing to Newton a doctrine to which the latter did not subscribe; far from it. The kinematic shift is simply Galileo's ship's cabin example writ large, and is a further illustration of the principle that Galileo was so keen to establish: the "inoperative quality of motion among the things which share it in common". Newton himself recognized the point in the *Principia*: "The motions of bodies included in a given space are the same among themselves, whether that space is at rest, or moves uniformly in a right line without any circular motion" (Corollary V to the laws of motion; Thayer 1974: 29). The import of this statement is far-reaching.

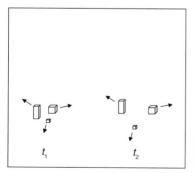 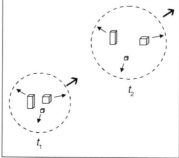

Figure 10.2 A universe in motion: the kinematic shift.

By "a given space" Newton means a *relative space*. The easiest way of defining a space of this sort is by taking one body as our origin – our "reference body" – and establishing three orthogonal axes (i.e. at right angles to each other). We then plot the movements of other bodies by reference to these coordinates. Corollary V is concerned with relative spaces of a particular sort: those established on reference bodies that are either at rest, or moving (but not rotating) with respect to one another at constant speeds and in straight lines. We can call relative spaces that meet these conditions "Galilean frames". Now, if like Newton we assume that absolute space exists, any Galilean frame has some state of absolute motion, or rest, depending on whether the relevant reference body is at absolute rest or not. On the face of it, the laws governing moving bodies could easily be such that bodies behave differently depending on their state of motion with respect to absolute space. For example, the force of gravity could be weaker for bodies in absolute motion. If this were the case, it would be possible – at least in principle – to discover whether or not a given Galilean frame is at absolute rest or not. Suppose that two stars, *Alpha* and *Beta*, are moving apart from one another at uniform speed in a straight line. Each star is at the centre of its own Galilean frame, *Frame-Alpha* and *Frame-Beta*, and both stars have planetary systems. By collecting data about the masses and orbits of the various planets, we could quickly ascertain which of these two frames has the greatest absolute velocity.

But, it turns out, no such test is possible. Newton's law of universal gravitation, for the simplest case, states that for two bodies of mass m_1 and m_2 separated by a distance d, each exerts on the other a force that is proportional to the product of the two masses and inversely proportional to the square of the distance between them, and acts along the straight line joining them (hence the equation $F = km_1m_2/d^2$, where k is the gravitational constant). There is no mention here of *absolute* quantities, the strength and direction of the force is determined by the relative configurations of bodies, and so is entirely independent of their position and velocity with respect to absolute space. The same applies to Newton's other laws: the forces between interacting bodies depend solely on their *relative* configurations and velocities. As a consequence, there is no way of determining the absolute state of motion of any Galilean frame. The bodies in Frame-Alpha and Frame-Beta will behave in exactly the same ways irrespective of any uniform absolute motion of these frames. This property is called "Galilean invariance".

As we can now see, Newton is plainly committed to the undetectability of both the static and kinematic shifts. Leibniz pounced on this, and with a series of arguments sought to establish that this commitment leads to absurd or objectionable consequences. Since these various difficulties all flow from the assumption that there is a distinction between absolute and relative motion, he argued that we must reject this assumption, and with it the notion that there is such a thing as absolute space.

10.5 The argument from indiscernibility

Leibniz was committed to the principle of the *identity of indiscernibles* (PII). When applied to objects, PII amounts to the following. Let S and T be names of objects. If S and T have exactly the same genuine properties then S and T are one and the same thing. Applied to possible *worlds*, PII entails that there cannot be two distinct worlds W_1 and W_2 that are exactly similar with respect to all genuine properties. If they are indiscernible in this way, W_1 and W_2 are one and the same world.[10] Here is how Leibniz deploys the principle:

> To suppose two things indiscernible, is to suppose the same thing under two names. And therefore to suppose that the universe could have had at first another position of time and place, than that which it actually had; and yet that all the parts of the universe should have had the same situation among themselves, as that which they actually had; such a supposition, I say, is an impossible fiction.
> (Alexander 1956: 37)

> To say that God can cause the whole universe to move forward in a right line, or in any other line, without making otherwise any alteration in it; is another chimerical supposition. For, two states indiscernible from each other, are the same state; and consequently, 'tis a change without a change.
> (Alexander 1956: 38)

The first passage is concerned with a static shift, the second a kinematic shift. Newton is committed to saying that both sorts of shift produce different states of affairs. Leibniz asserts that since in both sorts of case the pre-shift universe is indiscernible from the post-shift universe, given PII, the two supposedly distinct states of affairs are in fact one and the same. Newton's theory entails a result that is either absurd or false. If, on the other hand, we adopt the relational theory, and so hold that the only spatial relations that bodies have are relations with other bodies, then we get the right result: since in both cases the relative spatial relations are the same, the alleged "shifts" cannot have taken place.

The argument from PII is uncompelling. The principle is itself very controversial, and it is by no means the case that everyone accepts it.[11] But we need not enter into this dispute, for Leibniz's argument begs the question against Newton in a quite blatant fashion. The argument starts from the premise that the envisaged Leibniz alternatives are *indiscernible* with regard to all genuine properties. Since Newtonians hold that relations between material bodies and absolute space are real features of the world, they will simply deny the premise: the pre- and post-shift worlds *are* different, since the relations between objects and space are different in the two cases. It is true that these differences are not *detectable*, but the PII does not say "Two objects (or worlds) which do not differ in any detectable ways are numerically identical"; it says "Two objects (or worlds) which do not differ *with respect to any genuine properties* are numerically identical". Since Leibniz has not yet established that only empirically detectable differences in properties are *real* differences, his argument as it stands fails. The fact that "indiscernible" can be read as meaning "not detectably different" can easily obscure this point.

10.6 The argument from sufficient reason

Leibniz was also firmly committed to the "principle of sufficient reason" (PSR): "nothing happens without a sufficient reason why it should be so, rather than otherwise" (Alexander 1956: 25). This leads him to argue as follows:

Space is something absolutely uniform; and without the things placed in it, one point of space does not absolutely differ in any respect from another point in space. Now from hence it follows, (supposing space to be something in itself, besides the order of bodies among themselves,) that 'tis impossible there should be a reason why God, preserving the same situations of bodies among themselves, should have placed them in space after one certain particular manner, and not otherwise; why every thing was not placed the quite contrary way, for instance, by changing East into West. But if space is nothing else, but that in order or relation; and is nothing at all without bodies, but the possibility of placing them; then those two states, the one such as it is now, the other supposed to be the quite contrary way, would not at all differ from one another. Their difference therefore is only to be found in our chimerical supposition of the reality of space in itself.   (Alexander 1956: 26)

Leibniz seems here to be envisioning static shifts in the context of an initial act of creation: God had to plant the material bodies *somewhere* in space but, given the uniformity of space, what reason could he have for planting them in one collection of locations rather than another? If we assume that God does nothing for which he lacks a good reason, then we can be certain that God would not create a substantival space, for if he were to create such a space, he would be putting himself in a position where reasonless choices would have to be made, which he would never do. By *not* creating a Newtonian space, by creating only material bodies that are spatially related in certain ways to one another, the problem is avoided.

Do the same considerations apply to the kinematic shift? Suppose that someone were to argue thus:

Just as God could have created the world elsewhere than he did, he could have given it any number of states of motion relative to space itself. Just as there is no reason for creating the world in one place rather than another, there is no reason to give it one state of motion relative to space rather than another.

While there is no obvious reason why God should choose to locate the world in one part of absolute space rather than another, since by hypothesis this space is infinite and homogeneous, this same homogeneity suggests a solution to the motion problem. Since there is nothing to choose between any of the available directions, once God had decided *where* to place the world there would be no reason for him to set it in motion in one direction rather than another, so he would have good reason *not* to set it in motion: he would, therefore, create a cosmos whose overall centre of mass is at absolute rest.

Of course this leaves the "initial location" problem unsolved. However, whatever force this objection might have against theistic Newtonians, it has virtually none against the atheistic Newtonian. If we leave God out of the picture, the problematic choice never arises, and the issue of whether or not there could be rational grounds for making it is simply irrelevant.

Even if we work within Leibniz's framework, and assume that a rational God did create the universe, it is not at all clear that the argument from PSR is successful. The original source of the argument suggests that not all theistically minded philosophers would agree with Leibniz: it was *Clarke* who first introduced the shifts into the discussion, arguing that relationism is refuted by the "obvious" fact that God could have chosen to place the whole material universe somewhere else in absolute space while retaining all spatiotemporal relations between bodies.[12] As for finding a *reason* for putting the universe in one place rather than another, to suggest that the task is beyond God is to suggest that God is no wiser than Buridan's ass. Or as Clarke puts the point:

> when two ways of acting are equally and alike good . . . to affirm in such cases that God cannot act at all, or that 'tis no perfection in him to be able to act, because he can have no external reason to move him to act one way rather than the other, seems to be denying God to have in himself any original principle or power of beginning to act, but that he must needs (as it were mechanically) be always determined by things extrinsic.
>
> (Alexander 1956: 32–3)

It should be pointed out that PSR can be taken in a non-theistic way. The claim that "nothing happens without a sufficient reason why it is so" can be interpreted to mean "everything has a sufficient cause". If we read PSR in this way, how can the substantivalist respond to the question "Why is the world as a whole located where it is?" Sklar suggests that the substantivalist can answer thus:

> Because yesterday it was, as a whole, in this position and no forces arose to move it. So there is a sufficient reason for the present place of the material world in substantival space – its previous position and the forces that have acted in the meantime. After all, why is Jupiter where it is today relative to the Earth? Answer: Because of the positions and velocities they had yesterday and the forces acting upon them in the intermediate period.   (1977: 180)

Explaining the way things are at present in terms of causal laws and the way things were earlier is what *counts* as providing a "sufficient reason" in contemporary science. Leibniz would have refused to accept this sort of answer as providing a genuinely *sufficient* reason, since it explains one contingent state of affairs in terms of another contingent state of affairs, and he believed that ultimately nothing was contingent; everything is as it is because it *had* to be this way, as a matter of necessity. His argument thus rests on a particular version of PSR that, as Sklar notes, "most present-day philosophers and scientists would reject out of hand" (1977: 181).

10.7 The methodological argument

There is a further argument to be found in Leibniz, one that he did not greatly emphasize himself, but that has proved by far the most influential subsequently. In his "Fifth Paper" he writes:

> The author replies now, that the reality of motion does not depend upon being observed; and that a ship may go forward, and yet a man, who is in the ship, may not perceive it. I answer, motion does not indeed depend upon

being observed; but it does depend upon being possible to be observed. There is no motion, when there is no change that can be observed. And when there is no change that can be observed, there is no change at all.

(Alexander 1956: 74)

I noted earlier, in connection with the argument from PII, that Leibniz had failed to defend the claim that only empirically detectable differences are real differences. He is now claiming that although real changes can occur without being observed, any real change is such that it must be *possible* for someone to observe it.

Is this a plausible position? There are those who have found it so. The logical positivists subscribed to the "verification principle": for a claim about the world to be meaningful it must be empirically verifiable. This principle yields the result that claims about uniform absolute motion are meaningless, since none can be empirically verified. It is not at all clear that this is what Leibniz had in mind, and since few now are convinced that the verification principle is true, perhaps this is just as well. Scientists these days are quite prepared to posit unobservable entities, and thus unobservable changes, provided these unobservable goings on are among the postulates of a theory that explains and predicts what *can* be observed.

This last point takes us to the heart of the matter. Many of the entities postulated by contemporary scientists may not themselves be directly observable, but they do have *effects* that are observable. Sub-atomic particles are not visible to the naked eye, but not only are they the building blocks of the macroscopic objects that we can perceive, their movements can be (briefly) tracked by the detectors in particle accelerators. No scientist would be happy in accepting a theory that postulated entities that have *no* discernible empirical consequences. By way of an example, suppose that two theories of particle physics are on offer, P and Q. In many ways theory P is just like theory Q: both explain the behavior of all known particles and forces in equally elegant ways, but there is one difference. Theory P posits an additional type of particle – the *squillion* – which are swarming everywhere all the time. Theory Q does not posit squillions. Moreover, according to P, squillions are in principle unobservable, directly or indirectly: they never interact with any other particles. Clearly, we would have no reason to believe theory P, and so would reject the claim that squillions exist. According to Leibniz, Newton's absolute space is akin to the squillion. Not only is it unobservable, directly or indirectly, but we have an equally powerful theory of space, *relationism*, which is just as good as Newton's, but that does not posit any unobservable entities, or empirically indiscernible states of affairs.

This is a potentially compelling argument. I say *potentially* because Newton (as we shall see shortly) disputes Leibniz's claim that motion with respect to absolute space has no empirical consequences. Newton claims that *dynamic* (forced, accelerated) shifts *are* detectable.

11 Absolute motion

11.1 Inertial motion

Newton deploys two basic arguments against relationism. The best known is the *argument from inertial effects*: the notorious bucket argument. The second is the *argument for real inertial motion*. Both are connected with Newton's first law of motion, namely that any object will continue to move at a constant velocity (or remain at rest) unless acted on by a force. Before proceeding it will help to review some of the relevant terminology.

As we have already seen, Newton's first law is often called the *law of inertia*. *Inertial motion* is motion in the absence of impinging forces: a moving body will continue to move at the same speed in a straight line for ever if no forces act on it. *Inertial forces* are the forces experienced by a body that is undergoing acceleration.

We already know that in the context of Newton's physics a body is at *absolute rest* if it is stationary with respect to absolute space, and in *absolute motion* if is moving relative to absolute space. An object possesses *relative* velocity if it is moving at a certain speed and direction with respect to another object. Acceleration is rate of change of velocity. So an object is undergoing *absolute acceleration* when it is changing its velocity with respect to absolute space, whereas *relative acceleration* consists in a change in velocity with respect to some other bodies.

We can establish a *frame of reference* by centring a coordinate system on a particular body that we assume to be at rest. We can specify the motions of all other bodies relative to the reference body; these other bodies are at rest or in constant motion or accelerating, relative to this frame. So called *inertial frames of reference* are centred on bodies that are not undergoing absolute acceleration; these are the Galilean frames encountered in §10.4.

11.2 The argument for real inertial motions

We start with *De Gravitatione*, a work that remained unpublished until recently, but that contains important arguments not to be found in the Scholium. Newton's explicit target here is Cartesian physics, as formulated in Descartes's *Principles*, and he leaves rubble in his wake.

In §10.3 we saw that Descartes – in *Principles* if not *The World* – combined a commitment to the law of inertia with a relational conception of motion. As Newton is quick to point out, the two are incompatible. The law of inertia states that a body will continue to move in a straight line at constant speed unless acted on by a force. This entails that a body that is *not* moving in a straight line is being acted on by a force. Since the existence of forces is not a conventional matter, the law of inertia presupposes a real and objective difference between straight and curved paths. However, if like Descartes we hold that all motion is relative – to move is to have different distance relations to other bodies at different times – there is no such difference. Each body can legitimately be regarded as stationary, and so can be used to define a reference frame in terms of which the motions of other bodies can be determined. In Figure 11.1, A and B are two spaceships. In the situation depicted on the left, A takes itself to be moving in a straight line, and regards B as flying along a curved path. B does likewise, and regards A as flying along a curved path. For the relationist, neither perspective is privileged; both are equally valid. The relationist is thus committed to the view that every body can rightly be considered to have *many different* motions, in which case there is no objective fact of the matter as to what a body's "real" motion is, and hence there is no objective fact of the matter as to whether a body is moving in a straight line or not.

Consider another example involving competing reference frames. According to frame of reference F_1, the Earth is in orbit around the stationary Sun; according to frame of reference F_2, the Sun is in orbit around the stationary Earth. If we now suppose that the whole solar system is moving uniformly in a straight line, these two reference frames ascribe very different motions to the Earth. According to F_1, the Earth is moving in a spiral (since it rotates around the Sun, which is moving in a straight line), whereas according to F_2, the Earth is moving in a straight line, and the Sun is moving in a spiral. Bearing in mind that from a relationist point of view these two frames are equally valid, we now bring the law of inertia into the picture. According to F_1, the Earth is moving in a spiral, and so a force must be acting on it, dragging it away from its inertial straight-line path; by contrast, since the Sun is moving in a straight line, no forces are acting on it. But according to F_2, it is the Sun that is rotating around the Earth, whereas the Earth is following its straight-line inertial path, and so while there are no forces acting on the Earth, a force must be acting on the Sun. Clearly these two accounts are inconsistent. They ascribe differ-

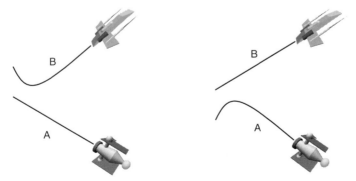

Figure 11.1 **Ships in flight – but along which paths?** If facts about motion are determined solely by the distances between objects at different times, the distinction between straight and curved paths becomes a matter of perspective.

ent forces to the same bodies in the same situation; both cannot be true, for the existence of a real force cannot depend on an arbitrary choice of reference frame. For a consistent physics we need a notion of *real* inertial motion: there must be an objective fact of the matter as to which motions really are inertial, and which are not. Newton's concept of absolute velocity fits the bill.

Newton follows up with a further assault on relationism. If an object's location consists solely in its distances from other objects at a given time, then if (as Descartes believed, and as seems largely true) bodies are continually changing their positions relative to other bodies, there are no facts about past locations, and this leads to disastrous consequences:

> For since, according to Descartes, place is nothing but the surface of surrounding bodies, or position among some other more distant bodies, it is impossible (according to his doctrine) that it should exist in nature any longer than those bodies maintain the same positions from which he takes the individual designation. And so, reasoning as in the question of Jupiter's position a year ago, it is clear that if one follows Cartesian doctrine, not even God himself could define the past position of any moving body accurately and geometrically now that a fresh state of things prevails, since in fact, due to the changed position of the bodies, the place does not exist in nature any longer.
>
> Now as it is impossible to pick out the place in which a motion began (that is, the beginning of the space passed over), for this place no longer exists after the motion is completed, so the space passed over, having no beginning, can have no length; and hence, since velocity depends upon the distance passed over in a given time, it follows that the moving body can have no velocity, just as I wished to prove at first. Moreover, what was said of the beginning of the space passed over should be applied to all intermediate points too; and thus as the space has no beginning nor intermediate parts it follows that there was no spaced passed over and thus no determinate motion, which was my second point. It follows indubitably that Cartesian motion is not motion, for it has no velocity, no definition, and there is no space or distance traversed by it. So it is necessary that the definition of places, and hence of local motion, be referred to some motionless thing such as extension alone or space in so far as it is seen to be truly distinct from bodies.   (Huggett 1999: 109–10)

Newton argues that, for there to be an objective fact of the matter as to how fast a body is moving at a given moment, and in which direction, there have to be objective facts concerning where it was located at previous times. If we define spatial locations by reference to material bodies, and if material bodies are constantly changing their spatial relations to one another, the required facts are simply not available. It would be different if we could take just one body as our standard: if (say) we knew the Sun to be motionless, we could plot the movements of Jupiter by measuring the distance between the two at different times. But this move is unavailable to the relationist. The situation is very different if we recognize absolute space, and hence absolute motion: there is now an objective fact as to the directions objects are moving in and the distances between them.

This argument is intriguing, but not decisive. There may be ways for relationists to meet Newton's challenge. But Newton is certainly right about there being a difficulty here, a difficulty to which his notion of absolute space supplies a solution.

11.3 The argument from inertial effects

In the light of the arguments considered thus far, we can provisionally grant Newton this much: the law of inertia does indeed presuppose a real difference between bodies that are in uniform straight-line motion and bodies that have some different form of motion (e.g. circular), a difference that the relationist who appeals only to distance relations between bodies cannot recognize. However, what is to prevent the relationist from rejecting the distinction in question? Is there anything to prevent Leibniz from holding that claims such as "the Sun is moving in a straight line and the Earth is rotating about it" and "the Earth is moving in a straight line and the Sun is rotating about it" are equally true? It is here that Newton's arguments in the Scholium come into play.[1]

While Newton fully accepted that absolute velocity has no detectable consequences – there are no experiments that will reveal how fast you are moving with respect to absolute space – he took pains to point out that the same does not apply for *changes* in velocity; that is, accelerations. We all know that forces are associated with acceleration. If you are in a quickly accelerating car you feel yourself being pulled back into your seat. This is a typical instance of so-called "inertial forces" making themselves felt; that is, the forces that are "experienced" by a body that is *not* following its inertial path through space as a result of some applied force, such as that supplied by a car's engine. The trouble is, just as velocity is frame-relative for the relationist, so is acceleration. If A and B are accelerating apart, then for the relationist there is no fact of the matter as to who is *really* accelerating: each of A and B can regard themselves as being at rest and the other as accelerating with respect to them. If I accelerate away from your car, which remains stationary at the pavement, then, *relatively speaking*, it is as true to say that I am accelerating away from you as it is to say that you are accelerating away from me. But it is clear that if we describe the situation in terms of relative accelerations we are omitting a crucial factor: it is only me who feels inertial forces, you do not. I am the one pulled back into my seat, not you. The obvious explanation of this is that only one of us is *really* accelerating. Hence it seems that an adequate dynamics requires a distinction between real and relative acceleration, for without it we cannot account for the presence/absence of inertial forces (i.e. those experienced by the accelerating body). Since the relationist has no means of grounding this distinction, his account of (accelerated) motion is inadequate. If we assume, as most do, that acceleration is a change of motion *with respect to something*, then what alternative is there to Newton's view that real acceleration is a change of motion relative to absolute space?

Once the existence of absolute space is established in this manner, we can deduce that material objects that are not accelerating must be either in uniform motion or at rest with respect to it, *even though there is no observational evidence that will tell us which state a particular body is in.*

In making these points in the Scholium Newton employed two thought-experiments. Neither involves linear accelerations of the sort that we have just been considering; rather they feature *rotations*. Newton claims:

> The effects which distinguish absolute from relative motion are, the forces of receding from the axis of circular motion. For there are no such forces in a circular motion purely relative, but in a true and absolute circular motion they are greater or less, according to the quantity of the motion.
>
> (Huggett 1999: 122)

And of course, rotation is itself a form of acceleration. Since acceleration is change in velocity, and velocity involves moving in a particular direction, a rotating object is continually changing its direction of movement, and hence is continually accelerating. In the case of a stone fixed to a string and whirled around one's head, at any instant the velocity of the stone is in a direction that is tangential to the circle it is moving around, the direction it would fly off in if released. The string has to exert a continual force on the stone to prevent it from pursuing this natural inertial motion. In the case of a body that follows a curved (but not circular) path, such as rocket ship B on the left in Figure 11.1, forces supplied, let us suppose by small rocket motors, must be applied to push the body off its inertial path. Since these motors apply force to the body of the ship, not to its occupants, it is left for the ship itself to force its occupants off their inertial path: unless strapped down, those on board would find themselves being pushed by the walls of the ship, just as people in a car turning a corner find themselves being pushed against the car doors. Similar effects, although in one case rather more complex, underpin Newton's examples.

The bucket

Suppose that there is a bucket full of water, suspended by a rope (or elastic cord) from the ceiling. Twist the bucket and let the rope wind up, then release the bucket. The following sequence of events will unfold.

1. The bucket and water are initially at rest with respect to one another and the surface of the water is flat.
2. After a short while, as the twisted rope starts to unwind, the bucket begins to rotate, but the water remains stationary and flat.
3. Then, a while later, as the rotation of the bucket is gradually communicated to the water via friction, the water starts to turn, and soon is rotating at exactly the same rate as the bucket; the water is no longer flat but concave, as it rises up the sides.

We are all familiar with the phenomenon: fluids rotating in containers rise up the sides. Who hasn't spilt their tea when stirring in milk? Newton talks of a tendency to "recede" from the axis of rotation, but this is not entirely accurate. The particles in a rotating fluid have a natural tendency to move *at a tangent* to the curvilinear path they are following. This is how they would move if unconstrained, just like a stone released from a sling. But since their natural (inertial) movement *is* constrained – by the other particles in the fluid, and ultimately the walls of the container – their actual movement takes the form of an outwardly expanding spiral. Since all the particles in the fluid move thus, there soon occurs a build-up around the sides of the container. (This build-up generates a marked increase in (downwards) fluid-pressure, which often prevents the liquid from overflowing, depending on the rate of rotation.)

The behaviour of the water in this very basic and familiar sort of case poses a severe problem for the relationist. Newton has an explanation for why the water in the third stage curves: it is moving with respect to absolute space. The relationist cannot adopt this explanation, and so has to say that the curvature of the water is due to its moving relative to some other material thing. But what? Descartes, it will be recalled, said that "real motion" takes place when a body moves in relation to its immediate surroundings. If this were right, then we should expect to find the curvature at a maximum when the relative motion of the water with respect to the bucket is at a maximum. But at this point, stage 2, the water is flat! The curvature is at a maximum at the stage 3, when there is *no* relative motion between the two, when both bucket and water are rotating at the same rate.

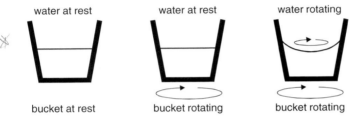

water at rest   water at rest   water rotating

bucket at rest   bucket rotating   bucket rotating

Figure 11.2
Newton's buckets.

If we cannot explain the inertial forces by appealing to the relative motion of the water with respect to the bucket, can we say that the water is rotating with respect to itself? No, for the same effect would be observed if none of the particles that make up the water rotated with respect to one another. Could the curvature be the product of motion with respect to we who are observing the bucket? Once again the answer is no. The curvature certainly occurs when the water is rotating relative to the onlooking observer, but it is clearly not the product of this relative motion. If the observer were to run around the bucket very quickly the water would remain perfectly flat (ignoring vibrations). The same would apply if we were to put the surrounding buildings on a giant turntable and whirl them around the motionless bucket. The experiment would be expensive to conduct but there is no reason to suppose that the water would be in the slightest way perturbed. Newton's explanation, that the inertial effect occurs when the water is rotating relative to absolute space, certainly has merit.

The globes

Newton does not explicitly consider whether the curvature of the water in the bucket case could be due to relative motion to some more distant bodies, such as the stars. Instead he provides a further example, one that is surely (in part) intended to refute the suggestion:

> if two globes, kept at a given distance one from the other by means of a cord that connects them, were revolved about their common centre of gravity, we

might, from the tension of the cord, discover the endeavor of the globes to recede from the axis of motion, and from thence we might compute the quantity of their circular motions.   (Huggett 1999: 123)

If the cord contained a device for measuring tension (such as an appropriately attached spring-balance), we could ascertain whether the globes are rotating or not simply by looking at it: if it shows a tension the globes are rotating. So far, so familiar. Newton now takes us a stage further. He points out that precisely the same would apply if the globes were not surrounded by stars and planets, but located in an "immense vacuum, where there was nothing external or sensible with which the globes could be compared" (Huggett 1999: 123). So imagine two possible worlds, W_1 and W_2. Each world contains a pair of globes connected by a cord, with a built-in tension meter, and nothing else. In W_1 the meter registers zero pressure and in W_2 the meter registers a significant pull. Newton can explain the difference: in W_2 the globes are rotating in absolute space, and in W_1 they are not. What can the relationist offer? Since there are no objects other than the globes and cord in either world, and the distance between the globes remains constant, from a relational standpoint there is no relative motion in either world, and so the globes in both W_1 and W_2 are stationary. It seems that the *only* difference between these worlds is the existence of the tension in the cord in W_2. The relationist is thus pushed into the uncomfortable position of having to view the tension in W_2 as an entirely inexplicable occurrence. Given the availability of an alternative explanation – Newton's – this position seems quite untenable.

Taken together, the bucket and globes amount to a formidable challenge to relationism. In both cases, accelerations give rise to observable forces; these accelerations are not, it seems, accelerations relative to ordinary physical bodies; but if acceleration is a change of velocity relative to something, what can this something be if not absolute space?[2] As Earman suggests (1989: 64), Newton's argument is best construed as (in part at least) an inference to the best explanation of the mechanical phenomena:

P1 The best explanation of mechanical phenomena in general (and rotating phenomena in particular) utilizes absolute acceleration (and absolute rotation in particular).
P2 Absolute acceleration (and absolute rotation in particular) must be understood as acceleration (and rotation) relative to absolute space.

If we accept the Newtonian account we have an explanation for why the connection between acceleration and force is so ubiquitous; why it exists at all places and all times. If all acceleration is a change in velocity with respect to absolute space, and absolute space is the same everywhere, it is to be expected that similar accelerations give rise to the same forces everywhere and everywhen, which as far as we know, they do. For as we have already seen, and as Newton was to go on to demonstrate in the rest of the *Principia*, inertial forces *are* ubiquitous, and are by no means confined to rotating bodies: every time you feel yourself pushing into the wall of a turning bus or car you are experiencing them.

11.4 Stalemate?

Although Newton's bucket amounts to a weighty argument, relationists did not simply surrender. This is not surprising, for although the argument from inertial effects is powerful, the relationist's stance is far from absurd. Indeed, the positions might seem perfectly balanced:

- Leibniz argued (and Newton accepted) that in Newtonian physics uniform absolute velocity has no empirical consequences. If you have an absolute velocity, it is impossible to tell what it is. If we assume that uniform absolute velocity is a suspect notion, since it has no empirical significance, then how can absolute *acceleration* have any empirical significance? After all, absolute acceleration is simply variation in absolute velocity.

- Newton argued in the other direction. Having good grounds for believing that absolute acceleration has significant empirical consequences, he concluded that there must be such a thing as absolute velocity, even though the latter is without physical consequences.

From this we see that Leibniz's kinematic shift continues to pose a problem for the Newtonian: the empirical insignificance of uniform motion with respect to absolute space renders the relevance of acceleration (relative to absolute space) questionable. This said, if relationists are to reject Newton's theory they must still come up with an alternative explanation of inertial effects. If these are not produced by absolute acceleration, construed as change in velocity with respect to absolute space, what does produce them? In the remainder of this chapter we will consider three lines of response.

11.5 The Leibnizian response

Although Leibniz was well aware of the bucket argument, and the general problem posed by acceleration for the relationist, he never got round to publishing a detailed response to it. However he did make some suggestive remarks. In his fifth letter to Clarke he wrote:

> 53. I find nothing in the Eighth Definition of the *Mathematical Principles of Nature*, nor in the Scholium belonging to it, that proves, or can prove, the reality of space in itself. However, I grant there is a difference between an absolute true motion of a body, and a mere relative change of situation with respect to another body. For when the immediate cause of the change is in the body, that body is truly in motion; and then the situation of other bodies, with respect to it, will be changed consequently, though the cause of the change is not in them. . . . Thus I have left nothing unanswered, of what has been alleged for the absolute reality of space.   (Alexander 1956: 74)

Unfortunately, Leibniz died before he could spell out exactly what he had in mind, but it may have been something along these lines. Since Newton's first law of motion tells us that absolute accelerations are correlated with forces, can't we say that inertial effects are produced not by change of motion with respect to absolute space, but rather the *forces* that Newton rightly recognizes are inseparably bound up with real as opposed to merely relative accelerations? Real accelerations are those in which the "immediate cause of the change is in the body".

This idea works well enough for some cases; for example, if two rockets in outer space are in relative acceleration it is easy to apply Leibniz's criterion to determine which is really accelerating – it is the one whose rocket motor is firing. However, not all cases of real acceleration are like this. We might extend Leibniz's argument to collisions, and say that (in a manner of speaking) when contact between moving bodies occurs, the cause of the change is "in the object". But there remain many instances of real accelerations that are not produced by external forces in this sort of way. First, there are the accelerations produced by forces that act at a distance, such as gravitational and magnetic attraction (at least in Newtonian physics). Secondly, as Friedman stresses:

> uniform rotation is *not* correlated with external forces in Newtonian theory. Even if we could replace the reference to absolute acceleration [in some cases] with a reference to force-generating causal processes, Newtonian theory does not allow us to replace the reference to absolute rotation in a similar manner. (1983: 229)

A rotating solid disc will continue to rotate at a constant angular velocity, without any need for any external force or torque to be applied: it is maintained in motion by the preservation of angular momentum (a basic principle of Newtonian physics, and confirmed by, for example, the Earth's daily rotation). And yet, if the disc rotates fast enough, the centrifugal forces generated by its rotation will tear it apart.

In other (earlier) Leibnizian writings a different approach can be discerned. In the "Specimen Dynamicum" he claims that all motion is in straight lines or compounded of straight lines, and then says:

> From these considerations it can be understood why I cannot support some of the philosophical opinions of certain great mathematicians . . . on this matter, who admit empty space and seem not to shrink from the theory of attraction, but also hold motion to be an absolute thing and claim to prove this from rotation and the centrifugal force arising from it. But since rotation arises only from a composition of rectilinear motions, it follows that if the equipollence of hypotheses is saved in rectilinear motions, however they are assumed, it will also be saved in curvilinear motions.
> (Loemker 1970: 449–50)

His idea may have been that although objects *look* as though they are rotating, in reality they are composed of microscopic particles that are moving in straight lines, and are zig-zagging back and forth. This is an intriguing speculation, but, as Earman points out (1989: 72), Leibniz's analysis is based on the assumption that there is a fact of the matter as to which trajectories are straight and which are not, and as we have already seen (§10.2), in a relational space a trajectory that is straight relative to one reference body will be curved relative to another, and each perspective is equally valid.

11.6 The Machian response

The nineteenth-century physicist Ernst Mach criticized Newton for positing absolute space on empiricist grounds. Mach was deeply suspicious of unobservable entities (such as absolute space), and believed that science should concern itself only with relationships between observable things and events. This general approach, which exerted a profound influence on Einstein, naturally leads to a relationist approach to space.

Like Berkeley, Mach criticized Newton for (as he saw it) going beyond the observational data. The globes argument is particularly vulnerable on this score. Newton claims that globes in an otherwise empty universe would behave in just the same way as globes in our universe; a tension would be created in the joining cord when the globes are rotating. Mach complains that this is an entirely groundless speculation. On the one hand, since the experiment has never been carried out we have no way of knowing how a pair of rotating globes would react if the rest of the matter in the universe were to disappear. On the other hand, the experiment is impossible to carry out. If the globes are the only objects in the universe, there couldn't possibly be anyone present to observe the results of their rotation. Newton is guilty of assuming the result he purports to be arguing for. Yes, given his theory, the globes would behave as he describes, *but* his theory is the only reason we have for believing that the globes *would* behave as he describes, hence the thought-experiment doesn't prove his theory true.

But what of the bucket? What of inertial effects that *are* observable in the actual universe? Since Mach is convinced that talk of absolute space is nonsense, and that the only motion it makes sense to talk about is relative motion, he argues that inertial effects must be the product of change relative to *other material bodies* in the universe.

How does this work in practice? Which other bodies? Well, for most practical purposes, from the standpoint of the Earth, the stars can be regarded as providing a fixed frame of reference. So instead of saying that the water in the rotating bucket rises because it is accelerating relative to absolute space, we say that it rises because it is rotating relative to the fixed stars. It is important to note that all the inertial effects we observe – on the Earth at least – *do* in fact arise when objects accelerate relative to the stars. Since we cannot observe motion relative to space, this relative acceleration is all we can directly detect. Mach is on firm ground here.

That there are predictive differences between Mach's account and Newton's is readily grasped. For Mach, since it is the relative motion between water and stars which causes the water to rise, it doesn't matter, or even make sense to say, which of the two is *really* rotating: if there is relative motion between water and stars the water will curve. For Newton, however, it does make sense to suppose that the water is really stationary, and if it is stationary (i.e. at rest relative to absolute space) it will remain perfectly flat no matter how fast the rest of the universe is rotating about it. Similarly, if there is *no* relative motion between water and stars, then according to Mach the water will remain flat. Not so for Newton. If the water and the stars are rotating at the same rate relative to absolute space, the water will rise, even though there is no relative motion between the two.

So far so good; now for the hard part. Mach has correctly pointed out that for most practical purposes on Earth we can regard inertial effects as arising when objects move relative to the stars, which we can regard as fixed. But this is only an approximation: in fact the stars are moving relative to one another, it is just that this motion is comparatively slow (at least when observed from Earth). So what *precise theory* of dynamics is Mach offering? If the stars are not fixed, and inertial effects are caused by relative motions, we need to specify what the relevant relative motions are. What must an object be accelerating with respect to in order to undergo inertial effects? What are the relevant laws of nature? What is the mechanism?

It is here that Mach is rather vague. The general character of the needed theory is plain. The distinction between inertial and non-inertial (accelerated) motion must be grounded in the overall arrangement of matter in the universe. It is because matter is configured in a certain way that certain paths are inertial and others are not. His one concrete proposal is that inertial forces are produced when a body accelerates with respect to the *centre of gravity of the entire universe*. By what mechanism, though, and why?

Presumably, there must be something akin to a force that links all bodies in the universe, and that is responsible for their moving as they do. This is a bold conjecture, which can seem implausible: "Think about it for a moment. Would you seriously entertain the possibility that the reason you get seasick is due to your motion relative to the stars? Is this notion any more plausible than astrology?" (Earman 1989: 211). Newton himself recognized the possibility of explaining the bucket in this sort of way but was sceptical: "But who will imagine that the parts of the [water] endeavor to recede from its centre on account of a force impressed only on the heavens?" (*De Gravitatione*; Huggett 1999: 108). It is not clear that this scorn is entirely justified: Newton's own theory of gravity, after all, posits a force of attraction acting at a distance that operates instantaneously and links every material body to every other material body! This is just the sort of force that Mach needs, but there is a key difference: Newton had a precise quantitative theory that matches known empirical data and whose predictions are well confirmed. Mach can offer only a sketch of a theory of comparable power. In the absence of a complete physical theory, the Machian approach cannot be judged successful.[3]

11.7 The Sklar response

In responding to Newton, the Leibnizian and the Machian both make the same assumption: they agree with Newton that inertial effects are due to acceleration, and that acceleration is change of velocity *with respect to something*. It is this last assumption, that change of motion has to be relative to something, that Sklar questions (1977: 229–32). Sklar pointed out that the relationist can simply abandon this assumption.

What follows if we do? Well, recall the problem. If we adopt a relationist view of acceleration, how can it be that when two bodies are undergoing acceleration relative to one another only one body experiences inertial effects? What account can the relationist give of this asymmetry? Sklar's suggestion is that there is a *primitive monadic property* that bodies moving on some trajectories possess, which comes in different quantities, and inertial effects are associated with the possession of this property. If we like we can call this property "absolute acceleration", but this is only a label; objects possessing this property should not be thought of as undergoing changes in motion relative to Newtonian absolute space, or Mach's centre of the universe. The property in question is primitive, and thus quite independent of absolute space, which we now say does not exist.

While Sklar's proposal on behalf of the relationist is interesting, in that it reveals a previously unsuspected region of logical space, in other respects it is not very appealing. The Sklarian relationist posits just what is needed for the relationist to solve the problem of inertial effects in a way that is immune to the problems facing the Leibnizian and the Machian, but the posited property has no independent motivation. In order to deliver quantitative predictions, the Sklarian will presumably employ all the normal machinery of Newtonian mechanics, and then re-interpret the results in his own terms; re-describing trajectories that Newtonians view as absolute accelerations in terms of varying quantities of the monadic property. The resulting picture is rather odd: there are bodies moving along various paths (construed in a relational way); only some of these objects possess the monadic property, and only those that do suffer inertial effects. Now, there is nothing in the objects *or* the paths they follow that distinguishes the objects that possess the monadic property from those that do not. The property seems to be distributed in a quite accidental or unprincipled fashion.

Of course the Sklarian can respond: "Not at all! The property is correlated with inertial effects. There is nothing unprincipled about its distribution at all." This is true, but does it help? The problem facing relationists is to explain how and why inertial effects arise as they do. Sklar is solving the problem by introducing a primitive property; the possession of this property is the proposed explanation of inertial effects. But, and this is the real problem, all we are told about this property is that it is "that property which is found wherever inertial effects are found". By comparison, the Newtonian offers a richer account: there is a difference between objects that suffer inertial effects and those that don't, namely, inertial effects are suffered by objects that are undergoing absolute acceleration. If we buy into the idea that some objects are moving uniformly with respect to absolute space and some are not, we have a principled explanation of what distinguishes those objects that suffer inertial effects from those that do not. The Sklarian cannot offer as much.

So at this stage of the game the Newtonian is clearly ahead on points, but the game is not yet over. We will next look at ways to improve the Newtonian account, but we shall also see that this revision gives rise to new problems and difficulties.

12 Motion in spacetime

12.1 Newtonian spacetime

Newton may be ahead on points, but his account is still blighted by a serious problem. To account for inertial effects Newton recognized absolute acceleration; having recognized absolute acceleration he also had to recognize absolute velocity (change of which constitutes absolute acceleration), despite the fact that absolute velocity has no physical consequences. In positing absolute velocities, Newtonians expose themselves to the claim that their theory gives rise to empirically indistinguishable states of affairs. Which of the infinite number of possible universes that can be generated by Leibniz's kinematic shift do we inhabit? Which corresponds to the actual world? If Newton's theory is correct, there is no empirical test that could be performed that would reveal the answer.

Since the work of Cartan and others in the 1920s and 1930s, Newtonians have come to see a better way to formulate Newtonian theory, a way that eradicates absolute velocities but retains (in a way) absolute accelerations. The improved formulation involves two steps: first, Newton's theory is formulated in terms of "spacetime"; then this *Newtonian spacetime* is discarded in favour of *neo-Newtonian* (or *Galilean*) spacetime.[1]

Before we can appreciate the difference between these two, we need to know what a *spacetime* is. We can then consider some of the ways that spacetimes can differ. Before proceeding there are some misconceptions to dispel. On hearing the word "spacetime" one naturally thinks of Einstein and relativity theory. But while Einstein's special and general theories of relativity are indeed formulated in terms of spacetime, they employ a particular form of spacetime (one in which spatial and temporal intervals are relativized to reference frames), but spacetime comes in many forms, and there are "classical" spacetimes that do not share this feature (see Earman (1989: Ch. 2) for a survey of classical spacetimes). A second misconception concerns the four-dimensional "block view" of the universe. It is certainly true that spacetimes are often, and understandably, construed in this way, but it would be a mistake to suppose that they *must* be. As we shall see in due course, there are alternatives. But in the meantime, to avoid overcomplicating, we can overlook this point. In what follows I will be assuming that all parts of a spacetime coexist in a four-dimensional ensemble.

There are two ways of approaching a spacetime: top-down and bottom-up. The top-down approach is to start with a complicated spacetime – Newtonian, for example – and then to subtract various elements of structure. The bottom-up approach is to start with the elementary components of a spacetime – a collection of distinct individual points – and then to add different types of structure. We will be following the top-down route.

Newton viewed space as an infinite and immutable three-dimensional Euclidean continuum of points, all of which endure through time (which is also infinite). To convert this space + time model into a spacetime we simply replace Newton's enduring spatial points with a succession of momentary and numerically distinct *spacetime points*. The switch in perspective is indicated in Figure 12.1. On the left are four spatial points persisting through time; on the right, these four persisting points have been replaced with an array of spacetime points. Newtonian space has been chopped into bits, in this case into forty numerically distinct points, although of course in a real spacetime there would be an infinity of points. Taken as a whole, the collection of spacetime points constitute a *four-dimensional* continuum: a spacetime.

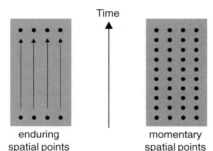

Time

enduring
spatial points

momentary
spatial points

Figure 12.1 From space to spacetime.
A spacetime continuum (or manifold) is composed of numerically distinct points-at-times, rather than the enduring points of classical space. While spacetime and the "block view" are certainly compatible, it is perfectly possible to combine the spacetime view with a dynamic conception of time. A growing block theorist, for example, will hold that spacetime increases in (temporal) size as layers of spacetime points come into being.

Just as a three-dimensional volume (think of a brick) can be thought of as being composed of a collection of two-dimensional planes, so a four-dimensional volume can be regarded as a collection of three-dimensional volumes, or *hyperplanes*. In Figure 12.2 the two cubes represent a Newtonian space at two moments of time (so each cube represents an infinite three-dimensional volume of space). A succession of these cubes composes a four-dimensional spacetime. By suppressing a spatial dimension, we can represent a four-dimensional spacetime by a succession of flat surfaces, as on the right. Each of these surfaces represents a three-dimensional hyperplane.

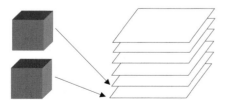

Figure 12.2 Four-dimensional spacetime as a sum (or "stack") of hyperplanes

In Newtonian spacetime every point is at a determinate spatial and temporal distance from every other point. Vertical lines running down through the planes indicate points at different times that are at zero distance from one another (and so, in effect, the "same place") (Figure 12.3).

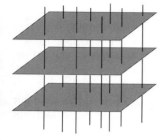

Figure 12.3 A Newtonian spacetime. Points in different hyperplanes (i.e. at different times) are spatially related. The diagram employs the convention that points directly above and below each other are at no spatial distance from one another; hence straight vertical lines represent the same place over time.

Distinguishing absolute velocity and absolute acceleration in Newtonian spacetime is easy. Persisting material objects are represented in spacetime diagrams by *worldlines* (or, in the case of objects larger than points, *world-tubes*, but I will ignore this complication for now). Each of these worldlines represents an object persisting through a succession of different spacetime points; the worldline in its entirety represents the complete history of the object in question. Objects that are at absolute rest have vertical worldlines; objects in uniform motion have straight worldlines that cut through the layer of grids at an angle; the greater the degree of deviation from the vertical, the faster the velocity. All curved lines passing through successions of hyperplanes represent the worldlines of bodies undergoing absolute acceleration; the steeper the curve, the greater the acceleration. In Figure 12.4 the worldlines of nine objects are depicted. Three of these objects are at absolute rest, three are in inertial (uniform straight line) motion, and three are accelerating, at different rates and in different directions. The objects in inertial motion are not moving relative to one another: the distance between their worldlines remains constant over time.

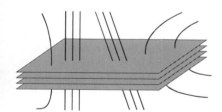

Figure 12.4 The states of rest, uniform motion and acceleration in Newtonian spacetime.

Looking at spacetime in a more general way, we can distinguish *inertial* and *non-inertial* trajectories within spacetime itself. These are continuous successions of spacetime points, occupied or unoccupied; straight paths are inertial, curved paths non-inertial. The straight paths are those that bodies will follow if they are either at rest or moving with a uniform velocity. All the curved paths represent accelerated motions. Again, the vertical lines represent absolute rest (each such line represents a different enduring point in three-dimensional space). Taken together, the collection of straight lines – vertical and non-vertical – is said to represent the *inertial structure* of the spacetime. This is quite natural: taken together, these lines represent all possible inertial trajectories that bodies could take.

12.2 Neo-Newtonian spacetime

To turn a Newtonian spacetime into a neo-Newtonian (or *Galilean*) spacetime is not difficult. In Newtonian spacetimes there are distance relations between points in different simultaneity hyperplanes; if we eliminate these distance relations, leaving everything else unchanged, we are left with a neo-Newtonian spacetime.

Once the distance relations between hyperplanes are abolished, it no longer makes sense to talk of how much distance separates a point in one hyperplane from a point in a different (earlier or later) hyperplane. Moreover, the concept "same place" cannot be applied over time, since points at different times are at no distance from one another; only points that are simultaneous are located at a spatial distance from one another.

Since all other aspects of Newtonian spacetime are preserved, the temporal relations between hyperplanes remain intact, and the points within a given hyperplane are all simultaneous. Crucially, the distinction between straight and curved paths is retained, despite the elimination of transtemporal spatial distance. Formally speaking, this is achieved by first establishing coordinates for each three-dimensional space (or hyperplane); this allows numbers to be assigned to points, and distances to be given a numerical value. The next step is to stipulate which sequences of points through successive three-dimensional spaces constitute continuous sequences. With these lines distinguished, we then specify which ones are curved and which are straight. In technical terms, the latter procedure is known as specifying the "affine structure" of the spacetime; the specification of straight-line paths is known as supplying the "connection". Since the straight lines are intended to represent the paths through spacetime that bodies would follow if no forces were acting on them, they represent the inertial paths, and the totality of straight lines represents the inertial (or affine) structure of the spacetime as a whole. The distinction between inertial and non-inertial motion is thus preserved: curved paths are followed by accelerating bodies, straight paths by non-accelerating bodies.[2]

To appreciate fully the difference between Newtonian and neo-Newtonian spacetime it will help to have a simple example on hand. Consider just two neo-Newtonian inertial frames, P and Q. As in the Newtonian case, P can be represented by a collection of parallel worldlines running through the simultaneity planes, and Q likewise. We start by assuming that the P-frame is at rest, and so depict its worldlines as vertical, and those in the Q-frame as slanting. In Figure 12.5 the rectangles represent the collections of lines corresponding to these two frames, but for the sake of clarity only a small proportion are shown. The depicted transformation is a re-alignment of the hyperplanes, which renders the lines of the Q-frame vertical (and hence "motionless"). A transformation of this kind is called an "inertial transformation", since it preserves inertial structure.

In neo-Newtonian spacetime *every* inertial frame can be turned into the zero-velocity frame by an inertial transformation – and every such transformation preserves the distinction between straight and curved lines, and hence between uniform and accelerated motion (see Figure 12.6).

Given the stipulation that there are no distance relations between distinct hyperplanes, there is no "right" way of aligning them, and the selection of one frame as "privileged" or motionless is entirely arbitrary.

Figure 12.5 An inertial transformation. P and Q represent two distinct inertial frames of reference; each frame consists of a family of (straight) parallel paths. On the left is a representation of spacetime that shows the P-frame to be the zero-velocity state. On the right is an alternative representation of the same spacetime – created by altering the distance relations between hyperplanes – in which the Q-frame is the zero-velocity state.

In Euclidean plane geometry, transformations that leave the structure of the space unchanged are called "symmetries". The symmetries of the plane are rotation about a point (by any angle), reflection about any axis, and translation (or displacement) in any direction by any distance. As shown in Figure 12.7, the geometry of a shape remains the same under these transformations.

For the same reason, inertial transformations are symmetries within a neo-Newtonian spacetime, for they leave spacetime structures unchanged. As we shall see later, different spacetimes have different symmetries (the Minkowski spacetime of special relativity (see §16.5) has "Lorentz transformations" as symmetries).

Velocity is distance travelled divided by time taken; since the notion of distance over time is undefined in neo-Newtonian spacetimes, there are no absolute velocities in such spacetimes. However, there are absolute accelerations, and these can easily be measured empirically. Suppose that we have an object that we have reason to believe is accelerating (e.g. it is being boosted by a rocket motor). Select

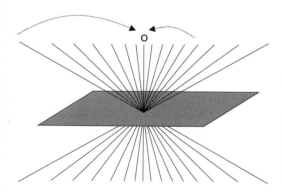

Figure 12.6 Neo-Newtonian spacetime. Every inertial frame can be transformed into the zero-velocity frame by an inertial transformation (or realignment of hyperplanes). Since there are no distance relations between hyperplanes, there is no fact of the matter as to which of these representations is correct. In a Newtonian spacetime there are distance relations between hyperplanes, and so one of these representations will reflect reality, the remainder will not.

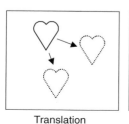

Translation

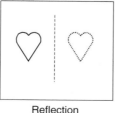

Reflection

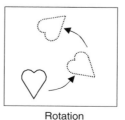

Rotation

Figure 12.7 Symmetries of the Euclidean plane.

some part of its trajectory as the "starting point", and select an inertial frame of reference for which the object is at rest at this time. Now select a later part of its trajectory as the "end point", and another inertial frame with respect to which the object is now at rest. All we need do now is ascertain the *relative* velocity of the second inertial frame with respect to the first. This gives the *absolute velocity change*, i.e. a change of velocity that observers in all inertial frames can agree on. If we now divide this absolute velocity change by the time separating the starting and end points, we derive the *average absolute acceleration* of the object over this part of its trajectory. The *instantaneous* absolute acceleration can be worked out using the usual limiting process (Sklar 1977: 205).

The following points may firm up your understanding of neo-Newtonian spacetime:[3]

- "This particle is moving at constant velocity" makes sense in neo-Newtonian spacetime: it simply means that the worldline is straight.
- "This particle has a speed of $10\,cm\,s^{-1}$" does not make sense. A particular speed requires a definite time between two events and a definite spatial distance. In neo-Newtonian spacetime the former is well defined but the latter is not. People in different reference frames will disagree about what speeds objects have, and there is no fact of the matter as to which groups are right or wrong.
- "This particle is at rest" does not make sense.
- "This particle collided with that one" makes sense: two particles collide if their worldlines intersect.
- "This rope is straight" makes sense: the claim is true if every temporal stage of the rope is straight in every three-dimensional space (or hyperplane).
- "This particle has travelled 10 cm" doesn't make sense: talk of spatial distances *over time* makes no sense.
- "Particle A is moving faster than particle B" does not make sense: in some frames A will be at rest and B moving, in others B will be at rest and A will be moving, and all the frames are equally legitimate.
- "Particle A is moving at $10\,cm\,s^{-1}$ relative to particle B" makes sense.

12.3 The only reasonable view?

Neo-Newtonian spacetime seems very odd initially, since we are certainly accustomed to assuming that where we are now is spatially related to where we were a little while ago, but is it really all that strange? Some writers suggest not. Geroch says:

> The Galilean [neo-Newtonian] view, although it takes a little getting used to, is really a simple and remarkably natural attitude about how space and time operate. It would be easy to be lulled into the position that it represents the only reasonable attitude one could take.   (1978: 52)

In a similar vein, Huggett writes:

> Again, it might seem strange that some points have a spatial distance between them when others do not, but points don't have to have every property that we can think of. For example, we could define some point to be the centre of

the Euclidean plane, so that every point would be a definite distance from the centre, but we don't. Similarly, we can define a spacetime in which there is no distance between some points. Indeed, it seems plausible that spacetime really is this way: How far in space are you from where you were in space a minute ago? 1m? 1km? 1,000km? It is reasonable to think that such a question has no answer. (1999: 194)

If you dwell for a few moments on the question "How far in space am I now from where I was a minute ago?", there is one thought that might occur to you that *would* seriously undermine any lingering conviction you might have that this question has an answer. The thought in question is this: "Suppose it is the case, as I sometimes think, that the past is unreal, completely and utterly gone from existence . . . If this is so, then of course it makes no sense to think that I am now some distance away from where I was." There is an amusing twist here, for you might well think "There is surely no metaphysic more hostile to this view of the past than the four-dimensionalist way of thinking about space and time!" However, while it may be natural to think like this, it is not necessarily accurate: a spacetime of the neo-Newtonian sort, equipped as it is with a well-defined simultaneity relation, is not in itself incompatible with the view that only the present is real.

It is interesting to note that the *Minkowski* spacetime of special relativity theory also has no place for the notion of absolute velocity, although it, too, retains the notion of absolute acceleration. But the reason is quite different. Whereas the neo-Newtonian keeps a rigid framework of temporal relations (simultaneity and succession) but abandons spatial distance over time, in Minkowski spacetime the same effect (abolition of absolute velocity) is achieved by a different means: abandoning absolute temporal intervals. Since velocity is the measure of how much space you are traversing in a given interval of time, the notion can be undermined from two directions, so to speak. This fact leads Geroch to say:

we observe that the transition from the Aristotelian [Newtonian] view to the Galilean view results in fewer things making sense. Although the evidence on this is perhaps a bit scanty, it seems to be the case that physics, at least in its fundamental aspects, always moves in this one direction. For example, relativity theory consists essentially of the realization that certain of the notions in the Galilean view don't make sense either . . . In quantum mechanics, to take another example, such notions as 'the position of a particle' or 'the speed of a particle' do not make sense. It may not be a bad rule of thumb to judge the importance of a new set of ideas in physics by the criterion of how many of the notions and relations that one feels to be necessary one is forced to give up. (1978: 52)

The claim that advances in physics result in "fewer things making sense" can be read in two ways:

- as physics advances it produces worldviews that are increasingly utterly unintelligible; and
- as physics advances, we find we can still make sense of the world despite the fact that some of our most basic assumptions about how things work have been refuted.

I leave it to the reader to judge which of these verdicts applies to neo-Newtonian spacetime.

12.4 A threat vanquished

Assuming that all this is at least reasonably clear (we shall be returning to the question of just how neo-Newtonian space should be interpreted), we can move on.

As should by now be plain, the move to neo-Newtonian spacetime solves the problem posed by the undetectablity of absolute velocity. Velocity is the measure of how much distance is traversed in a given time; since points within successive hyperplanes are not at any distance from one another, it no longer makes sense to talk of absolute velocity, of how fast a body is moving relative to absolute space.

Of course, in practice this is not what happens. In practice, we can set up a reference frame of bodies that are at rest with respect to one another, and from the point of view of this reference frame we can measure the velocity of other bodies. However, as we have just seen, there are other reference frames, comprising bodies that are not moving relative to one another, from the perspective of which *our* reference frame is moving. Given the existence of these other frames, *our* frame has indefinitely many velocities. Since there is no longer any such thing as *absolute* motion or rest, none of these inertial frames is privileged, and so there is no frame-independent fact of the matter about what our velocity really is. The only facts about velocities are facts about *relative* velocities.

So by situating Newtonian dynamics in neo-Newtonian spacetime the most serious Leibnizian objection is overcome. Since absolute velocities are no longer meaningful, the Newtonian is no longer guilty of positing innumerable physical states of affairs that are empirically undetectable and so indistinguishable. The move to neo-Newtonian spacetime does not solve the problem of the static shift. The Leibnizian could argue that, provided inter-object spatial relations are retained, the whole system of material bodies could, in each hyperplane, be located at a different and distant collection of spacetime locations. However, as we saw earlier, the static shift argument only has bite for those in thrall to the principle of sufficient reason, and these days few are.

However, we should pause before drawing the conclusion that Newton has been fully vindicated by the move from Newtonian to neo-Newtonian spacetime. The debate between the substantivalist (or absolutist) and relationist takes on a somewhat different complexion in the neo-Newtonian context, as we shall now see.

12.5 The charge of explanatory impotence

Teller (1991) questions the extent to which the substantivalist neo-Newtonian actually *explains* inertial effects. Teller suggests that in the context of standard Newtonian space the substantivalist's complaint against the relationist is best construed thus:

> We know that there are systematic connections between acceleration and inertial effects. With only relative accelerations to which to appeal, relationists

have no way of distinguishing between the relative accelerations which do and those which do not experience inertial effects. Thus they have no way of giving a systematic description of the relation between acceleration and inertial effects or of giving an explanation which such a systematic description might constitute. (1991: 370)

In response, the relationist can make Sklar's move and bite the bullet by maintaining that there is a primitive unexplained property (which we can call "absolute acceleration" if we like), which is systematically associated with inertial effects. This meets the challenge of the substantivalist. The relationist can say that the difference between relative accelerations that are, and are not, associated with inertial effects is that the former possess the primitive property while the latter do not. But this seems an unappealing move. The substantivalist will argue:

Sklar's move is *ad hoc*. To say that inertial effects are always accompanied by a primitive property, a property that is characterized solely as *the property that goes with inertial effects*, amounts to nothing more than saying that inertial effects are found where they are found. The substantivalist isn't restricted to this truism, for the substantivalist can say that absolute acceleration can be characterized independently of inertial effects, in terms of motion with respect to space (or spacetime). It is this independently characterized *absolute acceleration* that is connected with inertial effects.

Teller now states his claim. This plausible sketch of an argument relies on a slide from characterizing absolute acceleration relative to *space*, to characterizing absolute acceleration relative to *spacetime*. The argument is compelling in the context of absolute Newtonian space, but not in the context of spacetime, particularly neo-Newtonian spacetime. Since there is good reason for rejecting Newtonian space in favour of neo-Newtonian spacetime, the substantivalist's argument against Sklar has no force, for it turns out that the neo-Newtonian substantivalist is in a position that is precisely analogous to that of Sklar's primitive property theorist.

The justification for this claim runs as follows. In Newtonian space the distinction between absolute and merely relative acceleration is real and clear: if an object is located at the same point as time passes, it is at absolute rest; if an object is located at different points at different times, such that equal times are accompanied by equal changes in distance, and the changes in distance are always in the same direction, then the object is in uniform absolute motion; and all other motions constitute absolute accelerations. In neo-Newtonian spacetime the situation is quite different: absolute velocity is not well defined and absolute acceleration consists in having a curved worldline, rather than in change in absolute velocity (that is, change of rate of motion through persisting substantival spatial locations). So the notion of worldline curvature is key. But what does this involve? Let us recall the manner in which neo-Newtonians construct their spacetime.

We begin with a collection of spacetime points, organized into distinct Euclidean three-dimensional spaces, one for each moment of time, and impose additional structure:

- We specify which points are very close together; these count as "next to one another".

- We further specify what are to count as continuous sequences of spacetime points: these are the paths or trajectories through spacetime – continuous paths that objects could follow.
- We next specify the "connection" on the space; that is, we stipulate which trajectories are straight, and thus which are curved.

We can now identify straight trajectories with inertial paths. Once the connection is specified, we are provided with a distinction between absolute or true acceleration and merely relative or apparent acceleration. The problem now emerges. Teller suggests that the substantivalist:

> has an embarrassment analogous to Sklar's. For let us ask, Which trajectories are the ones to be called straight? That is, Which represent unaccelerated motion? Or again, Which, if followed by an object, would describe the object as unaccelerated? There is nothing in the collection of bare space-time points and their "next to" relations to answer this question. The facts presented so far do not enable a substantivalist to say more than that the trajectories to be labeled straight are just the ones which will exhibit no inertial effects (ones which, if followed by an object, describe motion free of inertial effects.) Such a substantivalist has no more of an account of inertial effects than Sklar's relationist.   (1991: 373)

This seems a sound point, for as Teller presents the situation, the *only* guide the neo-Newtonian has as to which trajectories are to be construed (or *made*) straight, and which curved, are which trajectories are associated with the production of inertial effects. So, just as the Sklarist relationist can do no more than tell us that the trajectories associated with inertial effects are the trajectories associated with inertial effects, so too for the substantivalist. Consequently, it seems plain that the neo-Newtonian substantivalist provides only a trivial explanation of the origin of inertial effects.

12.6 A rebuttal

The underlying worry here, I suspect, is that the move to neo-Newtonian spacetime seems to have fundamentally altered what absolute acceleration amounts to; since absolute velocity is no more, it can no longer consist in a change of rate of motion relative to an underlying substructure. As a consequence, our intuitive grasp on what absolute (as opposed to merely relative) acceleration consists in has all but vanished. It is not easy to see what role substantival spacetime is playing in the new scheme of things. However, while there is certainly some truth in this, Teller's argument is suspect.

As Teller presents the situation, the substantivalist *constructs* a spacetime to fit the bill (the observed interactions between bodies), and so in deciding which point-sequences to count as straight he looks to the inertial forces. He has, it seems, no other way of distinguishing spacetime point sequences that are straight or curved. But this way of looking at things is potentially misleading: isn't Teller overlooking the fact that for the substantivalist spacetime is a real entity, one possessing its own distinctive structure? In constructing a neo-Newtonian spacetime the physicist is constructing a mathematical model that is intended to represent this independently existing entity. The mathematical model doesn't constitute spacetime; it is merely a useful way of picturing it. From the standpoint of the substantivalist, *the physicist* doesn't introduce inertial structure into spacetime, rather this structure is already there, as it were, as a real feature of an autonomous physical entity. The substantivalist can thus claim:

> There is a real distinction in spacetime between straight and curved trajectories, and inertial effects are encountered when objects follow curved trajectories. So it is not the case that the substantivalist is in the same boat as Sklar. Sklar can do no more than say that inertial effects are always accompanied by a primitive property, a property that is characterized solely as *the property that goes with inertial effects*, which amounts to nothing more than saying that inertial effects are found where they are found. The substantivalist isn't restricted to this truism: absolute acceleration can be characterized independently of inertial effects, in terms of curved trajectories through neo-Newtonian spacetime.

While this reply may go some way towards dispelling the doubts raised by Teller, it doesn't go as far as it might. The question still remains, what exactly is the substantivalist offering by way of an *explanation* of inertial effects? One answer would run thus:

> An object that suffers inertial effects possesses a distinctive physical property; there is a law of nature that links the occurrence (and magnitude) of this physical property with curved spatiotemporal paths.

Now, there is nothing in principle wrong with an answer along these lines, in that it is certainly conceivable that there should be laws of this sort, laws that bring it about that objects following certain types of trajectory through spacetime come to possess intrinsic properties of a certain sort. However, I suspect that it would be a mistake to view matters thus: there is no reason to think that inertial effects involve *intrinsic* properties of a distinctive kind. Once this point is appreciated, it becomes clearer just what the neo-Newtonian is offering by way of an explanation of inertial forces.

What do inertial effects consist of? Typically, they consist of internal stresses between the component parts of a body. Objects follow inertial paths unless acted on by a force (which is created by gravity, magnetism, or an expenditure of energy). When objects are forced off their inertial path, this is typically achieved by applying force to one part of the object only, which sets up tensions *within* the object: that is, some parts of the object start exerting forces on other parts. In the rotating globes case, the tension is registered in the cord, which exerts a force on the globes (whose inertial motions are tangential to their circular motion). When a jet plane deviates from its inertial motion it pushes those parts of your body that are in contact with it off their inertial paths and you feel this as pressure on your skin. Since a similar motion is not imparted to the other parts of your body, a tension is created as your rear-parts push against your middle-parts (which is often accompanied by a sensation of queasiness). All this is explicable without positing any special properties. The neo-Newtonian's explanation of inertial effects is thus quite simple: the internal stresses arise when *absolute velocity differences* arise. As we saw above, to make sense of absolute velocity differences we have no need to posit absolute velocities. The absolute differences are those that observers in all inertial reference frames will agree on despite differing in their views as to who is moving and who is stationary.

The notion of *inertial paths* lies at the heart of this account. What does the neo-Newtonian have to say about these? Something like this:

- certain paths through spacetime are privileged by the natural laws which govern the relationship between material bodies and spacetime;
- the relevant natural law is that bodies continue to move along these paths unless acted on by a force exerted by another moving body, or a locally generated impulse (e.g. jet engine), or a distant attractive/repulsive impulse (gravitation, magnetism); and
- these nomologically privileged paths are the inertial paths.

This explanation seems non-trivial, and eminently reasonable.

12.7 Newtonian spacetime relationism

I want to conclude by looking at another way in which the debate between relationists and substantivalists may be affected by adopting the spacetime perspective. In a recent article Maudlin has defended the following interesting claims: if we adopt the spacetime perspective, in the context of *Newtonian* spacetime the relationist is in a very strong position, since the argument from inertial effects is completely without force; however, this changes once we move to *neo-Newtonian* spacetime, which is "an extraordinarily hostile environment for the relationist" (1993: 196).

Returning to Newton's buckets and globes and the argument from inertial effects, Maudlin poses this question: exactly which sorts of relationism does this argument work against? Well, suppose a relationist were to stipulate that the only spatiotemporal relations are simultaneity, temporal succession and spatial distance relations between simultaneous events. For such a relationist – a *Leibnizian* relationist (as Maudlin puts it) – the situations that the substantivalist describes as involving different rotations are all exactly alike physically. They must be. Since there is no spatial distance relation between states of affairs at different times, there is no way of distinguishing a pair of globes at rest from a pair of globes that are rotating. So, in a case in which the substantivalist explains different observed tensions by reference to different rates of rotation, the relationist can say nothing: there are simply the different tensions, which are just found to occur.

Introducing Newtonian absolute space solves the problem. Since absolute space consists of points that endure through time, it provides a reference frame that allows us to define distances between simultaneous *and* non-simultaneous events. And this is the key point. Newton's ability to account for inertial effects doesn't depend *immediately* on the positing of absolute space, but rather on the extension of the domain of the (spatial) distance relation from simultaneous to non-simultaneous occurrences. This suggests a new form of relationism, which Maudlin calls (for obvious reasons): Newtonian (spacetime) relationism.

The advocate of this doctrine rejects absolute space, and maintains that all spatiotemporal facts are facts about the relations between material bodies, but these relations now include a distance relation between non-simultaneous events. Material objects are now conceived in the four-dimensional spacetime way: an object consists of a succession of temporal parts (see §3.7). This change of view gives the relationist invaluable additional resources. The Newtonian relationist will say that a body is at *absolute rest* if the spatial distance between its successive temporal stages at different times is zero, and an object is in *absolute uniform motion* if its successive temporal stages are separated by the same spatial distance per unit time. The stages of an object undergoing linear acceleration are separated by different distances per unit time (think of a heap of placemats skewed to one side). Rotation poses no problem: the particles in a rotating sphere have different distance relations to one another at different times, even though they retain the same distance relations to one another at any given time.

With these augmentations to his armoury the Newtonian relationist has no trouble accommodating the bucket. In the stationary case, the successive bucket-stages are aligned on top of one another in spacetime. In the rotating case they are not: each stage is rotated to some degree with respect to its immediate predecessor.

Once all distance and direction relations between material objects are specified, the entire system can be regarded as situated or "embedded" in a full Newtonian spacetime, a four-dimensional continuum of points, some of which are occupied, and many not. The relationist will not regard this spacetime as a real entity, but as a fictional representation that is useful for purposes of calculation and prediction: for, once the relational system is so embedded, the entire mathematical apparatus that Newtonians have devised can be deployed by the relationist. The only *real* spacetime locations that the relationist will recognize are those occupied by material things and likewise for real spatial relations. The substantivalist, on the other hand, will regard the material bodies and their spatial relations as constituting just one part (or sub-model) of the totality of real spatiotemporal facts. This restriction provides the relationist with a solution to the static shift (the problem of "initial location"). For any given Newtonian spacetime S, regarded as a particular collection of (fictional) locations forming an infinite four-dimensinal continuum, there will be indefinitely many different ways to embed a given system M of material bodies into S that preserve all spatiotemporal relations between the bodies in M. For a start, M could be embedded at different locations in S. Alternatively, M can be set at rest relative to S or given a uniform velocity, of any magnitude.[4] However, since the embedding space is itself fictional, the relationist will not regard these embeddings as representing physically distinct states of affairs: each embedding is simply a different way of representing one and the same physical state.[5]

As Maudlin plausibly suggests, this option has not been noted by historical figures, because it requires a four-dimensional perspective and an event ontology, but it *is* a distinct position in its own right, and it shows that Newton's bucket argument is not effective against relationism *per se*. Maudlin concludes, "Relations that are sufficiently rich can provide the means for explaining inertial effects."

12.8 Neo-Newtonian spacetime relationism

Can the same trick be worked in the context of neo-Newtonian spacetime? Can the relationist devise a satisfactory theory of dynamics if all that exists are material objects whose successive temporal parts are linked by the spatiotemporal relations recognized by the neo-Newtonian? Maudlin suggests not:

> Neo-Newtonian spacetime does not have the complete four-dimensional metric provided by absolute space. It results rather from the addition to Leibnizian spacetime of an affine connection which (roughly) specifies which spacetime trajectories are straight, and hence inertial. So a reasonable version of neo-Newtonian relationism would add to the Leibnizian relations a three-place predicate col(x,y,z) which has as its extension all triples of non-simultaneous collinear events. That is, col(x,y,z) iff x, y and z all lie along some inertial trajectory. Specifying col(x,y,z) completely in a spacetime would fix the affine connection therein.
>
> The relationist, however, is given as data only the restriction of col(x,y,z) to the set of occupied points. This will, in general, be information too meagre to specify an embedding of the occupied trajectories into the full spacetime up to the freedom associated with choice of reference frame. For example, consider two particles rotating about their common centre of mass. Until the first rotation is complete, no triple of occupied event locations are collinear. Even after any number of rotations, the collinearity relations among occupied points will be consistent with any periodic rotation, uniform or nonuniform. So while the Newtonian relationist could make use of the full power of Newtonian absolute space, starting only with the relations between occupied points, embedding them in a fictional space, and deriving via Newtonian mechanics exhaustive predictions about the future relations between occupied points, the neo-Newtonian cannot perform the same trick. Knowing, for example, that two particles remain at a constant distance through some period of time and that no three points in their world-lines are collinear, the neo-Newtonian relationist could not predict when, if ever, a triple of points would be collinear. Nor could inertial effects be predicted or explained, since the absolute acceleration cannot be inferred from the data.
>
> (1993: 193–4)

To simplify, suppose that we are dealing with two rotating particles in an otherwise empty universe. Imagine the two particles as black dots on a thin sheet of glass. In describing the evolution of this system the relationist has to do two things: put the sheets into a particular order and provide them with an orientation with respect to one another. The temporal ordering is unproblematic (it is a given in neo-Newtonian contexts), but the orientation is not. If we draw an imaginary line between the particles, the relationist can align the successive sheets in any number of ways. If, say, the lines in successive sheets coincide vertically, the particles are at rest; if just 1 degree separates successive sheets, they are rotating but slowly; if the separation is 30 degrees, they are rotating more quickly, and so on. Now, since there are no distance relations between non-simultaneous spacetime locations, it seems that the relationist not only has no means of specifying what the correct orientations are, but he cannot even recognize that the different orientations are *physically distinct states of affairs*. The only fact about the relative orientations of the slices available to the relationist is collinearity: the relation that holds between particles that lie on the same inertial path. Given this, the relationist can tell when a rotation is *complete*, but there are no facts about how fast the particles have been moving in the intervening period. The Newtonian relationist is in a very different position. Since there are distance relations between particles at different times, there are facts about how fast the particles are moving from moment to moment, and the behaviour of the system through the relevant period of time is well defined. This is clearly not the case for the neo-Newtonian relationist.

The significance of this underdetermination is clearer for slightly more complex situations. When astronomers observe the behaviour of triple star systems they see the stars moving about in seemingly bizarre ways: the orbits are (seemingly) highly irregular, with stars quickly moving together and shooting off again. (Some "executive toys" approximate this behaviour, using magnetism rather than gravity.) Although the behaviour of the three bodies is seemingly random, in fact it is precisely predictable using Newtonian theory, provided one is given certain initial information, and this is when it becomes interesting.

As Poincaré showed, if we are given two snapshots of the system that show the relative distances between the bodies, then, provided we know both the time interval between the shots *and* the relative orientations of the snapshots, the future behaviour of the system is determined, and can be predicted. However, if we are not given the information about the orientation, we cannot predict how the system will evolve. Figure 12.8 shows just a few of the ways that such a system could develop; the different evolutions show the effects of different orientations. These different "orientations" correspond to different initial absolute accelerations. As these diagrams make plain, facts about absolute acceleration play a crucial role in the way a system of three mutually gravitating bodies behaves over time. The Newtonian relationist can view such systems in just the same way as the Newtonian substantivalist, since all the relevant distinctions are available to both. Not so for the neo-Newtonian relationist: in the absence of spatial distance relations between successive object-stages the absolute accelerations of objects is drastically underdefined.

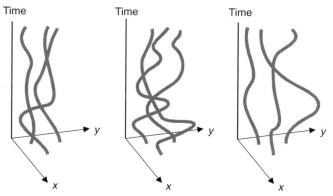

Figure 12.8 A system of three gravitationally interacting bodies. Different initial accelerations lead to dramatic differences later on.

In response to this problem the relationist could adopt an instrumentalist strategy. Is there anything to prevent the relationist from using the substantival neo-Newtonian apparatus for predicting the behaviour of moving bodies, but regarding this as nothing more than a fictional device useful for predictive purposes? Suppose that, when mapping his spacetime, the relationist "pretends" to be a substantivalist, and gets into the habit of using maps that display unoccupied locations. When the map is complete (or as complete as it can be given the observable facts), what is to prevent the relationist's deleting from his map all the unoccupied points and trajectories recognized by the substantivalist? What remains after the deletion are worldlines; that is, straight and curved four-dimensional material objects. At any given time, the spatial distances between the relevant temporal-stages of these objects are well defined. So, too, are the intrinsic curvatures: either zero, for straight worldlines, or non-zero for curved worldlines, with greater or lesser intrinsic curvatures for greater or lesser absolute accelerations. In trying to measure absolute accelerations the neo-Newtonian relationist has exactly the same observational data as the neo-Newtonian substantivalist: inertial effects found in accelerating bodies along with at-a-time distance relations. Moreover, the relationist can explain alterations in curvature in just the same ways as the substantivalist: in terms of collisions (worldlines that meet) and action-at-a-distance forces (e.g. gravity induces worldlines to curve towards one another). Hence it is not at all clear that the unoccupied positions and paths that the substantivalist (or spacetime plenist) posits are doing any real work. Look again at Figure 12.8. If we suppose that the relationist's world (a very simple one) is accurately represented by one of these spacetime diagrams, given the data he has, isn't it highly plausible that he could tell which one he lives in? More generally, it seems quite plausible to think that, given equal amounts of data, the relationist and substantivalist could work out equally accurate "maps" of their world.

This instrumentalist manoeuvre may seem innocuous enough, but closer scrutiny reveals it to be dubious. The substantivalist provides a precise explanation of the behaviour of the system in terms of Newtonian dynamics; according to this theory, every particle, at every moment of time, has a definite absolute acceleration, a feature that plays an indispensable role in determining how the dynamic system to which the particle belongs will evolve. Since the relationist does not recognize the same range of dynamic features as the substantivalist (in particular, the absolute accelerations of particles is often undefined), he cannot regard the Newtonian account as true. But since the relationist has nothing to offer by way of an *alternative theory* of remotely equivalent empirical power, he has failed to show that the dynamical features he rejects are irrelevant to the physics of moving bodies. Consequently, although the relationist is offering an ontologically more economical account, we have no reason at all to accept his account as true, and no reason to suppose that the discarded dynamical features do not exist.

12.9 Relationism redux

Is this the end of the story? Not necessarily, for as Maudlin himself notes in connection with Newtonian relationism "To defeat one tribe of relationists . . . is not to vanquish the whole nation" (1993: 187) for the simple reason that "the predictive and explanatory power of a relationist theory depends both on the nature of the spatiotemporal relations admitted and on the domain over which the relations are defined" (*ibid.*: 194). Newtonian relationism emerged as a serious rival to Newtonian substantivalism when the domain of the distance relation was extended to include temporal parts of the same objects at different times. Might it not be that the objections to neo-Newtonian relationism can be overcome if the relationist embraces a richer collection of spatiotemporal relations?

Consider again the system consisting of two rotating particles in an otherwise empty universe. It is quite true that in the absence of across-time spatial distances (we are working in a neo-Newtonian context) the relationist cannot distinguish between periodic rotations that are uniform and non-uniform, and so is incapable of discriminating states of affairs that are physically distinct. However, the relationist can solve the problem by introducing distance relations among simultaneous objects that have a specific *direction* as well as a size. There are two options: we can suppose either that distance relations have a *relative orientation* with respect to one another over time, or that distance relations have an *absolute orientation*. Since the latter eschews across-time comparisons it remains more faithful to the basic tenets of neo-Newtonianism, and for this reason it is this option that I will consider here.

For an intuitive picture of what is involved, start by imagining a motionless sphere in Newtonian spacetime. We now stipulate that every straight line connecting a point on the surface of the sphere to its centre both possesses and defines a different *intrinsic direction*. Focus on just one of these lines, and imagine many more with the same alignment, all originating at the centre of the sphere, but extending outwards for different distances, up to infinity; this collection of lines corresponds to all the potential spatial relations with the same intrinsic direction. A similar operation on the other radii generates the remaining spatial relations. Building a world is a matter of choosing different sorts of particles, plucking distance relations from the sphere, and fitting them all together. When a distance relation is removed from the sphere it retains its original direction and resists reorientation. Since we are operating in a spacetime framework, strictly speaking you are not connecting particles but rather particle-stages (momentary or very short-lived), and each configuration of particle-stages constitutes a temporal slice of your world.

Equipped with these newfangled distance relations we can reconsider the two-particle system. If the particles are not rotating, their neighbouring (but not simultaneous) temporal parts will be connected by a distance relation of the same size and the same intrinsic direction. If they are rotating, neighbouring temporal stages of the particles will be connected by distance relations of different intrinsic directions. Should the rate of rotation be uniform, the angles between the distance relations will be the same for unit intervals of time; in cases of non-uniform rotations, the angular variation will be different for unit times.

Intrinsically directed distance relations solve some problems but not all. In Figure 12.9, A and B are two bodies in an otherwise empty universe moving closer together in response to gravitational attraction; the curves represent the worldlines of these bodies, and B is more massive than A. R_1 and R_2 depict two different ways

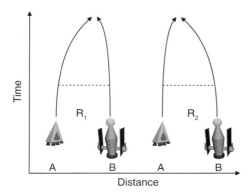

Figure 12.9 Distance and intrinsic direction are not enough.

these worldlines could converge. Since in actual fact A is the less massive of the two, it is accelerated more than B, and so its worldline will be curved to a higher degree. R_1 is therefore the accurate representation; R_2 misrepresents the situation, since it gives B the greater acceleration. However, while the difference between these two scenarios is both real and apparent on the page (which we can take to represent a Newtonian substantival space), the difference cannot be captured in a neo-Newtonian relational spacetime where only the following inter-object relations exist:

- inertial worldlines – the collinearity relation between particle-stages on inertial trajectories; and
- intrinsically directed distance relations between simultaneous spacetime particle-stages.

Neither of these objects is on an inertial trajectory, and at any time the distance relations between the object-stages is exactly the same (cf. the dotted lines). To capture the difference between R_1 and R_2 we need to introduce an additional property: different degrees of instantaneous intrinsic curvature.[6] As soon as the relationists augments his armoury with this property, the difference between states of affairs such as those depicted above is readily captured: the worldlines of A in R_1 and R_2 have different intrinsic curvatures at each moment of time.

I may have said enough to make this conjecture plausible: the neo-Newtonian relationist who embraces both intrinsic direction and different degrees of intrinsic worldline curvature has the resources to accommodate all the dynamically relevant features to be found in both relational and substantival Newtonian worlds. Of course, there is no need for the relationist to recognize every embedding of a given system into full Newtonian substantival space as physically distinct. For example, the "normal" and "upside-down" embedding of our world into Newtonian spacetime may well be physically indiscernible. Also, it may be that, with intrinsic curvature on board, the relationist can dispense with the services of intrinsically directed distance relations. Since a pair of particles rotating at a uniform rate have differently curved worldlines than a pair at rest, or a pair in non-uniform rotation, we only need directed distance relations if we want to be able to distinguish rotations of the same sort in different directions. We may not wish to regard such a difference as physically real and, even if we do, we may be able to capture it in a more economical fashion by positing inherently *directed* curvatures. But I will not attempt to explore these matters further. I introduce them only to illustrate that Maudlin may have underestimated the force of his own point: a relationist who is prepared to countenance sufficiently exotic species of spatial properties can reasonably hope to have available all the dynamical distinctions that are necessary for Newtonian theory.

Another lesson has emerged from the discussion thus far. Shifting to the standpoint of spacetime improves the prospects for the relationist; perhaps the geometrical way of conceiving motion from this standpoint makes it easier to develop viable forms of relationism. In any event, bearing in mind the resources available in the spacetime framework, one might be tempted to say that, for systems of material bodies conforming to Newtonian laws of motion, purely dynamical considerations do not weigh decisively in either direction: substantivalists and relationists may both be able to provide viable accounts of forced and unforced motions and their associated inertial effects. If this is so, then if the issue between the relationist and substantivalist is to be resolved, we must look further afield. Perhaps recent scientific discoveries fundamentally change the picture; even if the two conceptions of space (or spacetime) are equally viable in the context of Newtonian physics, the situation may well be very different when (say) electromagnetism or relativity theory enters the picture. Or then again, perhaps only purely metaphysical arguments can decide the issue, or at least tilt the balance. I have in mind questions such as these:

- What *are* these distance relations that the neo-Newtonian (and Newtonian) relationist simply assumes?
- How can these distance relations conspire to control the behaviour of bodies in inertial and non-inertial motions?
- Is a theory that posits vast numbers of relations any simpler than one that posits a substantival space or spacetime?

We will be turning to some of these questions shortly. However, lest it be thought that the interest of the issues we have been dealing with is entirely historical, the following point is worth bearing in mind. I noted earlier that the Minkowski spacetime of relativity theory is at once similar to and different from neo-Newtonian spacetime. It is similar because the concept of absolute velocity has no role, and different because of the reason for this: in neo-Newtonian spacetime there are no spatial distances between non-simultaneous points, and in Minkowski spacetime there is no absolute simultaneity. Recent writers have drawn attention to the fundamental incompatibilities between relativity and quantum theory. One such incompatibility is that quantum theory may be committed to absolute simultaneity, whereas relativity certainly is not. If, as many believe, the theory that synthesizes relativity and quantum theory will retain the basic character of a quantum theory, it may be that absolute simultaneity is about to make comeback.[7] Since there is no reason to believe that absolute velocity will do the same, might it not be that neo-Newtonian spacetime better reflects the actual character of *our* spacetime than any of its current competitors?

13 Curved space

13.1 New angles on old problems

We may have pursued the classical debate on space and motion a good way beyond its original boundaries, but there are further modifications to the classical worldview of a still more fundamental kind that we have yet to explore. Viewing motion as a process unfolding in spacetime does not require or involve the abandonment of the classical conception of space as a three-dimensional Euclidean structure, and the various spacetimes we have considered thus far retained this conception of space: there are no spatial distances over time in neo-Newtonian worlds, but the spaces-at-times that remain are entirely Euclidean. The spacetime perspective came into its own when Minkowski gave a spacetime interpretation of Einstein's special theory of relativity in 1908, but significant advances in the understanding of space – advances that play a crucial role in Einstein's general theory of relativity – had already occurred. By the middle of the nineteenth century it had become clear to mathematicians that space can take different forms and that the structure of Euclidean space is just one spatial structure among many.

This discovery gave rise to new questions: what sort of space do *we* live in and how can we find out? It also transforms the debate between substantivalists and relationists, as we shall see in Chapter 14. However, before entering these debates we need to know something about these strange non-classical, non-Euclidean *curved* spaces, and this chapter is devoted to this end (and so can safely be skipped by those already familiar with the basics).

13.2 Flat and curved spaces

In §9.2 we encountered Flatland, a two-dimensional plane inhabited by two-dimensional (but nonetheless intelligent) creatures. This imaginary world was used to undermine a deep-seated assumption that we naturally make and never normally have reason to question: that space, even nothingness itself, is necessarily three-dimensional. This same imaginary world is equally useful in undermining another basic assumption, namely that space is necessarily Euclidean. The Flatland space of §9.2 was truly *flat*: not only smooth and two-dimensional, but also lacking curvature. If by a "Flatland space" we mean any two-dimensional plane surface, not all Flatland spaces are flat in the latter sense. A tablecloth is a two-dimensional surface, and is flat when spread out over a table, but once picked up and rumpled there are any number of different and differently curved shapes that it can adopt. A Flatland space (or part of one) could have any of these more exotic *non-flat* shapes.

To make matters more concrete let us consider in more detail just one instance of a curved two-dimensional space, a Flatland shaped like the surface of a sphere. As previously, the Flatlanders live out their lives within their two-dimensional space and are entirely unaware of the third dimension, which they cannot move into (it is as if the interior of the sphere does not exist). Would the Flatlanders be aware that their space is curved? Not necessarily. Our Earth is curved, but because it is very large (in comparison with us) this isn't immediately obvious. Viewed from the surface it looks more or less flat. As ships sail away they do eventually disappear over the horizon, but it is easy to put this down to the fact that things look smaller as they get further away. However, this points to a potential giveaway. A Flatland explorer intent on discovering the edge of the world, who sets out in a particular direction and sticks to it, moving always in a straight line – never deviating to the right or to the left – would eventually find himself back where he started, having gone right around his space. After a number of such excursions, from different points of departure and in different directions, the Flatlanders would be in no doubt as to the shape of their world.

This result could be confirmed by other experiments. On a flat surface the internal angles of a triangle always add up to a total of 180 degrees. In the special case of a right-angled triangle, one angle is 90 degrees, and the other two are both of 45 degrees. As can be seen below, this is not the case for triangles inscribed on the surface of a sphere. If two Flatlanders were to set off from the North Pole travelling at right angles to each other, and continue until they reached the equator, and then turn another 90 degrees and walk along the equator until they met, they would trace out a triangle whose internal angles add up to a total of 270 degrees, since each internal angle is 90 degrees. A triangle of this sort cannot exist in flat two-dimensional space. The deviation from the 180 degree two-dimensional norm is not always so large. A triangle drawn within the boundaries of the one depicted below will have internal angles somewhere between 180 and 270 degrees, and the smaller the triangle the closer it will be to 180 degrees; a fact that illustrates the more general truth that the geometry of curved surfaces more closely resembles that of flat ones over short distances. Small areas of a sphere are *flatter* than larger areas. But there are also more extreme deviations. If the Flatlanders had set off from the North Pole walking in directions separated by 130 degrees, they would trace out a triangle whose internal angles add up to $130 + 90 + 90 = 310$ degrees.

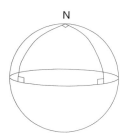

Figure 13.1 Sphere geometry. The internal angles of a triangle on a flat plane always add up to two right-angles (or 180 degrees). It is very different on the surface of a sphere.

There are further important differences between the geometry of spherical and flat two-dimensional spaces. The Flatlanders are, we can suppose, perfectly familiar with the geometry of circles on flat (or very nearly flat) surfaces: they know that the circumference of such a circle is its diameter multiplied by pi (π = 3.14159. . .). Suppose that one Flatlander decides to measure the circumference of Flatland space by travelling right around the equatorial circle depicted in Figure 13.1. After a trip lasting some years he returns to his point of departure. Another Flatlander decides to measure what is (in effect) the diameter of the equatorial circle. She starts from the same place as the first explorer, but heads directly north, crosses the North Pole, and continues in the same direction until she hits the equator again. She then returns to base, where she makes a curious discovery: on multiplying the length of the diameter by π, she obtains a result for the length of the circumference that is greater than its actual length (as measured by the other Flatlander). On reflection the reason for the discrepancy is obvious: the trip over the North Pole is considerably longer than the trip *through* the sphere to the same destination. Of course, the latter route, which corresponds to the diameter of the equatorial circle in flat two-dimensional space, does not exist in the curved space that we are currently considering. Hence the ratio of the diameter of a circle to its circumference on the surface of a sphere is always less than π, a figure that only gives the correct result for circles and diameters inscribed on flat surfaces. As with the internal angles of triangles, the discrepancy from the "flat norm" is not constant: it is smaller over smaller areas.

Parallel lines are straight lines that never meet, no matter how far they are extended. In a flat two-dimensional space, parallel lines not only exist, but there are plenty of them. For any given line, there is an infinity of others, at different distances both and above and below it, that are parallel with it. Does the same apply in our Flatland sphere-space? It might seem so. Are not all the (dotted line) circles in Figure 13.2a parallel with the equatorial circle? They certainly do not intersect the latter, and to this extent they share a feature of parallel lines on a plane. But parallels are *straight* lines that never meet, and there is a case for supposing that the dotted rings in Figure 13.2a are not straight lines in Flatland sphere-space.

Someone inclined to literal mindedness might reasonably argue:

The notion of a straight line is simply inapplicable in a curved Flatland: if we take "straight" to mean what it does in a flat space then *all* the lines on the surface of a sphere are curved to some extent, and so not straight. And since there are no straight lines, there cannot be any parallels either.

(a)

(b)

Figure 13.2 Unobvious parallels.

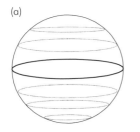

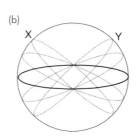

This may be going too far. Although there are no lines on a sphere that are *exactly* like their flat kin, might there not be some lines that are analogous to ordinary non-curved straight lines in important respects, and in virtue of this similarity deserve to be regarded as the counterparts of straight lines in sphere-space?

There are indeed. A straight line in flat space is the shortest possible distance between two points. If we apply this criterion to the surface of a sphere we find the following: for any two points, the shortest line between them is a line that forms part of a "great circle". Each of these great circles is a circumference of the sphere-space: the equator is one great circle, and so is a circle that joins the North and South Poles, but there are infinitely many others (see Figure 13.2b). A simple experiment reveals as much. A reliable way to find the shortest distance between two points on a sphere, X and Y, is to take a solid globe, pin a short length of rubber band to a point corresponding to the location of X and stretch the band to a point corresponding to the location of Y; the band will strive to find the shortest route between the two. Will it follow the horizontal ring? Well, try it (at least in your imagination). If you pull the band from X in the direction of Y horizontally, you will find that it immediately starts slipping upwards, only to stop when it reaches the North Pole: this great circle (running between the two poles) is the shortest route between X and Y. (This is why airliners flying between North America and Europe often take a northerly route rather than flying "directly" across the Atlantic – the route north is actually shorter.) We can now see that our literal minded objector's conclusion is true, even if not for the stated reason: there are indeed no parallel lines in our Flatland sphere-space, but not because there are no straight lines. There are lines that are analogous to straight lines in this space, but these are all paths along great circles, and when these paths are extended as far as possible, they all intersect, as can be seen in Figure 13.2b.

To sum up, a two-dimensional spherical space differs from a flat space in three key respects:

- the internal angles of triangles add up to more than 180 degrees;
- the ratio of the diameter of a circle to its circumference is less than π; and
- there are no parallel straight lines.

Interestingly, there is another sort of Flatland space where each of these results obtains in reverse, where:

- the internal angles of triangles add up to *less* than 180 degrees;
- the ratio of the diameter of a circle to its circumference is *more* than π; and
- there are *many more* parallels than there are in flat space.

One instance of a space with these properties is depicted in Figure 13.3. We will learn more about the characteristics of spaces of this type shortly, but as is intuitively

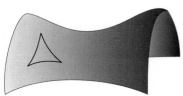

Figure 13.3 Negative curvature. The internal angles of a triangle add up to less than 180 degrees.

clear, a triangle inscribed on the two-dimensional surface of this object will have sides that bulge inwards (the opposite of what happens on the sphere-surface, where they swell outwards), and so its internal angles add up to less than 180 degrees.

In considering these curved Flatland spaces we have an advantage over the Flatlanders themselves: looking down from the vantage point provided by the third dimension we can *see* that the spaces are not flat. Confined as they are to their two-dimensional space, the Flatlanders cannot do the same. But as we have also seen, even so confined, there are ways the Flatlanders could learn about the shape of their space. Provided they know some of the basics of plane geometry, they could soon discover whether their space is flat or not by measuring distances and angles. There is nothing to prevent our doing the same, in order to find out if *our* space is curved. Once again we have not been considering Flatland for its own sake, but for the light it sheds on our predicament. Since the Flatlanders' visual field is restricted to a straight line, and their visual imaginations likewise, they can neither perceive nor imagine a curved two-dimensional space. In analogous fashion, we cannot visualize a curved *three*-dimensional space. Just try imagining a three-dimensional volume that extends the same distance in all directions but nonetheless curves back in on itself in the same way as the surface of a sphere. But a space of this sort is mathematically possible. In two-dimensional space there are circles, in three-dimensional space there are spheres, and in four-dimensional space there are hyperspheres, and the surface of a hypersphere is a three-dimensional "surface" that curves back on itself in just the same way as the two-dimensional surface of an ordinary sphere. If our space were of this form, a four-dimensional being could look down on us and *see* that our space is curved, in just the way that we look down on the spherical surface that constitutes Flatland.

There is, however, one important respect in which the Flatland cases can mislead. The curved Flatlands we have considered have all been embedded in three-dimensional space, but while viewing them thus makes their curvature undeniably apparent, it would be a mistake to suppose that curved spaces of n dimensions can only exist within a space of $n + 1$ dimensions. Mathematically speaking this is not so: the higher-order embedding space need not exist. It may be that a recognizably *physical* universe of just two dimensions, without any "thickness" at all, is an impossibility, in which case the embedding in three-dimensional space is not an optional extra. But there is no reason whatsoever to suppose that a three-dimensional space, whether curved or flat, can only exist when embedded in a four-dimensional space. This point, that a surface (or space) can be curved without having to be curved *in* anything else, is of crucial importance. It means we can make sense of the idea that our own space might be curved without having to suppose that our space exists as a "layer" in some higher-dimensional space. And the same applies to the notion of a curved *spacetime*. Although spacetime itself has four dimensions rather than three (the additional dimension being time), the curved spacetime of (general) relativity theory is regarded as a self-contained entity, one that is not embedded in any higher-order spatial (or spatiotemporal) medium. The sort of curvature that can exist independently of an embedding space is known as "intrinsic"; we will be looking at this notion in slightly more detail in §13.4.

Contemporary cosmologists often tell us that the universe is expanding, a claim that is not particularly puzzling in itself, but that quickly becomes puzzling when

Infinite because Flat [handwritten margin note]

they also tell us that not only is *space itself* expanding (as opposed to the objects within it), but also that irrespective of the direction in which astronomers point their telescopes they find that distant galaxies are invariably moving away from us, at speeds proportional to their distance (the more distant the galaxy, the faster the motion). It would be easy to see how this could be the case if we were located at the centre of the universe – the fragments from a bomb blast all move away from the site of the explosion – but the cosmologists also tell us that the Sun is situated in a nondescript suburb of the Milky Way, which is itself a perfectly ordinary galaxy whose location is in no way distinctive or distinguished. It is not obvious how to make sense of these cosmological claims in the context of a space that is both infinite and flat, but, as should by now be clear, they make perfect sense for a three-dimensional space that is finite and has a shape analogous to the surface of a two-dimensional sphere. Imagine a balloon covered in dots (the galaxies) which is expanding in size: from the vantage point of *each* dot all the others are receding, and the more distant the dot the faster the rate. If our universe is of this type, our three-dimensional space has the form of expanding hyperspherical surface.

This isn't infinite? [handwritten margin note]

13.3 The fifth postulate

These ideas may be mind-boggling when first encountered, but they are not particularly difficult to grasp, so it comes as something of a surprise to find out that it was only in the nineteenth century that physicists and mathematicians started to take seriously the idea that *our* space could be curved. The main reason for this lies in the history of geometry, and the spell cast down the centuries by the work of Euclid.

Many ancient peoples – in particular the Egyptians and Babylonians – had put together rules for calculating areas and volumes (on the basis of lengths), as these were handy for doing lots of practical tasks (such as allocating land and constructing buildings). The rules were *generalizations based on practical experience.* For example, the Egyptians knew that if you make a triangle out of pieces of wood of lengths 3, 4 and 5 units then the triangle will contain a right angle. Around 600BC, Thales of Miletus traveled to Egypt, learned something of their geometry, and did something quite new with it: he regarded geometric statements not as empirical generalizations, but as provable propositions, and showed that it was possible to logically deduce complex geometrical truths from more basic principles. Progress was swift (at least compared with the pace of later developments). Euclid's *Elements* were put together around 300BC, and contained a masterly synthesis of geometrical results, many discovered by others, some by Euclid himself. The hundreds of derived results fill thirteen books. What particularly impressed Euclid's contemporaries, and many subsequent generations, was the structure of the work: large swathes of plane (and some solid) geometry were logically deduced from a select handful of simple and largely self-evident axioms. Euclid himself may not have presented his work as having direct application to the physical world, but it was not long before people were working on the assumption that his geometry describes the structure of physical space. This view was far from unreasonable. Geometry is useful, in surveying, navigation, engineering and farming, and it provides us with knowledge about how the things in the world are related to one another by virtue of being located *in space*.

Viewed as an intellectual accomplishment, whether in science or pure mathematics, the *Elements* was unmatched in sophistication in its day and remained so long after. There are errors in Euclid's work, but these were mostly minor, and were detected and corrected over the subsequent centuries; the plane geometry taught in schools nowadays is essentially the same as in 300BC.[1] Later work in astronomy and mechanics – such as the *Almagest* of Ptolemy – presupposed Euclidean geometry; Newton himself compared the mechanics of his *Principia* to the *Elements*, suggesting that the two works were on a par in method, scope and significance. The influence on philosophy was equally profound. Plato had already taken geometry as a paradigm for all knowledge, and concluded not only that we could discover ultimate truths by reasoning alone, but that the objects of these truths would be perfect and immaterial entities, of which Euclid's circles and triangles can be seen as prime examples (Plato pointed out that a truly perfect circle is never to be found in the natural world). The rationalist foundationalism of Descartes is a more recent example of the same line of thought. Although Kant denied that we could know anything of reality in itself, questioning Euclid never occurred to him: he insisted that the principles of Euclidean geometry were hardwired into our minds (or whatever aspect of noumenal reality is associated with our capacities for thought and perception), so that even if space itself is not necessarily Euclidean we cannot possibly conceive of such a thing.

The status of the *Elements* as a cornerstone of Western intellectual life is one reason why it remained unquestioned for so long, but by no means the only one. The handful of axioms and postulates from which Euclid derived his geometrical results are such that it is not easy to see how any sane person could deny them, and, given the deductive character of the *Elements*, if the basic principles are true, so is the rest of the system. Euclid starts off with twenty-three definitions, none of which is controversial; for example, "1. A *point* is that which has no part. 2. A *line* is breadthless length. 3. The extremities of a line are points . . .". Following these are five axioms (or "common notions"), all of which seem self-evidently true:

1. Things equal to the same thing are equal to each other.
2. Equals added to equals yields equals.
3. Equals removed from equals yields equals.
4. Coincident figures are equal to one another in all respects.
5. A whole is greater than any of its parts.

Next come the five postulates:

1. Two points determine a straight line.
2. A straight line may be extended in either direction.
3. About any point a circle of a specified radius exists.
4. All right angles are equal.
5. From a point outside a given line, one and only one line can be drawn in a plane which does not intersect the given line, no matter how far it is extended.[2]

There is nothing obviously suspect here either but, if you grant Euclid this much, you must grant him the rest.

Of the five postulates the fifth, the so-called "parallel postulate", is by far the most complex. In Figure 13.4 the lower line is the "given" line, the X the point "outside" it, and the line with arrows, being perfectly parallel with the lower line, will never intersect it, no matter how far either is extended. How could there be *another* line running through X that is also parallel with the lower line?

Figure 13.4 Euclid's fifth postulate.

However, although no one doubted its truth, the fifth postulate was regarded with some wariness. Euclid himself only relied on it when he had to. He proved his first 26 propositions without using it at all. Noting its relative complexity, later geometers suspected that it might be redundant, and so tried to show that it could be deduced from the first four. All such attempts failed, and there were many such attempts.

In 1733 an Italian Jesuit, Girolamo Saccheri, tried to prove the parallel postulate by assuming it to be false, and deducing an absurdity from this assumption and the other four postulates (and the five axioms). This form of reasoning is familiar to us from logic's *reductio ad absurdum*: if you can deduce a contradiction from proposition P, then you can conclude that not-P is true (on the assumption that your other premises are true). Saccheri's proof had two parts. He first assumed that *no* parallel lines exist at all and deduced a contradiction; he then assumed that *more than one* parallel line exists and deduced a contradiction. He thought that he had vindicated Euclid and published the proof under the title *Euclid Freed from Every Flaw*. In fact his work had the opposite effect.

In the second part of his proof Saccheri demonstrated that a good many propositions that *seemed* false can be deduced when the "many parallels" postulate is assumed. When other mathematicians carefully examined the proof they found that, although Saccheri's propositions were counterintuitive, they did not in fact harbour a logical contradiction. Euclid was tottering. By the start of the nineteenth century several mathematicians – Gauss, Bolyai and Lobachevski – came to the same conclusion, more or less at the same time: if Euclid's fifth postulate is replaced by the alternative "many parallels" postulate, and combined with the other nine basic assumptions, it is possible to develop a consistent geometrical system. That there was an alternative to Euclidean geometry was a deeply shocking result, so much so that Gauss, the pre-eminent mathematician of his day, never published his results, fearing "the clamor and cry of the blockheads".[3] But it soon got worse. Just a few years later Riemann discovered that the assumption that there are *no* parallels to a given line also fails to generate a contradiction, and he developed yet another new geometry on this basis. Logically speaking, these alternative geometries were on a par with Euclid's. It couldn't be proved that any of the systems was perfectly consistent, so it remained possible that any one of them contains a concealed contradiction, but it was proved that *if any one of the geometries contains a contradiction, then they all do*. So, if a non-Euclidean system should yield a contradiction, then so will Euclid's. This "relative consistency proof" destroyed the logical priority of Euclidean geometry.

We have already encountered Riemann's non-Euclidean plane geometry. One possible realization of it is the Flatland space constituted by the surface of a sphere, so this geometry is often called "spherical" or (more often) "elliptical". Of the various theorems, the following are of special interest:

1. Through a given point external to a line no nonintersecting lines can be drawn.
2. The internal angles of a triangle add up to more than 180 degrees; as the area of a triangle approaches zero its internal angles approach 180 degrees.
3. The ratio of the circumference of a circle to its radius is less than 2π.
4. All straight lines are of the same finite length when fully extended.

The alternative *Lobachevskian* (or "hyperbolic") non-Euclidean geometry yields very different theorems:

1. Through a given point external to a line an *infinite* number of nonintersecting lines can be drawn.
2. The internal angles of a triangle add up to *less* than 180 degrees; as the area of a triangle approaches zero its internal angles approach 180 degrees.
3. The ratio of the circumference of a circle to its radius is *more* than 2π.
4. Straight lines are infinitely extendible.

One possible realization of a hyperbolic geometry is a flat but circular plane in which objects become smaller as they move further from the centre, shrinking to infinitesimal size as they approach the outermost boundary. Such a world would seem infinite to its inhabitants: they can never reach its edge, for as they approach it distances (as measured by them) get larger and larger. The geometrical properties of the space will be the same (when measured) by all occupants anywhere within the space. Straight lines are (in effect) curves that converge on the boundary at right angles. This "boundary" should not be thought of as being *in* the space at all, and neither should the Euclidean space surrounding the disc; it is only present as an artifact of our representation. (Some of Escher's "fish" pictures depict worlds with this geometry.) The surface of the object depicted in Figure 13.5 is another realization of a space possessing Lobachevskian geometry.

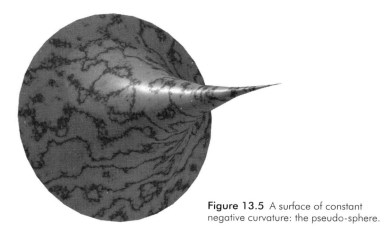

Figure 13.5 A surface of constant negative curvature: the pseudo-sphere.

13.4 Intrinsic curvature

In two papers of 1825 and 1827 Gauss showed how to measure the intrinsic curvature of a plane without referring to any external embedding space. Riemann subsequently extended the method, and laid the basis for our contemporary mathematical treatment of n-dimensional "manifolds", outlining his main ideas in the now famous 1854 Gottingen lecture "On the hypotheses that underlie geometry". The informal sketch of the Gaussian approach that follows is primarily intended to introduce some of the terminology used in the philosophical and scientific literature.

For any point P on a one-dimensional curved line it can be shown that there is a unique circle that most closely matches the curvature of the line at P. Circles with greater or smaller radii match the curvature of the line less well. The letter k stands for the Gaussian measure of the curvature of the line at P; the value of k is given by the reciprocal of the radius R of the best-matching circle: $k = 1/R$.

The same procedure can be extended to curved two-dimensional surfaces. It can be proved that all the lines running through a point P on such a surface have tangents that lie on the same plane: the so-called "tangent plane". Every point on a curved surface has a unique tangent plane, and each tangent plane has a unique line that is perpendicular to it and that passes through the relevant point on the surface: the so-called "normal line". Returning to our point P, any plane that includes P's normal line will be perpendicular to P's tangent plane, and slice through, or "section", the surface; this "normal-section" will be a curved one-dimensional line, and since we already know how to determine the curvature of such a line, we can determine the value of k for it. If we now take *all* the normal-sections (e.g. by rotating our initial section) through P, and find their curvatures, we will have completely determined the curvature of the surface at P. This is a complicated business, but Gauss showed how it could be simplified. He proved that for any point there will be two normal-section curves, one with a unique maximum curvature and the other with a unique minimum curvature, and moreover that the planes containing these sections will be at right angles to each other. These two "principal curvatures", k_1 and k_2, completely determine the intrinsic curvature of the surface at the point in question. The "Gaussian curvature", K, at P is given by the product of the principal curvatures: $K = k_1 k_2$. Although particular values for k_1 and k_2 depend on the units of measurement used, the magnitude of their product K does not; Gaussian curvature is thus an invariant and intrinsic feature of a surface.

We can now assign signs to our curvatures by arbitrarily calling one direction of a normal line N "positive" and the other "negative". A normal-section that is concave (or bends towards) the positive direction of N is *positively curved*, and a

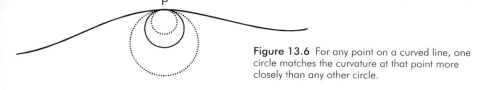

Figure 13.6 For any point on a curved line, one circle matches the curvature at that point more closely than any other circle.

normal-section that does the reverse is *negatively curved*. If k_1 and k_2 for a given point both have the same sign, whether positive or negative, their product, K, will be positive (a negative multiplied by a negative yields a positive), and the surface close to the point will be curved like a sphere or a bowl. If k_1 and k_2 have different signs, then K will have a negative value, and the surface close to the point will be curved like a saddle-back. If either one of k_1 or k_2 is zero, K will also be zero, and in this case the surface has zero curvature in the neighbourhood of the point in question, and the region has the same shape as the Euclidean plane.

These concepts allowed us to place the axiomatic geometries that we met earlier in a new perspective: Euclidean geometry is the geometry of a plane whose curvature is everywhere zero; hyperbolic (Lobachevsky–Bolyai) geometry is that of a surface of constant *negative* curvature; elliptical (Riemannian) geometry is that of a surface of constant *positive* curvature.

A few examples will help to make all this clearer. Think of a sphere, such as the Earth, and take a point P on the equator. Since the surface curves equally in all directions, we can take for our principal curves any great circles running along the surface that pass through the point, such as the equator itself, and the line running through P and the North and South poles. Call the radius of these circles R, then $K = 1/R \times 1/R$. Now imagine a sphere of *infinite* size. Its surface is effectively flat, and the radius of all principal curves on it will be infinite: $K = 1/\infty \times 1/\infty = 0$. This is why the Gaussian curvature of a flat Euclidean plane is zero. Now imagine a horizontal *torus* (e.g. a doughnut on a tabletop), and think of the surface inside the hole. There are two relevant curves here: one vertical (curving upward and downward from the central ring) and one horizontal (going round the inner surface). In this case the curves bend in opposite directions, and will have different signs, one positive, and one negative. $K = -k_1 \times +k_2$, and so is itself negative. Note also that the curvature on a torus is positive on its outer surface, since both circles bend in the same direction.

More intriguing is the case of a piece of paper that is bent into a gentle smooth curve (even a cylinder). Along one direction the paper is straight, and so the (maximum) radius of curvature is 0, but in the other direction the radius of curvature, R, will be finite. So the measure of curvature is $0 \times R = 0$. Hence we see that the Gaussian curvature of a tubular piece of paper is the same as a flat piece of paper! A Flatlander confined to this plane who went about measuring the internal angles of triangles would get exactly the same results as a Flatlander living on a flat plane. More generally spaces with the same Gaussian curvature can possess the same geometrical properties, even if intuitively the spaces are very different.

Figure 13.7 Appearances can mislead. The plane and the cylinder have surfaces with the same intrinsic geometry. Although the global properties of these spaces are different, this would not be apparent locally.

In more formal terms, the properties of a space, S, that could (in principle) be discovered by a geometer working within S, without any knowledge of the sort of space S is embedded in (if any) are *intrinsic* features, whereas those features of S that require knowledge of the embedding space are *extrinsic*. The case of the cylinder illustrates the fact that a two-dimensional surface with the same intrinsic features can be embedded in three-dimensional space in different ways; the difference is evident from the perspective of the third dimension, but invisible from within the space itself. As for curvature, while it is easier to grasp what K involves by considering two-dimensional planes from the perspective of the third dimension, which is what Gauss himself did, the higher-order embedding space is in fact inessential: the curvature of a given surface is exactly the same irrespective of whether the surface inhabits a space of higher dimensions. And although the value of K can be determined from the vantage point of a higher-order dimension (e.g. by positioning circles on the surface), it can also be determined from *within* the space itself, by measuring distances and angles.

Of course, a Flatlander living in a cylindrical space could discover that his space had a different overall structure from an infinite Euclidean plane by successfully travelling all the way around it in a straight line and arriving back where he set off. The points on a cylinder and a plane are path-wise connected in very different ways, and this is an intrinsic difference. However, it is a difference that can only be discovered by investigating the space as a whole; it cannot be discerned locally. Hence we need to distinguish between the *local* and the *global* properties of a space. Spaces with different global features can be locally indistinguishable.

The notion of intrinsic curvature was developed further in Riemann's extension of Gauss's methods, work that laid the foundations of modern differential geometry. Riemann provided the means for characterizing path-lengths and continuously varying curvatures in three-, four- and higher-dimensional spaces. Although he followed Gauss to the extent of supposing that the curvature in the vicinity of each point could be characterized by means of a Euclidean space of the required dimensionality, no actual embedding of the curved space in flat space is involved or presupposed: the claim is simply that the curved space around a specified point, provided the area involved is sufficiently small, can be approximated to any desired degree of precision, by a Euclidean space. The intrinsic curvature of more complex spaces cannot be captured by a single number such as Gauss's K. More sophisticated mathematical devices are required, such as the multi-component function now called the "Riemann metric tensor", g_{ik}, and the "Riemann curvature tensor", $R^i{}_{jkl}$. (A "tensor" is a mathematical operator, acting on vectors, which delivers numbers or vectors as output.)

Gauss had shown how to define the lengths of paths relative to an arbitrary coordinatization on a curved surface via his g-function, which assigns a collection of three numbers to each point on the surface; the values of these "metrical coefficients" vary from point to point, depending on the curvature (and so, in effect, these values register the effect of curvature on distance). Given the coordinates of two points, the lengths of all the different paths between them can be calculated, provided the values of g-functions are known for the points in the space. Riemann generalized this approach, and showed how it could be applied in spaces of higher dimensions. For a four-dimensional space, ten metrical coefficients are required, represented by the symbols $g_{11}, g_{12}, g_{13}, g_{14}, g_{21}, \ldots, g_{34}, g_{44}$ (these can be organized into a matrix of sixteen members, but six of these are redundant since $g_{21} = g_{12}$, and so on). Together these go to form the metric tensor. Twenty functions produced from certain combinations of the metrical coefficients together determine the curvature tensor, R^i_{jkl}, which provides a full description of the geometrical properties of the space around the point in question. The twenty numbers (or components) of this tensor are needed to capture the way a four-dimensional space bends and twists in different directions. Generally speaking, the higher the value of the metric tensor, the more curved the space at the point in question. In a flat n-dimensional space all of its components are zero. This "vanishing" of the metric and curvature tensors at every point in a space is what distinguishes n-dimensional Euclidean spaces. For a complete representation of the curvature of an entire space the value of these tensors for every point in a space must be supplied. (Such collections of tensors are known as a "tensor fields".)

Using the tensor apparatus to represent the properties of a space has a useful feature: the quantities thus represented (if not the actual numbers employed) are independent of the system of coordinates used to refer to points in the space. Coordinate independence (or "covariance") is a criterion of a quantity being physically real as opposed to an artifact of a particular mode of representation. The mass of a truck does not depend on whether it is measured in pounds or kilograms, and just as pounds and kilograms are interconvertible, so, too, are the tensor representations of geometrical quantities based on different systems of spatial coordinates.[4] It was this property that led Einstein to seek out the tensor apparatus when formulating the general theory of relativity.

One further piece of terminology will prove useful later on: in n-dimensional variably curved spaces, the counterparts of straight lines in Euclidean space – paths of shortest distance – are called "geodesics"; these are also the lines of least curvature.

13.5 Topology

As well as developing methods for representing differently curved spaces, nineteenth-century mathematicians moved on to investigate the properties of spaces at a still more abstract level, and topology was born. The topological properties of an object are insensitive to continuous transformations. Intuitively,

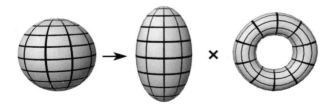

Figure 13.8 Continuous transformations and their limits. Topologically speaking, the sphere and the ovoid (rugby ball) are the same shape – a continuous transformation leads from one to the other (and back). But no such transformation (which permits stretching and squeezing but no tearing) can turn either of these shapes into a torus. The sphere and the torus are therefore topologically distinct.

the operations of bending, stretching and squeezing (e.g. as applied to a piece of soft rubber) count as "continuous transformations", provided no cuts, perforations or joining of edges takes place. Hence a teacup and a doughnut are topologically indistinguishable – stretching without tearing can take you from one to the other – but both are topologically distinct from a sphere, which is itself topologically indistinguishable from a rugby ball and a cube. There are various interesting concepts in topology, but I will mention just one, which will prove relevant later.[5]

Manifolds that (like the torus) are "holed" (or that possess a "handle" like a teacup) are said to be "multi-connected", whereas manifolds lacking holes (such as the surface of a sphere, or the Euclidean plane) are said to be "simply-connected". If a surface is simply-connected, every loop on it can be shrunk down to a point; on multi-connected surfaces, the presence of one or more holes means there will be some loops that cannot be shrunk down to points. Although it becomes harder to define, this distinction applies to all spaces, irrespective of their dimensionality. From a purely mathematical perspective multi-connected spaces are far from uncommon; indeed, they tend to outnumber their simply-connected counterparts.[6]

A torus can be made from a rectangular rubber sheet: first take two opposite edges and curl the sheet into a tube, then take the two ends of the tube and bend them towards each other until they join, and apply glue. Such operations cannot be performed on a topological manifold (since no joining of edges is allowed), but we can achieve much the same result by supposing that the points along the opposing edges of the rectangle are numerically identical. A Flatlander on the surface of a torus who moves straight up or down (imagine that the torus is horizontal, like a doughnut lying flat on a plate) will soon end up where they started, likewise if they keep moving forwards, thus taking the longer of the two routes around their curved space. The dotted line connecting X and X* in Figure 13.9 represents a "short" route, whereas the line between Y and Y* represents one of the "long" routes. Anyone who has played the (now venerable) video game "Asteroids" will have enountered a toroidal space of this sort. Objects moving across the screen behave in the same way as the aircraft shown in Figure 13.9.

The torus created by identifying the opposite edges of a rectangle is geometrically distinct from the more familiar torus depicted in Figure 13.8: unlike the latter it is *entirely flat* everywhere (and so is called the "flat" or "normal" torus), and in this respect resembles the surface of a cylinder (which is also locally

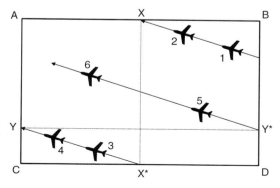

Figure 13.9 A flight through toroidal space. Sides AB and CD are identical, as are AC and BD. To an observer ignorant of the topology of this space, the plane would seem to leap (instantaneously) across space as it flies through X, before resuming on its original heading at X*. But in fact, since X and X* are the same point, no such leap occurs. Similarly, when the plane arrives at Y on AC, it appears immediately at Y* on BD, and continues on its way. Some video games employ spaces of this type.

Euclidean). The surface of the ordinary torus, by contrast, curves both positively and negatively. Unlike its curved counterpart, the flat torus cannot exist in three-dimensional space.

We construct a flat *hypertorus* by identifying the opposite faces of a cube; the surface of the hypertorus constitutes a three-dimensional multi-connected space that is locally Euclidean. The rectangle and cube constitute the "fundamental domains" of the (flat) torus and hypertorus. Many other geometrical figures can serve as fundamental domains for more complicated multi-connected spaces. In the three-dimensional case, the fundamental domains are polyhedra whose faces (or facets) are pairwise identified, sometimes after being rotated by specific amounts relative to one another (see Figure 13.10). Some of these spaces are flat but most are curved, either positively or negatively (usually the latter). Some are closed (and so have finite volumes), and others are open. The number of multi-connected three-dimensional spaces is infinite, and mathematicians have yet to complete their classification.

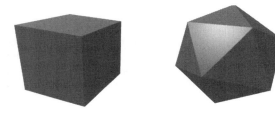

Figure 13.10 The fundamental domains of three-dimensional multi-connected spaces are polyhedra. The hypertorus is created by identifying the opposite faces of a cube.

Detecting multi-connectedness in a space can be quite easy. If both you and the contents of your bedroom were miraculously transported to a hypertoroidal space, whose fundamental domain is a cube with sides a hundred yards long, the unusual character of your space would be immediately obvious. Looking around, you would see something remarkable: repeated replicas of yourself and your bedroom furniture, appearing from different angles, stretching off into the distance on all sides. It would be rather like being in a room whose walls, ceiling and floor are all mirrors. These images are due to light making several "passes" through your space (in the manner of the aircraft in Figure 13.9) before arriving at your eyes. If you were to wait long enough, you would eventually be able to see every object in your room from every angle. More generally, the apparent structure of a multi-connected space is a function of the fundamental domain; an observer in such a space will see a "tiling" of endlessly repeated images of the relevant polyhedron stretching out, seemingly to infinity on all sides. Figure 13.11 shows the effect for a closed Seifert–Weber space.[7] Of course, if the fundamental domain is very large, these effects will not be so easily discerned; if the fundamental domain is larger than the visible universe, they will not be detectable at all. But more on this later (§18.7). It is time to consider some of the philosophical implications of these geometrical discoveries.

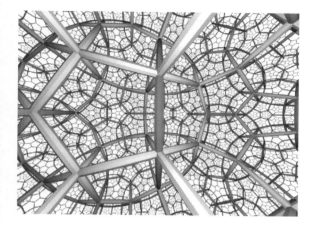

Figure 13.11 A closed, multi-connected, hyperbolic space viewed from within. The tiling of the fundamental polyhedron – a dodecahedron – appears to stretch off to infinity. (Courtesy of the Geometry Centre, image created by Charlie Gunn.)

13.6 Conventionalism

Most of the early pioneers of non-Euclidean geometry had this thought: "Since there are alternatives to Euclidean geometry, we cannot be certain that *our* space is Euclidean", and went on to devise tests that they hoped would reveal the truth. Gauss used ordinary surveying equipment to measure the internal angles of a sizeable triangle (69 km × 85 km × 107 km) formed by the Brocken, Hoher Hagen and Inselberg mountains. Lobachevsky worked on the astronomical scale, and conducted measurements on a triangle that had the star Sirius as apex and one of the diameters of the Earth's orbit as base. Both were disappointed to discover that there were no discernible deviations from what one would expect if space were Euclidean. However, they also realized that this proved nothing. Very gradual spatial curvature would only reveal itself over very large scales indeed, of a sort that were beyond the measuring techniques available to Gauss and Lobachevsky. The existence of other galaxies was unknown at the time (indeed, their existence was only firmly established in the 1920s).

At the turn of the twentieth century, shortly before the advent of Einsteinian relativity theory, Poincaré put forward an argument that, in the words of one commentator, "demonstrated once and for all the futility of this controversy and the fallacy of any attempt to discover by experiment which of the mutually exclusive geometries applies to real space" (Jammer 1993: 165). The "conventionalist" approach to space that Poincaré advocated initiated much subsequent discussion (see Sklar 1977: 88–147 for a thorough overview), and was an important influence on neo-positivist thinkers such as Reichenbach and Grunbaum.

Poincaré's starting point is uncontroversial. Since Gaussian methods for ascertaining intrinsic curvature involve measuring distances and angles, before we can even start to investigate the geometry of physical space we have to give *physical meaning* to basic geometrical concepts; we must settle on some practical ways of measuring distances, straight lines and angles. This was not usually seen as problematic. Short distances can be measured by putting rigid rods end to end, or with tape or string; for longer distances it suffices to measure the time taken for a

fast-moving signal with known speed, such as light or sound, to cross the intervening space. Light rays also provide a measure of straightness, at least on the
assumption that light always travels in straight lines through a vacuum.[8] Poincaré's
basic point was simple. If we conduct measurements that suggest that our space is
non-Euclidean we cannot immediately draw the conclusion that our space *is* non-
Euclidean, for there is always an alternative hypothesis, namely that our space is in
fact *Euclidean* but (for one reason or another) our measuring equipment is delivering inaccurate results.

Reichenbach provides a good example of how this might come about (1958:
§1.3). Imagine a Flatland inhabited by two-dimensional creatures who move about
making measurements as they attempt a large-scale geographical survey of their
world. They find that most of their space has Euclidean characteristics (with respect
to triangles, circles, etc.), but there is one region where this isn't the case. As they
proceed through this region their measurements suggest that space is becoming first
negatively curved (a two-dimensional saddle-space), then positively curved (like the
surface of a sphere) and then negatively curved again, before flattening out. They
conclude (reasonably enough) that their space is non-Euclidean: although mostly
flat it contains a hump-shaped region.

These beings live in world A. Imagine now beings who live in world B. Although
world B is completely flat, the observable behaviour of physical bodies in world B
is exactly the same as in world A: when moved across a certain region of space they
change shape. The relationship between the two is shown in Figure 13.12.
Whereas the rod in the curved region of world A retains the same length but
becomes curved, the corresponding rod in world B remains straight, but contracts
in length. Since *all* bodies behave in this way, expanding and contracting, the
inhabitants of world-B would be utterly unaware of this behaviour, at least until
they conducted detailed measurements. We further suppose that all processes that
can be used to measure geometrical properties (such as light rays) act in ways that
render the size changes unobvious. When they do conduct detailed surveys and
measurements, the inhabitants of world B soon discover that one region is
different from others: at the centre of this region parallel lines converge, but at the
interchange region many parallels exist; they find similarly odd results for
measurements on triangles and circles.

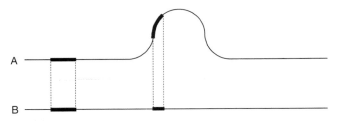

Figure 13.2 Reichenbach's world.

Since the results obtained using measuring rods and protractors are just as in
world A, the inhabitants of world B could postulate that their space is non-Euclidean,
and has a hump. But as it happens, they don't. The geometers of world B argue thus:

There is a region in our world where measuring rods *shrink and expand* in various odd ways. These changes in shape are such that someone who didn't realize that they were occurring could easily be led to conclude that our space contains a region curved like a hump, for if the measurements were taken to be accurate, this would be the case. But since the equipment is in fact distorted, this is not so: our space is flat.

Not only does this hypothesis seem just as reasonable as the one proposed by the geometers in world A, but if it were put forward by a dissenting geometer in world A there is no observation that could directly refute it. Even so, a proponent of curved space might reasonably argue like this:

Our account has the merit of explaining exactly *why* we get the measurements we do: the medium in which we are conducting the measurements is curved in ways that we can specify very precisely. By contrast, the postulated shape changes are brute facts that are entirely unexplained.

This objection can be countered, but the reply reveals an important feature of the alternative theory.

To explain the changes in shape we can posit a special kind of force-field emanating from the centre of the relevant spatial region, and the measurable alterations in shape can be explained in terms of variations within this field. Since physicists often postulate force-fields to explain observed effects, this general strategy is perfectly respectable, but the *type* of field required in this case is somewhat unusual. Since metal bars expand when heated and contract when cooled, it might be thought that something like heat (or temperature variation) is causing the size change of measuring rods in world B. But heat has *differential effects*, in that different kinds of substance expand at different rates in response to the same applied heat. Put a glass rod and an iron rod into the same oven and the latter will emerge longer than the former. The contraction in world B is a *universal effect*, which changes all kinds of body in the same way and to the same degree, depending only on their location. If the force-field affected some bodies but not others, its influence could be detected, and world B would differ from world A in observable ways.

When the flat-space hypothesis is supplemented by a detailed account of the location and action of the universal force, it offers as powerful an explanation of the observational data as the curved-space hypothesis. So what should the inhabitants of these worlds conclude about the geometry of their space? Poincaré suggests that since they cannot *know* the shape of their space, the preference of one hypothesis over the other can amount to nothing more than a *choice of convention*. The inhabitants of world A have decided that they prefer to think of their space as curved, whereas the world B inhabitants prefer the flat-space hypothesis. This is just one example, but it is easy to devise others (Poincaré himself used an imaginary disc-world whose inhabitants shrink as they move from the centre), and the lesson applies in three dimensions as much as in two. No observational evidence could decisively refute the hypothesis that our space is Euclidean since the relevant deviations can always be explained by postulating a universal force that produces precisely the same observational results as curved space. More generally, since for any physical universe the observational data will always be equally compatible with any number of hypotheses concerning the geometry of physical space, the preference for one hypothesis over the others is not justified by the empirical evidence. In actual fact, Poincaré suggested, we will always favour the simplest available hypothesis, and since Euclidean geometry is simpler than the alternatives, we will never adopt a theory that ascribes a non-Euclidean geometry to our space. There is, however, no reason to think that our preferred theory reflects the real structure of space; what it reflects, rather, is our preference for theories we find simple and natural.

13.7 Realism versus anti-realism

In one respect Poincaré was soon proved wrong: by the 1920s most physicists had come to accept Einstein's general theory of relativity, according to which gravitational effects are due to mass-induced curvatures in spacetime. Scientists do have a preference for simpler theories, and Euclidean geometry is simpler (in some respects) than non-Euclidean, but what Einstein saw (and Poincaré did not) is that an entire theoretical package of the form {non-Euclidean space + laws + forces + entities} could provide a simpler account of the observational data than {Euclidean space + laws + forces + entities}. Spatial geometry is just one aspect of a physical theory, and the complexity that comes with non-Euclidean geometries can be more than compensated for by simplifications elsewhere. Einstein showed that by introducing variable curvature into spacetime we can economize on forces: we no longer need to posit gravity as an attractive force (more on this later).

However, Poincaré's failure to appreciate this point does not undermine his argument as a whole. Before going into the reasons why this is so it is important to be clear on what the argument purports to establish. The thrust of Poincaré's conventionalism is *epistemological* rather than metaphysical. He is not claiming that the very idea of a substantival space is incoherent, he is not denying that such a space exists. The claim, rather, is that, even if a substantival space does exist, we could never be justified in thinking that we know its true geometry.

The reasoning that Poincaré uses to establish this conclusion is a particular application of a familiar anti-realist strategy in the philosophy of science. The anti-realism in question is directed against theories that posit unobservable entities, and the claim is that, even when such a theory provides a plausible explanatory account of the observable data, and delivers accurate predictions as to how observables will behave, we should not believe that the posited unobservable entities exist. The reason for this is that quite generally scientific theories are *underdetermined* by observational data. The relevant data does not logically entail the truth of any one theory, since there will always be alternative theories – theories that posit different unobservable entities – that explain the same data equally well. These alternatives may not strike us as being equally plausible, indeed it will often be the case that one theory seems to explain things better than the others, but it would be a mistake to think that this means the theory in question has a greater chance of being *true*. Owing to the quirks of human psychology we find some explanations "better" or "more satisfying" than others; similarly, the theories that we find "plausible" are those we find easy to believe. But what reason do we have for thinking that our preferences in these matters are reliable guides to the structure and functioning of reality? Assuming that we evolved through natural selection, our minds are far more likely to be geared to solving the problems encountered by our ancestors struggling to survive in the hostile environments in which they found themselves, rather than the invisible deep structures and basic ingredients of physical reality. Given this, we should adopt an attitude of cautious agnosticism toward even our best theories concerning these structures and ingredients.

There are those who are convinced by this argument, but, equally, there are many who are not. Realists point out that the best explanation for why a theory generates tremendously accurate predictions across a vast range of cases is that the theory is true, or at least approximately so. As for the charge that the criteria we use in evaluating theories are subjective (at least at the species-level), realists can accept that, while there is some truth in this, it would be wrong to think that our criteria are entirely fixed. As science has progressed (and most anti-realists accept that progress has been made), our notions about what counts as a "plausible theory" or "good explanation" have evolved a good deal – the standards of plausibility in current physics are not Aristotle's – and it is not unreasonable to suppose that the criteria we now employ better reflect the functioning of nature itself.

This is not the place for a thorough treatment of this important issue. For our purposes it will suffice to note the implications of the general realist–anti-realist debate in the philosophy of science for the substantivalist–relationist dispute. If it turns out that our best theory of matter and motion requires a substantival space, realists will be substantivalists; but if our best theory takes a relational form, realists will be relationists. Those of anti-realist inclinations will keep an open mind on the ontological issue. No matter what theoretical simplifications accrue from positing a substantival (or relational) space, anti-realists will remain uncommitted on the question of whether space really is substantival (or relational).

14 Tangible space

14.1 Manifestations of curvature

Now that we have a reasonable grasp of the distinction between flat and curved spaces we can turn to our main business, and consider the significance of the distinction for the reality of space. That the distinction *does* impact on the debate between substantivalists and relationists has been forcefully argued by Nerlich, on various occasions (1979, 1991, 1994b: §1.7). I will start by looking at some of the ways that spatial curvature can manifest itself in discernible ways, and them move on to consider the use that Nerlich makes of these manifestations.

Leibniz's various anti-substantivalist arguments all rely on the *innocuousness* of space: there would be no discernible or significant difference if the physical universe as a whole were shifted with respect to space, either by having a different velocity, or by being located elsewhere. A further variant along these general lines that Leibniz could have employed is *nocturnal doubling*. If everything were to double in size overnight, would we be able to tell? You would wake up in the morning twice your previous size, but since your bed would also have doubled, this would not be immediately apparent. Similarly, your room would also have doubled in size, as would your house, your town and everyone in it. Distances between houses and towns are now twice what they were, but since all measuring implements have also doubled in size, there is no way that this can be detected. Your trip to work takes no longer than usual. To make the example pedant-proof we must further suppose that the laws of nature have been altered to conceal the doubling: light travels twice as fast; the mass of an object is increased in proportion to its length rather than its volume and so on. Let us simply assume that the alterations required for the doubling to be indiscernible have all occurred.

If this all seems very plausible it is because a tacit assumption is being made: that the doubling occurs within a flat Euclidean space. Were the doubling to occur within a *non*-Euclidean space it would be very different. This is most easily seen for a space of constant positive curvature in two dimensions, the sort of space inhabited by Flatlanders living on a spherical surface. (Precisely analogous consequences obtain in curved three-dimensional spaces of the same type.) One of the distinguishing properties of spaces of this sort is that the internal angles of triangles add up to more than 180 degrees. Moreover, this divergence from the Euclidean norm is size-dependent: the smaller a triangle is the more closely it approximates to the Euclidean, and the larger it is the more it deviates. As we saw in Chapter 13, triangles that stretch from the pole to the equator can easily have internal angles greater than 270 degrees.

As a consequence, a doubling in size in such a space would be easily detectable, since it would result in easily measured changes in the "shapes" formed by objects separated from one another by discernible distance. Nerlich goes on to point out that it is not just triangles that are affected. Since any polygon can be decomposed into triangles, and any closed curve can be approximated by some polygon, shapes of all kinds will change on being doubled. (In a space of constant negative curvature there are changes too, but of the opposite kind: the internal angles of triangles will *decrease* with the increase in size.)

There is the potential for more dramatic consequences. As the two-dimensional spherical surface illustrates, a space of constant positive curvature can be finite, and as far as doubling is concerned this makes a definite difference. If our Flatlander's space is more than half full of matter (think of a planet whose continents cover more than half its surface area), there won't be enough *room* for them all to double in size. In such cases, nocturnal doubling is simply impossible. Shifting up to three dimensions, if our universe were both finite and positively curved, and the stars are uniformly spread through space, it makes no sense whatsoever to suppose that the distances between the stars could all double.

As might be expected, in spaces of variable curvature the effects of doubling are more variable. Again, consider the two-dimensional case. Our Flatland now undulates, and contains mountains, valleys and plains; although in some places the curvature is zero, in others it is positive and in others negative. The changes that would be produced by a doubling in size now depend on where we envisage the doubling starting, for whether or not an object gets pushed from a valley on to a plain, or from a mountain top into a saddle-back, depends where we "centre" the doubling; different centres yield different results. As Nerlich says:

> there are uncountably many ways of performing the thought experiment and uncountably many distinct thing–thing relations resulting from them. So doubling-in-thought experiments depend on which thing–place relations of spatial occupancy (position, place) are preserved, if any, and how others are altered.
>
> (1991: 174)

Spaces that vary markedly in curvature also cast new light on Leibniz's static and kinematic shifts. In forwarding these arguments Leibniz assumed Newtonian space to be uniform. Whatever force they have in the context of uniform space does not survive in the context of non-uniform spaces. It may make no discernible difference where God chooses to locate the material universe in a flat space, but it does in a variably curved space: the measurable (Gaussian) geometry of space will be different in differently curved regions of space. The same applies to the kinematic shift. If the material universe were moving through the three-dimensional analogue of a Flatland mountain range, the discernible geometry would vary over time in quite dramatic ways, and the patterns of variation would vary, depending on the speed and direction taken.

In extreme cases, the changes in shapes and alterations in distances wrought by spatial curvature would be both tangible and visible in ways not yet considered, as Nerlich's example of the *non-Euclidean hole* clearly shows (1994a: 38; 1994b: §7). In this imaginary (but logically possible) case there is a region of space in the room you are now in, about the size of a football, where space is drastically curved, but both positively and negatively; space elsewhere in the room is flat. (The "hole" in question is not an *absence* of space (construed substantivally), but simply a hole-shaped region where space is curved.) When light rays approach the hole they are moving in parallel straight (Euclidean) lines; on entering the hole they continue to follow geodesic paths, but by virtue of this region's intrinsic curvature they first all converge and then all diverge again before leaving. The fact that light rays change direction means that objects viewed through the hole will look different, just as they do when viewed through a flaw in a windowpane, or through a heat haze. Wandering around the room, observing how different things look distorted when looked at from different angles, would soon give you a firm impression of the hole's location, shape and size. Indeed, as Nerlich points out, "If all this came about, would we say, not just that we see *that* there is such a hole, but that we *see the hole itself*, just as we see warps in clear glass? I think we well might" (1994a: 39, original emphasis). In cases such as this, space itself becomes visible, or at least regions of it do.

This is not all: these regions would be *tangible*. Like rays of light, moving particles follow geodesics unless acted on by a force. Suppose you take a sizeable cube of soft foam rubber and move it into the hole. What would happen? Would it pass freely through? There are no material barriers to prevent it, just air and empty space. In fact, you would very likely feel a resistance. On entering the hole, the particles in the rubber will try to follow geodesic paths, and so initially they will converge, and then diverge. But this convergence will be resisted by the inter-particle bonds: you can imagine these electromagnetic forces as akin to elastic bands connecting the particles; as the particles try to converge, the bands have to stretch. Consequently, for the rubber cube to succeed in entering the hole a force needs to be supplied – to stretch the bonds – and hence you feel resistance when pushing the cube forwards. What if *you* were to walk through the hole? The same would apply. You would have to push yourself forwards, you would feel resistance, and since stress tensions would be generated among your middle parts you would feel a distinctly queasy sensation. You would *feel* the curvature of space.[1]

Why doesn't the previous Flatland example apply p216

14.2 The detachment thesis

We will be returning to the tangible manifestations of spatial curvature, but we have seen enough for the purposes of Nerlich's pro-substantivalist argument. To start with, recall the general form of Leibniz's case against Newtonian space. Underpinning the argument is an assumption about what we should regard as real. Leibniz assumes that an entity or property that can change without making any observable difference to anything is not a genuine ingredient in or feature of the world. Given that unobservable entities now play a more prominent role in scientific theories than they did in Leibniz's day, this criterion can be broadened to something like this: only entities that are needed to explain and predict the observable behaviour of objects should be regarded as real.

On the face of it the substantivalist's space satisfies this criterion. An object's size is the quantity of space that it fills, its location is the place in space that it occupies, the distances between objects are the lengths of the paths through space that connect their locations, and, quite generally "the spatial relations between things ... are all mediated by relations of things to space" (Nerlich 1991: 171). Leibniz's anti-substantivalist argument can now be viewed as akin to a *reductio*: "Let's suppose that this substantival space *does* exist." The doubling case (or one of the other shifts) is then introduced, and straight away the initially plausible assumption concerning the role of space is undermined. Since all spatial relations between things can remain observably and measurably the same through dramatic changes in the relations between things and space, it seems that substantival space does no useful work after all; it is an idle and inconsequential posit. To put it more succinctly, all *thing–space* relations can change while all *thing–thing* relations remain the same ("thing–thing" relations are spatial relations that are empirically detectable).

Nerlich now suggests that Leibniz's argument relies on the following general principle:

> *Detachment thesis*: thing–thing spatial relations are logically independent of thing–space relations. (1991: 172)

The viability of relationism presupposes the detachment thesis. Relationists hold that only material objects and spatial relations exist. Someone could object, "But spatial relations require the existence of *space*! Things are spatially related when some path through space connects them." For relationists to avoid this charge of circularity they must establish that spatial relations *can* exist independently of space; hence the need for the detachment thesis.

So the question to be addressed is whether the detachment thesis is true. Nerlich's central contention is simple: the detachment thesis is only plausible for the special case of Euclidean space, it is manifestly false for spaces that are not Euclidean. We have already seen why. A universal doubling may have no observable consequences in a flat space, but it has in curved space, and likewise for the kinematic and static shifts. Since in curved spaces varying the thing–space relations can produce dramatic variations in *observable* thing–thing relations, the claim that thing-space relations are inconsequential is straightforwardly false, likewise the claim that thing–thing relations are logically independent of thing–space relations.

Of course at the time Leibniz formulated his arguments for the detachment thesis, the existence of non-Euclidean spaces was not even suspected, let alone established, so he can scarcely be blamed for not noticing the flaw in his case. But since his arguments *do* have some force in the context of Euclidean space, a relationist might argue thus:

> Even if I grant for the time being that the detachment thesis is false for certain non-Euclidean spaces, rather than rejecting the thesis entirely, we should restrict its scope to spaces of zero curvature. If it then turns out that *our* space is flat, the detachment thesis is true for our world and the argument against substantivalism retains its force.

Not so argues Nerlich:

> the dependence of the outcome of doubling on the nature of space applies with equal force to the special and actually indiscernible case of doubling in

Euclidean space. It is not only when doubling will change shapes that we need to specify in which space we are performing the thought experiment. That the doubling in Euclidean space has no thing-thing consequences depends on specific symmetries unique to the structure of that space. These symmetries pick out the similarity transformations from the rest. They do not show that there is no structure and no space. They mean that there are specific structures in a specific space.   (1991: 175)

The various changes in distances and shapes that universal doubling produces in non-Euclidean spaces are to be explained by the specific features of the space concerned: different curvatures produce different changes. But equally, the fact that operations such as doubling, translation and relocation produce *no* variations in shapes and distances in Euclidean space is to be explained by the specific features of this space; it is only because Euclidean space possesses its distinctive symmetries that these operations leave no discernible traces. Euclidean space is no less influential than other sorts of space, but by virtue of its distinctive geometry its influence is uniquely discrete: by virtue of its symmetries linear translations of any size and in any direction do not change the distance relations between a system's component parts, and doubling (or trebling, or halving) operations do not change the relative sizes or distances between objects in a system.

As Nerlich goes on to argue, once the active role of Euclidean space is appreciated, the substantivalist's position looks stronger than ever. Given the Euclidean symmetries, the hypothesis that we live in a substantival space whose geometry is Euclidean explains why it is that the difference between rest and uniform motion, doubling and constant size cannot be observed:

The realist owns the symmetries, not the relationist. It is the realist's realism that tells him (and tells us) why we cannot care about the difference between one of these states and another. In fact, it is the relationist who has no clear, coherent, non-circular account of the relativity of motion.   (1991: 181–2)

And this means that a standard objection against the substantivalist is without force. There is no reason for the substantivalist to be concerned about the fact that various translations through space have no observable consequences, for this is precisely what his own hypothesis (space is Euclidean) predicts.

In §9.2 I pointed out that since a pure void has no determinate structure or dimensionality, the void-conception of space leaves something unexplained – why it is that our movements are constrained in the ways they are – and I called space the "unseen constrainer". The point still applies but, as is now clear, this particular label is particularly appropriate in the case of Euclidean space.

14.3 The explanatory challenge

Nerlich's argument can usefully be recast in the form of a challenge to the relationist. Objects moving through non-Euclidean spaces – especially spaces of extreme and variable curvature – will behave oddly: their shapes will vary in observable ways; they will move closer together and further apart; internal stresses will be generated within them. Since no known forces (e.g. gravity, magnetism) can explain these movements, we are left with a puzzle. The substantivalist has a simple and economical explanation for the strange behaviour: a substantival space exists; this space is curved; and it is a law of nature that objects follow geodesics through this space unless acted on by some external force. For this explanation to work the substantivalist must be able to fill in the details, by providing at least an approximate mapping of the actual shape of space, and show that this mapping suffices to explain the observed motion of bodies when combined with data about the known forces that are in play. Providing these details is a non-trivial task, but if it can be done the substantivalist's explanation certainly deserves to be taken seriously. What can the relationist offer in its stead?

In some cases it is very hard to see how the relationist could come up with a remotely plausible explanation. To see this it suffices to imagine a world that contains non-Euclidean holes of the type introduced above. From the substantivalist's standpoint, the space of this world contains various stationary regions of tightly curved space, and these regions have discernible effects on moving bodies, light rays, and so on. How can the relationist explain these various effects?

Consider first a very simple case that the substantivalist would describe in these terms. The system contains one star, one orbiting planet and one hole. The curvature of the hole is symmetrical around its centre; the planet's orbit is elliptical and regular, and it passes through the centre of the hole at precisely the same time each year. Since the relationist denies the existence of both substantival space and the hole, the various discernible effects that the substantivalist puts down to passing through the hole must be explained solely in terms of the relationship between the planet and its star. Can we say that the effects are correlated with the planet being at a certain distance from the star? No, since the planet is at the same distance from its star twice per orbit, and the problematic effects only occur once per orbit. But another simple option is available: correlate the effects with *time*. Since the orbit is perfectly regular, the time between the onset of the effects is regular too – they happen exactly once each year – so the relationist can explain the effects by positing a law that states that the various effects occur at yearly intervals. Laws of this form are odd, but perhaps not impossible; there might be worlds where all green things change to red for one day at yearly intervals. Even so, there is a sense in which the relationist's explanation is less complete than the substantivalist's. The latter can (in principle at least) explain the precise form the effects take, in terms of differing degrees of positive and negative curvature within the hole, and why these effects only occur on the planet but not the Sun. The relationist must take these to be nomologically basic features of the world.

In this simple case the discrepancy between what the relationist and substantivalist can offer is not dramatic, but this quickly changes if we introduce more complexity into the system. Suppose that we introduce just one more planet on a different orbital plane. The two planets are attracted to one another by both gravity and magnetism in such a way that their orbits are complex and aperiodic, but still deterministic and predictable. Both planets pass through the hole region at varying intervals, and along varying trajectories. Since they don't always pass through the centre of the hole the effects are usually only felt in some regions but not others. Since the relationist can no longer explain the effects by the simple time law that worked in the previous example, the only remaining option is to seek an explanation in terms of the relative positions of the two planets and the star. But given the irregular orbits of the planetary bodies, and the different ways the effects are felt by the planets on different occasions, there will be no simple correlation between relative orbital configurations and effects. At best the relationist will be able to come up with a law of this type: $\{C_1\text{-}E_1, C_2\text{-}E_2, C_3\text{-}E_3, \ldots\}$, where each 'C' stands for a different orbital configuration and each 'E' for a different pattern of effects. Not only is each distinct pairing a brute nomological fact, but the list may well be infinitely long. Opposing this hideously complex gerrymandered effort is the simple and powerful account offered by the substantivalist. There is simply no competition. And all this holds for a very simple system; if a few more holes and planets are introduced, if the star moves, the relationist's predicament swiftly becomes utterly hopeless.

However, if the sort of universe just considered is extremely hostile to the relationist, others are far less so. To take another simple example, consider a world consisting of just three equally spaced non-interacting point-particles moving away from each other in straight lines at uniform speeds. In a spacetime diagram of this world, the particles' worldlines resemble an inverted tripod. Needless to say, the substantivalist can account for this pattern of movement in terms of geodesics through flat space, but this account is not very economical, featuring as it does vast (perhaps infinite) tracts of empty substantival space, the vast bulk of which serves no explanatory purpose. The relationist's account is just as simple and far less profligate. At any instant the universe contains just three objects and three spatial relations, and the evolution of the universe is controlled by a simple law: the distance relations between the objects increase at a constant rate over time. More complex worlds present the relationist with more complex challenges. As we saw in the course of our earlier explorations of Newtonian dynamics, there are worlds where the relationist needs to distinguish between inertial and non-inertial paths. But, as we also saw, there are ways that the relationist can hope to do this.

As for the conclusions that we can draw from all this, I suspect that there is just one: there are no easy answers. There are, however, three lessons that have emerged. First, even in worlds where there is no evidence of the sorts of behaviour that can be explained in terms of spatial curvature, there is still *something* to be explained, namely why the movements of material bodies conform to the geometry of Euclidean space. Since there are many other possible geometries, this is a fact that needs explaining, and the explanation advanced by substantivalists is simple and effective. Unless the relationist can provide an account of comparable merit we have no reason to reject the account offered by the substantivalist. Secondly, in worlds where the substantivalist finds it natural to posit a variably curved space to explain the observed motions of bodies, it is quite likely that the relationist will be unable to offer an explanation of comparable simplicity. Finally, however, there are worlds where the relationist *may* be able to provide accounts of observed motions that rival or even surpass the substantivalist's accounts in simplicity and economy. In cases such as these we cannot pronounce a priori on the likely outcome; all depends on the details of the specific theories that are proposed.

We will be returning to these issues in Chapters 18 and 19, when we consider Einstein's general theory of relativity; a theory that posits a variably curved spacetime. We shall see that while a substantivalist interpretation of this theory is hard to avoid, it is also far from unproblematic. But before moving on to these real-life scientific cases, there is a famous a priori argument for spatial substantivalism that we have not yet examined.

14.4 A solitary hand

At various times dating back to the 1768 essay "Concerning the Ultimate Foundation of the Differentiation of Regions in Space", Kant used "incongruent counterparts" in a novel attempt to refute the relationist conception of space. Sometimes he used the argument to reject relationism in favour of Newtonian absolutism, and at others he used it to defend the idea that *our concept* of space was of a structured entity, distinct and independent of the things in space. This is all he could say in his anti-realist phase. Although the argument (and the phenomenon in question) have generated a considerable literature, I will not be entering into the debate very far, but since Kant's argument does cast light on the main theme of this chapter from a new angle, this is the place to consider it.

The discussion will move more swiftly if some relevant terminology is introduced at the outset. Objects are *enantiomorphs* or *incongruent counterparts* if they are related to one another in the same way as mirror images. Although in one sense they are of exactly the same shape, in that their parts are of the same size, distance and angular separation, they cannot be brought into congruence with each other (i.e. made to occupy the same part of space) by any amount of sliding and turning. The letters **b** and **d** are two-dimensional examples (at least in certain fonts). If both letters are confined to the same two-dimensional plane, they cannot be made to occupy the same space. There are three-dimensional examples too, such as the left and right gloves in a pair. Of course, if the letters were made of clay, the **b** could easily be squished into a **d** shape, but in the present context, bending, stretching and turning inside-out are not allowed. The permitted spatial translations are called *continuous rigid motions*, or crms. Any two objects whose parts are of the same size and disposition, but that cannot be brought into congruence by a crm are enantiomorphic; objects are *homomorphic* if they *can* be made congruent by crms.

Kant noticed the peculiarity of handed objects: "What can be more similar in every respect and in every part more alike to my hand and to my ear than their images in a mirror? And yet I cannot put such a hand as seen in the glass in the place of its archetype" (Kant 1996: 55). Hands and ears are prime examples of three-dimensional incongruent counterparts. No measurement of size or angle will reveal a geometrical difference between the left and right ears of someone who is perfectly symmetrical, and yet the two are clearly *different*, since they cannot be made to occupy the same space. Kant saw that this difference could be very troublesome indeed for the relationist. To bring this out he used a memorable thought-experiment.

Imagine that the universe is empty except for a single human hand. Is it a right hand or a left hand? Since there are no intrinsic differences between enantiomorphic objects, we have no basis for calling the hand left or right. Of course, if you imagine yourself looking at the hand floating in otherwise empty space, you will see it as a right hand or a left hand, but in imagining this you are putting yourself into the space of the hand, together with your sense of right- and left-handedness. If you perform the thought-experiment rigorously, and don't include yourself in the picture, it seems clear that we would have no more basis for calling the hand left or right than saying that it is pointing *up* or *down*. Now suppose that a human body – yours – materializes in space near the hand. You have with you a pair of gloves. Clearly only one of these gloves will match the floating hand. Let us suppose that the glove for the left hand fits. Since we can stipulate that your appearance on the scene does not change the hand in any way, it seems clear that the hand must *already* have been a left hand *before* you appeared in the space. In which case, there must have been a fact of the matter as to the handedness of the hand even when it was the sole occupant of the universe. The key question is whether the relationist can explain this fact, and it seems not. Since by definition all the spatial relations (distances, angles) among the parts of a pair of incongruent counterparts are identical, it seems that from a purely relational perspective there can be no difference whatsoever between a right hand and a left hand. Kant concluded that since there clearly *is* a difference between them, and this difference cannot be explained relationally, it must somehow be due to the relationship between the hand and substantival space.

14.5 Global structures

Unfortunately, Kant was not very explicit as to how introducing substantival space into the picture explains the difference between oppositely handed objects. Later substantivalists have tried to fill this lacuna, and their efforts have helped clarify the phenomenon of handedness. I will summarize their arguments, and then consider the relationist replies.

Some have tried to refute Kant by arguing thus:

Kant's argument rests on the assumption that it makes sense to say that a *solitary* hand is left-handed or right-handed. In fact this doesn't make sense. To see why this is so, think of a solitary (and asymmetric) flat hand on the plane surface of Flatland. Does it make sense to say that this flat hand is right-handed or left-handed in the absence of any Flatlanders? The answer is no. From our three-dimensional perspective, the hand is *either* a right hand or a left hand, depending on the direction that we view the plane from (e.g. seen from above it is a left hand, and from below it is a right hand). So it seems that it doesn't make sense to say that the hand *is* a right hand or a left hand when it is all alone, because what we say it is depends on the perspective that we view it from. One and the same hand is *both* a right hand and a left hand, depending on how we view it, and so in itself it is neither.

Suppose that we now introduce a Flatlander into the world. She has two gloves, one for a right hand and one for a left hand, labelled R and L. Obviously only one of these gloves will fit the hand. Let us suppose that the left one does. Does it mean that the hand was a left hand *before* we introduced the Flatlander? Not at all, because the glove that fits depends on *how we introduce* the Flatlander; place her in Flatland one way, and she finds that the hand fits the left-hand glove; turn her over and put her down, and she finds that the hand fits the right-hand glove. In turning the Flatlander over in three-dimensional space, we don't change anything about her or the hand in Flatland; all we change is how they are spatially related to each other. As is clear, precisely the same applies to us in three-dimensional space. We can imagine materializing next to the floating hand, and finding that our left-hand glove fits it; but this doesn't mean that the hand was a left hand before we arrived, because if we had been rotated in four-dimensional space we would have found that the hand fitted the right-hand glove. Nothing in the *hand itself* determines in advance which glove will fit.

The Flatlanders look at their pairs of gloves and wonder "how can two things be so alike and yet so different?" We three-dimensional beings can understand. The Flatlander's hands *are* exactly alike; it is only because they are trapped, physically and imaginatively, in their two-dimensional space that they think the gloves are different. Likewise, a pair of three-dimensional gloves are exactly the same shape, since a four-dimensional being could make them match exactly; they only appear different because of the way they are implanted in three-dimensional space.

This line of argument may seem to provide strong grounds for supposing that enantiomorphs are not intrinsically different, and so it might seem to undermine Kant's argument. However, the substantivalist can reply as follows:

> Far from refuting Kant, the argument from Flatland strengthens his conclusion. It is true that one and the same two-dimensional hand can be a right hand or a left hand, but, as Kant foresaw, the handedness it has depends on *how it is inserted into space*. Turning a Flatlander's left-hand glove into a right-hand glove by rotating it in the third dimension amounts to inserting the glove into space in a different way.
>
> We can now see that the very possibility of enantiomorphism depends on the properties of space itself. A pair of two-dimensional glove shapes can *only* be enantiomorphs in two-dimensional space; in a three-dimensional space they are homomorphs, since there is always a crm that will make the hands congruent. Quite generally, an asymmetrical n-dimensional object is enantiomorphic in a space of n dimensions, but not in a space of $n + 1$ dimensions. So we can see that the phenomenon of enantiomorphism depends on two factors: the shape of a particular object, and the dimensionality of the space in which the object exists.[2]

This conclusion can be strengthened, for there is another general topological property of space that is relevant to enantiomorphism: *orientability*. The technical definitions of "orientable" and "non-orientable" spaces are quite complex, but what the distinction amounts to is easily grasped.

Suppose that space is an infinite two-dimensional plane. Take two shapes, such as **b** and **d**, which cannot be smoothly slid so that one coincides perfectly with the other (contrast with **b** and **q**). The **b** and **d** are enantiomorphic both locally and globally: no matter how far you move the two apart (and then bring them together again) you will be unable to make them congruent. A space like this is *orientable*. Now consider **b** and **d** on a Möbius strip.[3] In the immediate neighbourhood of **b** and **d** it is impossible to get the two letters to perfectly coincide: no crm will render them congruent. However, if you slide **b** right around the strip, when it returns it will look just like **d**; the trip through this two-dimensional space has transformed the **b** into a **d**. A space where this can happen is *non-orientable*. (In effect, such a space has a kinked structure when viewed globally.) In an orientable space, any two figures that can be brought into congruence by "global" continuous transformations can be brought into congruence "locally", too; in a non-orientable space, objects can be locally enantiomorphic without being globally enantiomorphic. If we define enantiomorphism in terms of the non-existence of *any* crms that will render two asymmetrical objects congruent, then in a non-orientable space of *n* dimensions, all *n* dimensional objects are homomorphic.

The substantivalist seems to have a strong case: whether an object is enantiomorphic or homomorphic depends on *the global properties of space itself*, the relevant properties being dimensionality and orientability. However, in the light of the discussion earlier in this chapter, it should be clear how relationists must respond. They must show that we can make sense of the relevant features of space in purely relationist terms, mentioning only material bodies, spatial relations, and their nomologically possible modes of combination. Once again, the prospects for this seem mixed.

Dimensionality poses no problem; the relationist can explain it in terms of natural laws governing the angles or directions in which objects can be spatially related. Orientability, however, is a very different kettle of fish. In a simple world containing two objects we might perhaps posit a law that states that the objects undergo a Möbius-like "flip" each time they move a certain distance apart, and remain in the same orientation unless they first move closer together and then move further apart again. But what of a universe that consists of numerous objects all moving relative to one another that behave as if they exist in a static non-orientable space? As with the case of multiple non-Euclidean holes, it is very likely that in such cases any laws that the relationist can devise to explain the observed behaviour of bodies will be hideously complex and unwieldy.

It seems that the phenomenon of enantiomorphism yields precisely the same lessons as spatial curvature. When all possible spatially structured worlds are taken into consideration, substantivalism will often be the only viable option. In the vast majority of cases the relationist will be unable to provide an explanatory account of why objects behave as they do that is of comparable power and simplicity. However, as we have also seen, some spaces are more dispensable than others. In

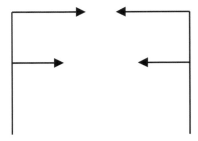

Figure 14.1 Relations with attitude. A relationist who posits distance relations possessing intrinsic directions has an account of how objects such as those depicted here differ from each other, irrespective of how they can move relative to one another.

worlds where the motions of objects are not such as to warrant the postulation of one of the more exotic forms of space, worlds where a substantival space – if it existed – would be comparatively boring and bland, the relationist's prospects of providing a satisfactory account of why things move as they do are greatly increased.

Having conceded a good deal to the substantivalist, it should be noted that the shoe is sometimes to be found on the other foot. While we can explain enantiomorphism in terms of the global structure of a substantival space, the relationist who is prepared to recognize inherently directed vector-like spatial relations can recognize handedness as an *intrinsic* property, as Figure 14.1 illustrates.

15 Spatial anti-realism

15.1 Foster on matter and space

While the conceptions of the large-scale composition of the physical universe offered by substantivalists and relationists are undeniably very different, there is also a sense in which they do not differ at all: both camps accept that we inhabit a spatial world. The disagreement is limited to the question of how space (or spacetime) should be characterized. Despite their differences, substantivalists and relationists both count as spatial *realists*. In this chapter we will be examining a purely metaphysical argument for spatial *anti-realism*; the claim that our universe is not in fact spatial, despite appearances that suggest the contrary. The claim that space is unreal may well seem as absurd as McTaggart's assertion that time is unreal, but just as McTaggart's argument (or at least a close relative of it) proved to be both non-trivial and independent of idle sceptical considerations, so too will the argument we will be considering here.

In *The Case for Idealism* (1982) Foster develops an argument intended to establish that the physical world is not (and could not) be a part of what he calls "ultimate reality", and the key step in reaching this conclusion is an argument for spatial anti-realism. Unfortunately, despite its interest, Foster's argument is complex, multi-faceted and often difficult; to do it full justice would require a lengthy exposition, and this is not the place for it. Consequently I will be selective in my approach, and confine my attentions to what he has to say about spaces that are substantival in character and taken to be ontologically basic components of the physical world. Foster develops similar (and equally potent) arguments against relationism, but since these depend more heavily on the sometimes exotic and unfamiliar framework employed in his book I will leave them for interested readers to explore for themselves. (In his more recent "The succinct case for idealism" (1993) Foster himself employs just this strategy.)

Although the import of Foster's argument is ontological whereas that of Poincaré's conventionalism is epistemological, in certain respects the routes they take are similar. Both arguments are concerned with spatial geometry, and there are similarities in approach: both employ thought-experiments in which natural laws conspire to conceal the real structure of the underlying (spatial) reality. The similarities do not end there. As the following sketch of an argument shows, Poincaré's conventionalism might easily be taken to have ontological implications:

1. Physical space has the geometry specified by our best physical theory.
2. This geometry is a matter of convention.

3. The character of objective reality is not a matter of convention.
4. Hence the geometry of physical space does not belong to objective reality.
5. As the bearer of this geometry, physical space does not belong to objective reality either.

Whatever its merits, this reasoning will not convince those who adopt a realist stance towards scientific theories, for anyone falling into this camp will reject conventionalism, and hence point 2. Foster's argument poses a threat even for realists. He begins by assuming that there is a physical world, and that our best scientific theories provide accurate guides to its laws and structure. Then, having thus assumed that the scientific picture of the world is broadly correct, he goes on to show that this assumption is incompatible with adopting a realistic stance with respect to physical space.

Although I will be concentrating on Foster's spatial anti-realism, rather than his more general physical anti-realism, a brief comment on the transition from the latter to the former is in order. It may not be immediately apparent why an argument that threatens the existence of physical space is a threat to the existence of the physical world *per se*. Supposing that such an argument were successful, shouldn't we conclude not that the physical world doesn't exist, but rather that it consists solely of material objects? Bearing in mind what was said in Chapter 9 about the constraining and connecting role that space plays, this is not a plausible option. Even if we could make sense of individual physical objects existing in the absence of space, it is surely uncontroversial that for a physical *world* to exist these objects must be spatially connected. In the absence of space (whether construed relationally or substantively) this is simply impossible.

15.2 The intrinsic and the inscrutable

We can start with a doctrine that plays an important role in what is to come. Foster sometimes calls it "the inscrutability of physical content". It can be broken down into a conjunction of claims, one of which has two components:

1. The *content thesis*: at least some material things possess their own intrinsic natures.
2. The *inscrutability thesis*: (i) it is impossible for us to know anything about these intrinsic natures; (ii) these intrinsic natures do not fall within the scope of physical science.

If "intrinsic nature" is taken to mean "intrinsic property", 1 is clearly true, but 2 is equally clearly false. Shape, size, mass, colour, impenetrability and charge are commonly taken to be intrinsic properties of material bodies, but all are eminently scrutable. However, in talking of "intrinsic natures" Foster is not talking of "intrinsic properties" in this familiar sense, as the following passage makes plain:

> from our observations, and from the way these support certain kinds of explanatory theory, we may be able to establish the existence of an external space with a certain geometrical structure . . . we can never find out *what* . . . *the space is like in itself* . . . Likewise, while, by the same empirical means, we

may be able to establish the existence of external objects located in this space
... we can never discover the ultimate nature of their space-filling content
... we can never discover, beyond a knowledge of their spatio-temporal
properties, what the simplest objects (the elementary particles) *are like in themselves* ...   (Foster 1993: 295)

The idea that there is something that elementary particles "are like in themselves" has to be taken in the right way. The claim is not that "there is something that it is like to *be* an elementary particle"; Foster is not saying that elementary particles are *conscious* entities. The claim is simply that the most basic physical things have some intrinsic characteristics over and above their spatiotemporal and causal properties, over and above *all* the properties we ever actually ascribe to physical things. Put in this way, the content thesis may well seem to lack motivation (to put it mildly). Why believe that these additional properties exist? In fact, as quickly becomes apparent, the content thesis has a good deal of plausibility.

Consider the intrinsic properties we do ascribe to ordinary physical objects. Some are spatiotemporal: size, shape and motion fall into this category – shape, size and mode of arrangement are sometimes called "structural" properties, for obvious reasons.

As for the others, it may not be immediately obvious, but they are all causal powers: that is, capacities to produce certain specific types of change in (or to be changed *by*) other objects. This is often obviously the case. Objects that are *rigid* have the capacity to retain their shape under pressure. Objects that are *brittle* have a tendency to shatter when struck. Electric charge is associated with a variety of causal properties, such as the capacity to attract or repel other charged objects and the capacity to interact with magnetic fields. Mass is a causal property, too: more massive things are harder to stop, and require more pushing to get them moving. Likewise impenetrability: if we know that two particles are impenetrable with respect to each other, we know that neither can occupy the space the other occupies, that each has the capacity to repel the other at its borders, as it were. Less obviously, physical colour is a causal capacity, too. Or at least, it is if we follow Locke, Foster and many others in rejecting "naïve realism" and subscribe to a "representative" theory of perception, according to which, to say that a physical object has a certain colour amounts to saying that the object is causally disposed to produce in perceivers a certain sort of visual experience in certain sorts of circumstances (e.g. normal daytime lighting conditions). The same applies to all the other secondary qualities, such as warmth, texture, and solidity (as these feature in our experience).

The question now arises as to whether all non-structural intrinsic properties could be causal (or dispositional). If so, then the physical world as a whole might consist of nothing but "bare powers" or "bare potentialities"; *bare* because the potentialities in question are not grounded in anything that is not itself a potentiality. Foster maintains that this conception of physical reality is incoherent: the causal potentialities that physical things possess must be grounded in (i.e. possessed by) something that is not itself nothing but a potentiality. Hence physical things must possess *some* properties that are *not* mere potentialities. These non-dispositional (or "categorical") properties are the "intrinsic natures" referred to in the content thesis.

Foster is certainly right about one thing: the idea of a world consisting of *nothing but* bare powers is a very odd one indeed. In a very real sense nothing can ever happen in such a world. Such changes as there are amount only to alterations in some of the various potentialities that exist; but since everything that ever exists is a potentiality, any activation of a power only results in an alteration in some other potentiality, and this potentiality in turn can never result in any actuality. Is a world without *anything* that is more than a potentiality, a world in which the activation of any potentiality can only bring about alterations in other potentialities, really possible? I suspect not.

Consider a simple world, consisting of just two atoms moving in space. Atom A_1 has the power to repel another atom A_2 should it ever approach, and A_2 has the power to repel A_1. We can now ask, what is atom A_1? Given that it is nothing but a spatially located mobile disposition, we can only answer that A_1 is the capacity to repel A_2. But what is A_2, if not the power to repel A_1? This seems problematic, since in order to make sense of what A_1 is, we need to make sense of *what A_1 is a power to do*, since it is nothing but a causal power, but all that A_1 can do is repel something which itself is nothing more than a power to repel A_1, and of course the same applies to A_2, which is nothing more than a power to repel something which itself is nothing more than a power to repel A_2. It seems that, no matter where we begin to consider the situation, we are left with nothing. Before we can take A_1 to be something that exists at all, we need to be able to specify what it is able to do; but we cannot do this, because *ex hypothesi* all it can do is repel A_2, and the status of A_2 as something that exists is only secure when we have stated what *it* can do, but this cannot be done, since *ex hypothesi*, all it can do is repel A_1, whose status as an extant thing we have yet to secure. So it seems that it is impossible for the universe we are envisaging to contain anything at all, contrary to the initial stipulation that it contains two atoms.

The argument above concerned only two atoms and a single causal power. But it is clear that adding other atoms (consisting of nothing but the power to repel other atoms) will not alter the situation at all. Moreover, adding further causal powers will not do any good, unless they are powers to affect something intrinsic (i.e. non-dispositional), which is something we have ruled out. Going back to our (allegedly) two-atom universe, let us suppose that each atom has the power to attract the other, in addition to their powers of repulsion. As soon as we try to spell out what each atom's attractive power is the power to do, we run into the same problem as before: A_1 is the power to attract A_2, which is itself nothing more than the power to attract A_1, and so on. Adding to the specification of A_2 that it possesses the power to repel A_1 achieves nothing, since we already know that the power of mutual repulsion is insufficient to secure the existence of either atom. Once again, bringing additional purely dispositional atoms into the picture does nothing to improve matters (see Foster 1982: 67–70).

The "bare power" ontology may be incoherent, but does it follow that *all* physical things must possess an intrinsic nature in addition to their causal powers? Probably not. But if some physical things are nothing more than mobile clusters of causal potentiality, they can only exist because there are other physical things that do possess intrinsic natures. As for which physical things fall into the latter category, Foster is adamant on one point: a substantival space must possess an intrinsic nature of its own. And he is surely right about this, at least on the assumption that a substantival space, as a fundamental ingredient of physical reality, does not depend for its existence on any other type of physical thing.

Let us turn from the content thesis to the inscrutability thesis. Now that we have a better idea as to what Foster means by "intrinsic nature", it is hard to disagree with claim (i), which states that we know nothing about them. We know that (some) physical things *have* intrinsic natures, but this is all we know: the precise character of these intrinsic natures is a complete mystery. Science tells us a great deal about the intrinsic properties that physical things possess, but as we have already seen, since these properties are entirely causal or structural in character, they tell us nothing about the intrinsic *natures* of the things that have these properties.

But Foster goes further: the inscrutability thesis also includes claim (ii), according to which intrinsic natures do not even fall within the *scope* of physical science. As for why this is so, Foster argues that physical concepts – the concepts that it is legitimate to apply to the physical world in scientific theorizing – are without exception entirely neutral on the question of intrinsic natures. This "topic neutrality" thesis is a conceptual truth, or so Foster maintains: genuine physical concepts are restricted to those that can feature in an empirically testable theory, and claims about intrinsic natures cannot feature in such theories.

An example will help bring these claims into clearer focus. Imagine a world superficially very similar to our own, but consisting of a substantival space and a single type of atom – a simple and indivisible particle of small but finite size. Physicists have a well confirmed and seemingly complete theory of atomic interactions and motions, and this theory – call it T – is topic neutral. Now, there are any number of competitor theories that are *not* topic neutral. These theories are the same as T in all respects save one: they attribute intrinsic natures to both atoms and space. Our purposes will be served by considering just two of these competitors. According to T_1, atoms are spheres of an intrinsic quality Q_1, whereas space consists of an infinite expanse of an intrinsic quality Q_2. According to T_2, space is filled with Q_1, and atoms are pervaded by Q_2.

Now, Foster does not claim that we can make no sense whatsoever of theories such as T_1 and T_2, for we can, especially if we give the posited qualities a positive interpretation in phenomenal terms (e.g. by thinking of Q_1 as *red* and Q_2 as *blue*, as these colours feature in our experience). But it is one thing for a claim to be intelligible, and another for it to be an intelligible *scientific* claim. Real-life physicists do not produce theories such as T_1 and T_2, and any physicist who did would not be taken seriously; intrinsic natures, in Foster's sense, play no role in

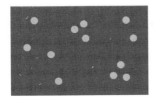
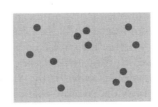

 Q_1 Q_2

Figure 15.1 Two competing accounts of the distribution of intrinsic qualities.

actual scientific theorizing. Although the difference between T_1 and T_2 is easy to depict (and imagine), it would be impossible to detect in practice. Since both T_1 and T_2 ascribe exactly the same causal powers to atoms and space as T, they make no novel or testable predictions: objects composed of Q_1-type atoms look and behave in just the same ways as objects composed of Q_2-type atoms. Quite generally, the only properties that science recognizes (or should recognize) are those that contribute to explaining and predicting the observable features of the physical world. All such properties are causal in character or have causal implication in the manner of structural properties; it is precisely because of their causal character that they *can* contribute to explaining the observed behaviour of material things. Since intrinsic natures are properties that things possess over and above all their causal and structural characteristics, they do not fall within the proper purview of science. Hence Foster's claim: the concepts that can legitimately feature in science are topic neutral.

15.3 Modes of deviancy

Substantivalism, in its traditional guise, is the doctrine that space is an entity in its own right, possessing its own inherent structure. The existence and structure of such a space is (to some degree at least) independent of the existence, distribution and behaviour of any material bodies that it may happen to contain. If we follow Foster, and subscribe to the content thesis, then a substantival space will also possess an intrinsic qualitative nature of its own. Given the inscrutability thesis, we cannot hope to discover what the intrinsic nature of our own space is. Nonetheless, it is easy to imagine how a space *could* have its own intrinsic nature. A Flatland space could take the form of a homogeneous circular blue plane. To imagine a space with the same structure but a different intrinsic nature, simply picture a homogeneous circular *red* plane. The surface of a red sphere represents a Flatland space with the same intrinsic nature as the latter but a different structure. Three-dimensional analogues of flat two-dimensional spaces are easily imagined (just picture a colour-pervaded volume); three-dimensional analogues of curved Flatland spaces are beyond our imagining, but certainly possible. As these examples make plain, the geometry of such spaces and their intrinsic natures are linked, but the determination goes only one way: the geometry of a space is determined by the structure of its qualitative "filling", but spaces of different qualitative types can have the same geometrical structure.

Foster's argument is general in scope, but easier to expound in the context of simple spaces. So let us suppose that our space (or spacetime) – call it S – is Newtonian in character, and that material bodies are composed of a single kind of atom. To make matters more concrete, we can imagine these atoms as minuscule mobile spheres of red, and S as an unbounded volume of uniform grey. Thinking of these items thus is a useful way of reminding ourselves that they possess intrinsic natures, but it should not be misconstrued: these "colours" are not visible to us. S does not impede light, and so is entirely transparent, whereas material objects reflect light in different ways (and so have different observable colours) depending on the way their surface atoms are configured.

Since the material things we perceive are composed of atoms moving through S, it is natural to think that S and physical space – call it P – are one and the same. Foster's key claim is that S and P cannot possibly be one and the same: they are, and must be, numerically distinct entities. The argument for this conclusion relies on the possibility of certain modes of deviancy, ways in which P and S might differ from one another.

One mode of deviancy features variations in the intrinsic natures of S and its occupants. For example, S could contain a cube-shaped region, R, which is systematically linked (through a law of nature) to variations in the intrinsic natures of atoms: as atoms enter R they change from "red" to "blue", and remain "blue" until they leave (see Figure 15.2b). Both sorts of atom have exactly the same causal powers, so they behave in exactly the same ways. Alternatively, we can suppose that R itself has a different qualitative nature than the rest of S; for example, it is a different shade of "grey", as depicted in Figure 15.2c. Possibilities such as these pose no threat to the status of the physical world; there is no obstacle to our holding that S and P are one and the same space. Nonetheless, they are a useful way of bringing to the fore the implications of the inscrutability thesis.

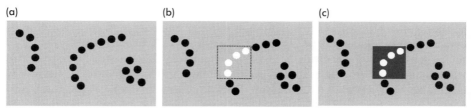

Figure 15.2 Deviant and non-deviant distributions.

Since the qualitative changes exist solely at the level of the intrinsic natures of the items concerned, given the inscrutability thesis, they are not empirically detectable. Hence the qualitative variation (or deviancy) in S found in the case in Figure 15.2c would not register in our physical theorizing: our best physical theories would posit a homogeneous space, and a single kind of atom. Now, a natural response to this situation would be to suppose that these theories provide us with a largely accurate but *incomplete* account of the physical world, but this is a mistake. Not only are the changes in intrinsic nature undetectable, but they are not *physical* changes at all. If, as the inscrutability thesis states, physical concepts are without exception topic neutral, then the worlds in Figure 15.2 fall under precisely the same physical descriptions, and so are physically indiscernible in all respects. The worlds are different, but not physically different, and the relevant physical theories are not incomplete.

The mode of deviancy that Foster is most concerned with – a mode that poses more of a threat to the status of physical space – is geometrical rather than qualitative. His primary example (he provides plenty of others) consists of a case in which the laws governing the behaviour of material things conspire to conceal their real spatial locations. Although the mechanism involved is very simple, its dramatic consequences may not be immediately apparent.

Once again S is a three-dimensional Newtonian space. Although S itself is homogeneous, there are two spherical regions within it, R_1 and R_2, some distance apart, whose boundaries have unusual effects on any material objects that cross them, effects due to the relevant laws of nature. Any object that enters R_1 emerges in R_2 and continues on its way, with the same speed and in the same direction; similarly, an object entering R_2 emerges in R_1, and continues on its way, with the same speed and in the same direction. An object that leaves R_1 appears at the boundary of R_2, and continues on its way, and in a similar fashion, an object that leaves R_2 appears at the boundary of R_1, and continues on its way. All of these "transitions" occur instantaneously, and all forms of matter and energy (light included) are affected. These deviant laws have the effect of making it appear as though R_1 is located where R_2 actually is, and vice versa. A concrete example will help make this clear.

Let us suppose that R_1 is a spherical region of ten miles diameter in Cambridgeshire, and R_2 is a similarly sized and shaped region in Oxfordshire. We go back several thousand years, to a time when these areas were just starting to be occupied by the first human settlers. A particular settler, walking through the region we now know as Oxfordshire, finds that he wants to be alone, and so decides to build a hut a few miles beyond his tribe's most outlying village, which – it so happens – is close to the boundary of R_2. (Since this boundary is itself invisible, and has no observable effects, no one realizes it is there.) Our settler walks five miles further on, and gets to work. He builds a small hut. After a few months of solitude, our settler decides he wouldn't mind some company after all, and it is not long before a small village has formed, centred on his original hut. From this modest beginning, the city we now know as Oxford developed. However, although Oxford appears to be within the Oxfordshire countryside, in reality it is situated in *Cambridgeshire*, in region R_1. As our founding settler first crossed the boundary of R_2 he was re-located inside R_1, and since R_1 is in Cambridgeshire, so too is the hut that he built, and the village that developed around it. When the villagers went in search of suitable stones to use in their constructions, they frequently crossed the boundary of R_1, and emerged in Oxfordshire; the material they brought back to the village – after crossing the R_2 boundary – all end up inside Cambridgeshire, which is why the resulting village is constructed of what seem to be "local" materials (of the sort found in Oxfordshire), even though it is actually situated in Cambridgeshire. All subsequent movements – of animals, people, matter, light – undergo the same instantaneous translations. We can tell a similar story for a settler walking through the Cambridgeshire fenlands: on crossing the boundary of R_1, this settler emerges in Oxfordshire, and founds the village that develops into the city we know as Cambridge.

So we see that, in this imaginary scenario, the actual locations of Oxford and Cambridge are the opposite of their apparent locations: Oxford is in Cambridgeshire, and Cambridge is in Oxfordshire, but given the immediate re-locations that occur when things cross the boundaries of R_1 and R_2, no one realizes this. Figure 15.3 (cf. Foster 2000: 268) illustrates the journey of someone travelling from Oxford to Cambridge by car. As can be seen, the journey through S (space as it really is) and the apparent journey, are very different.[1]

Since light is similarly affected, anyone flying high over Oxfordshire who looks down on to R_2 will see light emanating from R_1 (as depicted in Figure 15.4), and so will actually perceive Oxford, even though Oxford in fact lies a good many miles to the east.

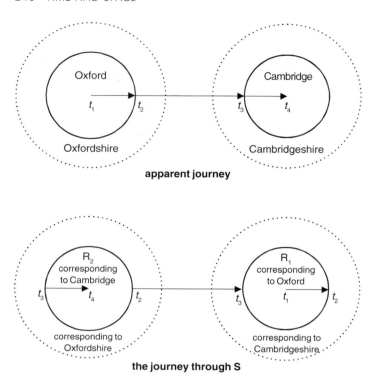

apparent journey

the journey through S

Figure 15.3 The Oxford–Cambridge case.

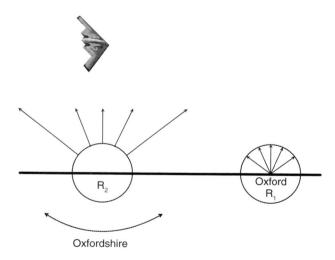

Figure 15.4 **The consequences of deviancy.** Light from a spherical region of S in Cambridgeshire (where Oxford is situated) emerges in Oxfordshire, leaving the pilot of a plane flying over Oxfordshire with the impression that Oxford lies directly below.

It might be thought that none of this makes any sense: if all physical processes are nomologically organized in exactly the same manner as they would be if Oxford were really in Oxfordshire, and Cambridge in Cambridgeshire, what sense is there in the idea that this is not the case? It is here that the underlying (but invisible) space S plays a crucial role. With respect to this space, which we are assuming is perfectly real, Oxford really is in Cambridgeshire, and Cambridge really is in Oxfordshire. As noted earlier, if *per impossible*, we could see S, it would be glaringly obvious that Oxford is surrounded by the Cambridgeshire countryside, and Cambridge by the Oxfordshire countryside.

15.4 Intrinsic versus functional geometry

Let us say that the laws governing the behaviour of atoms in space are "nomologically uniform" if they put the same constraints on the behaviour of atoms at all points in space, and, in particular, the laws apply to all the geometrical properties of atoms in a uniform way throughout space (and time), where by the "geometrical property" of an atom we mean its shape and possible spatial trajectories. It is clear that the laws operative in S are not nomologically uniform: they require atoms to move continuously in regions outside R_1 and R_2, but discontinuously across the boundaries of these two regions, and a spherical atom will remain spherical as it moves through space except when it crosses a boundary of R_1 or R_2, at which point it ceases to be spherical, and for a while is, in effect, divided into two parts.

In addition to not being nomologically uniform, the laws operative in S have a distinctive feature: they are such that, if the regions R_1 and R_2 were interchanged, then nomological uniformity would be restored. Since the imagined interchange would result in R_1 (which contains Oxford) being located in Oxfordshire, and R_2 (which contains Cambridge) being located in Cambridgeshire, the envisaged laws no longer force objects to follow discontinuous paths. A car driven through Oxfordshire along a road that leads to Oxford will pass through Oxford (not Cambridge): as the car crosses the critical boundary it "emerges" in R_1, just as before, but since R_1 is now in Oxfordshire rather than Cambridgeshire, the car's path through S is continuous.

Taking nomological uniformity to be the norm as far as the nomological organization of a world is concerned, Foster calls worlds that are nomologically non-uniform and also possess this distinctive feature "nomologically deviant". A nomologically deviant physical world is one that is not nomologically uniform, but in which nomological uniformity would be brought about by some alteration in the geometry of the world. In the case we have been considering, the required alteration is comparatively mild – the points in two small regions have simply to be switched around – but Foster describes more radical cases where the deviance is at the level of dimensionality (the laws range over three spatial dimensions, but the underlying space is only two-dimensional).

Foster next draws a distinction between what he calls *functional* and *intrinsic* geometry. The intrinsic geometry of a space is its real geometry (the "colour-determined" geometry in the case of S). The functional geometry is the geometrical structure that empirical theorizing (whether at the level of common sense or advanced science) will lead us to ascribe to the physical world. Since space itself is invisible, we have no direct empirical access to its real geometry, and so functional geometry is ascribed on the basis of the observed behaviour of material bodies together with the assumption that the laws of nature are uniform (always the default assumption in the absence of evidence to the contrary). In short, functional geometry is the *apparent* geometry of a world, the geometry the world in question would seem to have to beings such as ourselves – beings with no direct empirical access to the real (intrinsic) geometry of the underlying substantival space. In the Oxford–Cambridge case, the functional and intrinsic geometries are clearly different. Let us call the intrinsic geometry in this scenario "G_i" and the functional geometry "G_f". According to G_f, the points in R_1 (those underlying the various material processes in and around the city of Oxford) are close to, and surrounded by, points in Oxfordshire, whereas, according to G_i, the points underlying Oxford are close to, and surrounded by, points in Cambridgeshire; the same applies, *mutatis mutandis*, for Cambridge and Oxfordshire. G_f and G_i are thus distinct, since they assign different metrical and topological properties to points and regions associated with the same physical objects and processes. The question of quite how this situation should be interpreted is a difficult one, and we will be turning to it shortly.

It is worth noting that Foster's distinction between intrinsic and functional geometry is not the same as the usual distinction between pure and applied (or physical) geometry. Both functional and intrinsic geometries are applied; the difference between them lies in what they are applied to. The intrinsic geometry of a space is the one we would ascribe to it if we could look upon the space as it really is, from an all-revealing transcendental perspective (Foster sometimes uses the term *transcendentally natural intrinsic* (or TNI) *geometry*). Construing a space as a single colour-field gives us a concrete way of understanding what this means. The functional geometry of a space, on the other hand, is the geometry the space would possess if the laws controlling the behaviour of its occupants were nomologically uniform. Foster's contention is that the two could differ. Provided we grant his assumption that a genuine substantival space possesses both an intrinsic nature and a spatial structure that is independent of the behaviour of its occupants, then he is surely right.

15.5 The nomological thesis

Foster now goes on to argue that it is logically impossible for any *physical* space to be nomologically deviant, since it is a conceptual truth that physical geometry coincides with functional geometry. He calls this claim the "nomological thesis", since in effect it posits a conceptual connection between the structure of physical space and the laws governing the behaviour of its occupants, these same laws being the basis upon which functional geometry is constructed. The nomological thesis is crucial to the anti-realist argument, and its defence occupies several chapters of *The Case for Idealism*.[2]

The first point to note is that there could never be any empirical evidence in favour of the hypothesis that a physical space is nomologically deviant. Empirical evidence points, however unreliably, towards functional geometry alone. By definition, the hypothesis that the geometry of physical space is its functional geometry will be simpler than, and to this extent preferable to, any alternative geometrical hypothesis (which does not preserve uniformity), provided we take nomological uniformity to be our criterion of simplicity, which of course we do. Consider, for example, what we would make of the hypothesis that Oxford is really in Cambridgeshire, when all the observational evidence suggests that Oxford is in Oxfordshire. Although someone could construct a theory involving non-uniform laws that explains why it seems that Oxford is in Oxfordshire when really it is in Cambridgeshire, the needless complexity of this theory means we would never be inclined to take it seriously.

But Foster's claim is stronger than this. He holds that not only could we never have any scientific grounds for accepting the hypothesis that our physical space is nomologically deviant, but the hypothesis itself is incoherent. Consider again the Oxford and Cambridge case, in which the intrinsic geometry of the external colour-field (S) does not correspond with the functional geometry. What can we say here about the physical world and its geometrical structure? Three positions are open to us:

1. The physical world does not exist.
2. The physical world exists, and its geometry is the intrinsic geometry of the underlying space S.
3. The physical world exists, and its geometry is its functional geometry.

Foster suggests that 1 is not a genuine option: "It is just obvious that the local organizational quirk with respect to R_1 and R_2, and the resulting effects on empirical appearance, do not suffice to eliminate the physical world altogether" (2000: 269). Since this is clearly right, the issue is between 2 and 3.

In regard to 2, Foster's claim is that it simply makes no sense to think that the physical world could have a structure that is both quite different from its apparent structure and wholly concealed from us as a consequence of natural law. However, as Foster acknowledges, it is far from obvious that this is so, and he defends his claim with a variety of arguments.

Where the deviation between functional and intrinsic geometry is as slight as in the Oxford and Cambridge case, it might be objected:

> Surely in this case it does make sense to hold that the structure of physical space might be that of the intrinsic geometry, even if we could never discover what this geometry is. After all, our conception of the physical leaves room for some discrepancy between how the physical world really is, and how our best scientific theory represents it as being.

The force of this point can be countered by considering cases in which the functional and intrinsic geometries deviate more wildly. Foster supplies two such scenarios (in enough detail to show that they are workable). In the first, the underlying geometry is two-dimensional, whereas the functional geometry of the resulting physical world (i.e. the world to which we have access) is three-dimensional. In the second case, a single physical space is sustained by events in two quite separate spatial systems, so there is spatial unity at the functional level, but disunity at the intrinsic level. Since in both cases we can suppose that the empirically manifest world is exactly like the physical world we are actually acquainted with, it would be very implausible to claim that the physical world in either scenario *really* possesses the structure of the underlying space (a structure to whose existence we are entirely oblivious). A more plausible position would be that physical space exists in such cases, as a unitary three-dimensional continuum, but given the discrepancy between the geometries of this space and the underlying reality, we cannot regard physical space as being metaphysically fundamental.

What Foster says might be correct for cases in which there are radical discrepancies between intrinsic and functional geometries, and we must admit that such cases are logically possible, but what of those cases in which intrinsic and functional geometries exactly coincide, or nearly coincide (as in the Oxford and Cambridge case)? Can't we say that in these sorts of case the physical geometry is the intrinsic geometry, or at least, that it *makes sense* to think this? Foster rejects this. What grounds could there be for adopting two different methodological principles for selecting a geometry for physical space? If we admit that the functional criterion is appropriate in cases of radical divergence, then it must be appropriate in *all* cases, since even when functional and intrinsic geometries coincide, our epistemological predicament is exactly the same as when the geometries are radically different. We *never* have any evidence that functional and intrinsic geometries are similar, and so never have any grounds for employing any selection criterion other than the functional one.

But, an objector might continue:

Isn't the hypothesis that the functional geometry coincides with the intrinsic geometry the simplest hypothesis? Doesn't this hypothesis yield a simpler set of physical laws? If we take nomological simplicity to be an explanatory virtue, doesn't this hypothesis provide the best explanation for the world's being as it seems to be? If so, then the hypothesis that physical and intrinsic geometries coincide at least makes sense, even if it is not true.

The inscrutability thesis comes into play here. Foster does not deny that simplicity considerations have a role to play in the selection of an explanatory hypothesis concerning the structure and laws of the physical world. But since physical theory is topic neutral, simplicity considerations can only enter into the empirical justification of theories that make no claims about the *intrinsic* nature of physical reality. The hypothesis that intrinsic and functional geometries coincide clearly isn't topic neutral, and so is not a hypothesis to which the usual canons of empirical reasoning can apply.

This point seems decisive, and doesn't rest simply on an appeal to intuitions about the sorts of reality it is appropriate to call "physical". The intrinsic geometry of the external item underlying the physical world is grounded in the intrinsic nature of this item; for example, the intrinsic geometry of the colour-field S is fixed by the trajectories and distance relations between the different regions of this field. The same applies in other cases, no matter what the intrinsic nature of the external item is (we can't of course *imagine* in any way the nature of this item if we suppose it to be non-phenomenal). Now, *if* we accept the inscrutability thesis, it follows that considerations concerning the intrinsic nature of the external item cannot play any role at all in the considerations that lead to the adoption of a physical geometry. The only considerations that do have a role to play are those concerning the behaviour of those aspects of the external item to which we have empirical access. The empirical evidence may underdetermine physical spatial geometry – there may be a family of competing hypotheses each assigning to the physical world a different combination of geometries and laws, each of which is equally compatible with all possible empirical evidence – but *every* member of this family will be an acceptable hypothesis only because it helps explain the observed behaviour of physical objects. The intrinsic geometry of the underlying external reality does not, and cannot, enter into the selection principles we employ in choosing a physical geometry.

If we grant this much, the case for the nomological thesis seems compelling: it is a conceptual truth that the physical geometry of a space is its functional geometry.

15.6 Nomological contingency

One further general assumption is relevant to the main argument. Foster assumes that the physically relevant geometry of S is contingent. I shall refer to this as *the nomological contingency assumption*. The nomological contingency assumption amounts to the claim that the physically relevant laws governing S and its occupants could be different from how they actually are. Given its importance to his central arguments, it is somewhat surprising to find that Foster never provides any argument for the nomological contingency assumption; he simply takes it to be self-evidently true. However, one does not need to reflect for very long before finding oneself quite convinced that the assumption in question is eminently reasonable.

Suppose that S is a colour-field, and its occupants are tiny mobile spheres of (a different) colour moving about within it. It is easy to imagine these spheres conforming to any number of different laws of motion without the structure of S changing. The intrinsic geometry of S is no more affected by the behaviour of the atoms that move through it than is the geometry of a cinema screen by the play of light across its surface during the projection of a film. Although this sort of exercise lends the nomological contingency assumption considerable prima-facie plausibility, we are dealing here only with a model. Might it not be that the laws governing real-life equivalents of S and its occupants are necessary rather than contingent?

If the "bare powers" metaphysic applied – that is, if a purely causal–dispositional account of the constituents of the physical world were viable – then it would be very plausible to maintain that physical things possess their causal powers essentially, since such things are nothing but clusters of causal powers. But given the content thesis, S and (at least some of) its occupants must possess space-filling intrinsic natures of *some* kind, so this hypothesis can be ruled out. Consequently, the question we need to focus on is whether there could be necessary connections between intrinsic natures and their causal powers. Since the intrinsic natures of S and its occupants are entirely inscrutable, it is hard to see how the claim that a particular intrinsic nature is necessarily associated with a particular set of causal characteristics could ever be substantiated. A non-contingent connection between an intrinsic nature and its causal characteristics cannot be analytic: since *ex hypothesi* we have no conception of the intrinsic nature we are considering, no amount of conceptual analysis will do the job. But nor can the connection be a posteriori since, given the inscrutability thesis, no amount of empirical investigation will shed even a glimmer of light on the intrinsic nature of something external to our minds. The claim that water is necessarily H_2O may be an a posteriori truth, but this truth does not reveal any connection between an intrinsic nature and a set of causal properties, simply because we have no idea of the intrinsic nature of hydrogen and oxygen atoms (or their constituents).

The nomological contingency assumption is also supported by purely metaphysical considerations. We are working on the assumption that S is the metaphysically basic item. If we held that S itself could not exist if its occupants obeyed different laws, we would be saying that S depended for its existence on its occupants behaving in a specific way, a claim that is manifestly incompatible with the assumption that S is ontologically basic, for an ontologically basic item is one that is dependent on nothing else for its existence.

15.7 Realism rejected

We can now, finally, turn to the crucial part of Foster's case: the argument that yields the conclusion that physical space cannot be a part of what is metaphysically basic, or ultimately real. To simplify, we again assume that the external reality corresponding to the physical world is a giant colour-field, S. We also assume that the physical world exists, and that one of its metaphysically basic constituents is physical space, P. Facts about physical objects supervene upon facts about the distribution of atoms through S. We have seen already that there is a sense in which S can possess two geometries: its own intrinsic geometry, and the functional geometry that applies to its occupants. Since the physically relevant geometry is the functional geometry, the geometry of P is the functional geometry of S. The question now arises as to the relationship between these two spatial items, P and S. Foster's claim is that S and P are, as a matter of logical necessity, numerically distinct entities.

His argument rests on some widely accepted assumptions concerning the identity-conditions for spaces and their constituent points. The most general level at which the geometrical properties of a space can be characterized is topological. Spaces that are topologically indistinguishable may be different in other geometrical respects (e.g. as to whether or not a Euclidean metric applies). Foster makes the minimal assumption that the topology of a space is essential to it.[3] So for a given space S with topology T, in any possible world in which S exists, it possesses T. We are assuming here, of course, that S is a concrete item, a particular, not merely a mathematical structure (which will "exist" in all logically possible worlds). Foster further assumes that spatial points owe their individuality to the spaces to which they belong: whatever geometrical features are essential to a space are also essential to the points that constitute the space.

Now, from this it follows that there are logically possible worlds that outwardly resemble ours in all respects, but where S and P are numerically distinct. Foster provides several examples of worlds where S and P are topologically different – the Oxford–Cambridge scenario is one such – and in these cases not only must S and P be numerically distinct spaces, but none of the constituent points in S exist in P, and vice-versa. But, as Foster is fully prepared to concede, we have no reason whatsoever to suppose that our world is like this. Suppose that in our world there is no discrepancy or deviance; the laws are uniform and S and P have the same topology. In this instance, is there any obstacle to holding that S and P are one and the same space? Foster maintains that there is: S and P may share their actual properties, but they still differ in their *modal* properties.

If P is a genuine space, as we are currently assuming, it must possess its topology essentially. As it happens, S shares this topology. But does it do so essentially? No. Given the nomological contingency assumption, there are possible worlds where S has exactly the same intrinsic geometry as in the actual world, but where the laws governing its occupants are such that it has a very different functional (physically relevant) geometry. In some of these worlds the functional topology of S is different from that of P. In such worlds P does not exist (in its place is a quite distinct physical space). Hence there are possible worlds where S exists but P does not. If we assume that identities are necessary (i.e., that if $x = y$, then $x = y$ in every possible world where either exists), then S and P cannot be numerically identical.

The same conclusion can be reached by a slightly different route. Leibniz's principle of the identity of indiscernibles may be highly controversial, but the converse – the *indiscernibility of identicals* – is not. According to this principle, if x and y are numerically identical then x and y must possess the same genuine properties. Needless to say, if "x" and "y" are merely different names for the same object, it is hard to see how this principle could be false. Now, S has a physically relevant geometry with a certain topology; call this topology T. P has precisely the same topology. But there is a difference: S possesses T only contingently, but P does not. Although in actual fact both S and P have the same physically relevant topology, it is possible for S to exist without this topology (if the occupants of S obeyed different laws of motion, S's physically relevant topology would be different), but P cannot do likewise (on the assumption that P is a genuine space in its own right, it possesses its topology – T – essentially). Let us give the modal property "could exist without T" a convenient label, X. The argument that P and S are numerically distinct is simple:

1. S possesses property X.
2. P does not possess property X.
3. So, by the indiscernibility of identicals, S cannot be numerically identical with P.

Provided that X is a genuine property, which it certainly seems to be, the argument is sound.[4]

Foster's argument bears a striking resemblance to a somewhat more familiar argument concerning the relationship between ordinary physical objects, such as statues and plastic dishpans, and the matter out of which they are composed. The idea that two physical objects could exactly coincide in space throughout their entire careers, but nonetheless be numerically distinct may seem odd (just as odd as Foster's claim that S and P can have the same geometry, but nonetheless be numerically distinct spaces). But there is a powerful argument that leads to precisely this conclusion.

To take the best-known example, consider *Lumpl* and *Goliath* (Gibbard: 1975). Goliath is a statue, and Lumpl is the lump of clay from which Goliath is made (we can treat both "Lumpl" and "Goliath" as proper names, since we *can* name lumps of clay if we so choose). Goliath and Lumpl come into existence at exactly the same time, and both cease to exist at exactly the same time. (Take a "lump" of clay to be a collection of bits of clay that are stuck together. There were first two lumps of clay. The first is modelled into an upper-body shape, the second into a lower-body and these two lumps are then stuck together at the waist, bringing into being both a new lump of clay, Lumpl, and a new statue, Goliath. At a later date, Lumpl-Goliath is smashed to bits.) Since Lumpl and Goliath are composed of exactly the same material constituents throughout their respective careers, and likewise share exactly the same physical properties (at least of the kind recognized by physics), it could be argued that there is only a single *physical object* here, not two. However, there is an obstacle to this identification.

Even though Lumpl and Goliath coincide (occupy the same volume of space at each moment) throughout their spatiotemporal careers, it is possible that they might not have done. Shortly after completing Goliath the artist might have become bored with his work, and squeezed Lumpl into a ball. Provided that no bits of clay drop off during the squeezing, Lumpl survives; but Goliath clearly does not. This doesn't happen, but it might have. In which case, the modal property "might have been squeezed into a ball without being destroyed" is possessed by Lumpl but not by Goliath. Hence, by the law of the indiscernibility of identicals, Lumpl and Goliath cannot be numerically identical. Anyone who accepts this conclusion (and many do) must also accept the validity of Foster's modal argument. If they wish to resist his conclusion, they must do so by rejecting his premises.

If we do accept Foster's conclusion, that S and P cannot be numerically identical, the only remaining task is to settle on the ontological status of physical space. Since physical space exists only in virtue of the particular contingent laws that apply to the underlying external reality, and the latter could exist without sustaining any physical space at all, it is clear that physical space has an ontologically dependent status, contrary to our initial assumption. Furthermore, if we take the view that physical space is *the* metaphysically basic item in the physical world, and that physical objects depend for their existence on physical space, then given our conclusion that physical space is necessarily numerically distinct from the space of the underlying reality (should this reality be spatial at all), the underlying reality is, necessarily, *wholly* non-physical. Even if we discard this assumption, and take the view that material things are not ontologically dependent on physical space, we are left with the conclusion that a substantival space cannot be an ontologically basic ingredient of reality. Spatial realism in its most interesting form is thus undermined.

One last point. Although he argues that we should be anti-realists with respect to physical space, Foster is not saying that statements such as "Paris is closer to London than to New York" cannot be true. His position is analogous to that of the B-theorist, who recognizes tensed truth while maintaining that reality itself is not tensed: the truthmaker of a tensed statement such as "World War One is past" is the tenseless fact that this token statement occurred at a time later than the ending of the First World War. In a similar vein Foster would argue that statements about the spatial properties of physical objects have as truthmakers states of affairs that are entirely non-physical. Foster is perfectly happy to concede that there are physical facts, but only in the trivial sense that there is a fact corresponding to every true statement. Since statements about the physical world are made true by non-physical facts, physical facts are not basic; rather they supervene upon, or are *logically sustained* by, non-physical facts. What holds for facts also holds for objects. Not all objects are basic: chairs exist, but – in our world at least – chairs are always composed of various particles, and any particular chair supervenes on, or is the *logical creation* of, facts concerning these more basic entities. For Foster a fact is basic if it is not logically sustained by any other fact or facts, and an entity is basic if it is not the logical creation of any fact or facts. What he calls "ultimate reality" is the totality of basic entities and facts. Formulated in these terms, Foster's anti-realism amounts to the claim that ultimate reality is wholly non-physical. So, on the assumption that there are physical entities and facts, these must be logically dependent (sustained or created) by something else, some reality that is non-physical. Foster suggests that this ultimate reality is mental, hence his idealism. But whatever its nature, it is this *non-physical* reality that is responsible for our sensory experience being as it is.

15.8 Geometrical pluralism

Any argument as complex as Foster's can be challenged in different ways. For example, the argument presupposes the necessity of identity and "geometrical essentialism" – the view that the geometry of a space is essential to the identity of all its points – and although both doctrines are widely accepted, they could certainly be questioned.[5] Indeed, if there were no alternative, many might prefer to abandon one of these doctrines rather than accept Foster's anti-realist conclusion. But a better response is available, and emerges as soon as we question another basic assumption about the nature of space, an assumption that is by no means unique to Foster, but that his own argument shows to be questionable.

Not only is Foster committed to geometrical essentialism, but he is committed to what we can call "geometrical monism". His anti-realist argument is based on the unstated assumption that a genuine space can only have a single geometrical structure. It is taken as obvious that when P and S have different geometries they must be numerically distinct; even when their geometries coincide the possibility that they could diverge ensures their distinctness. But if a genuine space is *not* restricted to a single geometry – if *geometrical pluralism* obtains – we have the option of regarding the divergent geometries of S and P as belonging to one and the same space. If a case for pluralism can be sustained, realism can be restored without rejecting the other essentialist doctrines that Foster's argument requires.

We saw in §9.6 that there are two viable accounts of distance relations within a space. Intrinsic distance consists of a direct spatial relation between a pair of points; Gaussian distance is measured by the shortest continuous path through space. Although the two accounts yield the same results for classical Euclidean spaces, they can easily diverge; for example, creating a hole in a space leaves the intrinsic distances between points on either side unaltered, but increases their Gaussian distances. When presenting these accounts I tacitly assumed that the points in any given space could be related by only one sort of distance relation, but there is no obvious reason why this must be so. Consider the case of a *corrugated* space: an infinite two-dimensional plane that undulates in a regular way, its crests and troughs taking the form of alternating half-cylinders (see Figure 15.5). Call this space C and compare it with the flat Euclidean plane E. When a flat surface is bent without any tearing or stretching, the lengths of paths confined to the surface do not change, so from the Gaussian perspective C and E are geometrically equivalent (both have zero intrinsic curvature). But clearly, if we suppose that the points in C are connected by intrinsic distance relations – indicated by the arrow-headed lines in Figure 15.5 – the spaces are not geometrically equivalent. Corrugating a space changes the distances between points; for example, the intrinsic distances between points on successive mounds (and valleys) are shortened as they are moved together.

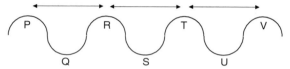

Figure 15.5 Gaussian versus intrinsic distances in a corrugated space. In a flat space, intrinsic and Gaussian distances diverge when points are deleted from the manifold. In a curved space there are other ways for divergence to occur.

On the assumption that the intrinsic distance relations are genuine properties of the points in C it is tempting to suppose that C is *really* a bent space, that its true geometry is the one captured by the intrinsic measure of distance. But the Gaussian conception of distance remains perfectly valid; the facts about the continuous path-lengths between points within C are not affected by the existence or non-existence of intrinsic distance relations. The significance of this fact is easily obscured by what happens when we try to picture C: we will embed it within three-dimensional space and imagine ourselves looking down on a surface that visibly undulates. If we can *see* the bends in C, isn't it obvious that this space is *really* bent? However, the "obviousness" of the ripples in C evaporates when we abolish the embedding space and the Gaussian paths through three-dimensional space connecting its peaks and troughs. If we were Flatlanders living in C, no ordinary measuring devices would reveal the ripples in our space and their existence would not be obvious at all.

This case could be construed in a Fosterian way. The true geometry of C is its intrinsic geometry (as determined by intrinsic distances), for a being equipped with transcendent God-like powers of observation would discern the intrinsic distance relations between the points of C, and hence the ripples. The Gaussian geometry of C is clearly its *physical* geometry, since the Gaussian properties of a space are those that can be discovered by geometers working within the space using rods, strings and so forth. Since C is intrinsically corrugated but physically flat, physical space is numerically distinct from C and, given that C really exists, physical space itself cannot be ultimately real. But this interpretation of the case is far from compelling. For, as is clear, the Gaussian geometry is a real feature of C itself. Gaussian distances comprise the shortest continuous path-lengths through a space, and so are determined by the presence or absence of points within C: if a hole in C is created, path-lengths connecting points at opposite sides of this region are increased. Rather than populating our ontology with additional (and not fully real) spaces, why not simply accept that C possesses two equally valid geometrical structures simultaneously, each structure depending on real features of C itself – Gaussian and intrinsic distances.

The discrepancy between physical and intrinsic geometry in C is of a comparatively mild form. The sort of deviancy exhibited in Foster's Oxford and Cambridge example is not only more extreme geometrically, but a product of a form of nomological deviance that is absent from C. What it reveals is different, too: not the possibility of Gaussian and intrinsic distances coexisting while diverging, but rather the existence of two distinct forms of Gaussian or path-based distance, which likewise can coexist without coinciding.

We can distinguish two varieties of "shortest path-length", which I will call "intrinsic geodesics" and "material geodesics". The material geodesics of a space are the paths of shortest distance as measured by some specified physical process. A natural (and standard) way of giving this notion substance is to say that inertial paths (those followed by bodies that are not being acted on by a force) are geodesics. Other definitions are perfectly conceivable, but for present purposes the inertial account will serve. Intrinsic geodesics, on the other hand, are defined relative to the structure and composition of space itself, and so independently of the behaviour of the material (or other) contents of a space. On the assumption (which we are currently making) that space is substantival, and so a real entity in its own right, the idea that space *has* intrinsic features that possess a specific geometrical structure seems eminently reasonable (all the more so given the strength of Foster's case for the content thesis). There is no denying that an entity shaped like the surface of a sphere is geometrically distinct from one shaped like a flat piece of paper; likewise, a flat Euclidean plane with a hole and an otherwise similar plane without a hole are geometrically different. Quite *what* geometrical features a given space possesses will depend on the nature of the spatial fabric concerned, which may come in very different forms. It is sufficient for present purposes that we accept the hypothesis that a substantival space can possess geometrical features that are intrinsic to it, and this is something most substantivalists will be prepared to accept.

We can now draw a further distinction, between *naïve* and *circumspect* substantivalism. The naïve substantivalist assumes without question that material and intrinsic geodesics are one and the same; the idea that they could be different is never considered or taken seriously. The circumspect substantivalist is willing to countenance at least the possibility that the two might not always coincide. On finding a region of space where bodies in inertial motion do not move in (Euclidean) straight lines, and where this deviation cannot be explained in terms of known forces pulling or pushing the bodies, the naïve substantivalist will immediately conclude that a non-Euclidean hole has been discovered (a "hole" in Nerlich's sense, a region of strong local curvature in an otherwise flat space, rather than an absence of points). While circumspect substantivalists will accept that this is one possible explanation for the peculiar motions, they will be prepared to consider an alternative: that the intrinsic geometry of space in the relevant region is flat, but the nomological organization of the world is such that it seems holed; bodies behave *as if* they are moving along geodesics in a holed space, but they are not. In short, in this region the material geodesics and the intrinsic geodesics fail to coincide; perhaps when the universe was created occasional lapses in nomological uniformity occurred. Unless this possibility can be ruled out, nor can the alternative explanation, or so circumspect subtantivalists will argue.

The naïve substantivalist could try arguing thus:

> You should reconsider your position. If you are prepared to recognize this sort of explanation for the deviant behaviour, you are committed to recognizing another possibility: there could be variations in the intrinsic geometry of space that are *concealed* by nomologically deviant behaviour. There could be regions of space that are intrinsically curved, but where the material geodesics are Euclidean, thanks to the laws governing space–body interactions that apply in these regions. Isn't it absurd to posit physical structures that are by their very nature empirically undetectable?

But it is by no means clear that this objection should sway the circumspect substantivalist, who can offer this response:

> It would be nice to rule out even the possibility of material and intrinsic geodesics coming apart, but this is simply not an option. There is no reason to suppose that it is metaphysically necessary for bodies in force-free inertial motion to follow intrinsic geodesics. If they do follow them, this is only because of the particular nomological organization that obtains in this world; other nomological regimes are perfectly conceivable. Once the possibility of geodesic dislocations is recognized, the possibility of *concealed* dislocations must also be recognized. This said, there is no reason whatsoever to suppose that such dislocations are commonplace, or even exist at all: for like you I believe we have good reason to suppose that our world is governed by basic physical laws that are simple and uniform. *Unlike* you, if we did discover a region where (for no apparent reason) the material geodesics are curved, I am prepared to countenance the possibility that this deviation from the norm is due to unusual laws rather than variations in the shape of space itself.

In the absence of any reason for thinking that dislocations between intrinsic and material geodesics are absolutely impossible, I think we should conclude that substantivalists should be circumspect rather than naïve.

In our earlier case of geometrical conflict (intrinsic distance versus Gaussian distance) I suggested that the substantivalist should embrace the option of geometrical pluralism, and hold that both geometries are simultaneously possessed by the space in question. The same applies when material and intrinsic geodesics fail to coincide. Here, too, pluralism is a compelling option: since material geodesics are paths through *physical space* that are privileged by *physical law*, there is every reason for the substantivalist to accept them as genuine geometrical features. On the assumption that intrinsic geodesics also exist, pluralism is unavoidable.

It might be claimed that intrinsic geodesics are more basic from a metaphysical perspective than their material counterparts. This claim has some justification: there are certainly counterfactual situations in which a space retains its intrinsic geodesics but has different material geodesics, thanks to alterations in the relevant natural laws. But this fact does nothing to impugn the status of material geodesics as geometrical features of a substantival physical space; after all, not all the features of a physical entity are essential to it. Indeed, if we focus on the physical world as a whole, rather than its space, it is the material geodesics that have most prominence: as the paths privileged by the laws connecting material objects with space, they provide us with the only clues we have as to the structure of our space. These amount to strong grounds for taking both geodesic structures as essential to the identity of a space, at least for worlds where they come apart.

Where does this leave us *vis-à-vis* Foster's case for anti-realism? Let us grant that in worlds where the discrepancy between the real and apparent spatial structure of the world is large, physical space is as Foster maintains: an indispensable theoretical construct that supervenes on a non-physical reality. (I have in mind Foster's more radical examples, where the physical world partly depends on each of two distinct external spaces, or a two-dimensional external space.) Since, as Foster himself concedes, we have no reason to believe that our world is anything like this, the more interesting possibilities are where the discrepancy between real and apparent spatial structure is comparatively small or non-existent. In these instances, by rejecting geometrical monism, and accepting that a substantival physical space can have a multiplicity of different geometrical structures at the same time, we can accept that the scenarios envisaged by Foster are possible without rejecting realism. In the Oxford and Cambridge example there are extreme but highly localized divergences between material and intrinsic geodesics, and the circumspect substantivalist will regard both geometries as real features of the same spatial entity: physical space. No less importantly, in the case of the actual world, where, as Foster willingly concedes, we have no reason to suppose that material and intrinsic geometries diverge, the mere possibility that they could gives us no reason to reject realism, for the realist can take the view that the material geodesics are geometrical features of a physical space that are possessed only contingently. But perhaps the realist does not need to concede even this much. If we choose to individuate physical spaces by reference to *both* intrinsic and material geodesics, it no longer makes sense to suppose it is possible for the two to diverge, if in actual fact they coincide.

16 Special relativity

16.1 Time, space and Einstein

Our investigations up to this point have not been in vain – they have certainly led us to a deeper understanding of the complexities of the issues we are concerned with – but they have also been inconclusive. Very different accounts of the large-scale structure of time seem both metaphysically viable and compatible with the character of our everyday experience. Space has proved an equally stubborn topic. Substantivalism may be ahead on points in certain respects, but there have been few clear-cut victories. But for those looking for definite answers all is not yet lost. We have yet to consider the impact of more recent scientific developments, and some believe that answers to the questions we have been considering can be found here. Of course, these answers will only concern the space and time of our world, rather than space and time in general, but being so restricted would scarcely reduce their interest.

However, although contemporary science has a great deal to teach us, anyone looking for *definitive* answers to our questions will be disappointed. Our current best theory of the very small – quantum theory – and our current best theory of the very large – Einstein's general theory of relativity (GTR) – have yet to be reconciled, and there is no consensus among the physicists working in these areas as to what the overall character of theory that synthesizes the two, "quantum gravity", will be. Some believe that the successor theory will retain the most distinctive features of both relativity and quantum theory; others believe that at least some of these features will disappear. The waters are further muddied by the fact that there are continuing disputes over the correct interpretation of relativity and quantum theory themselves. As a consequence, much of what we can learn about our space and time from current physics can only have a provisional status. This said, the results that can be gleaned from physics are vitally important, for, no matter how unreliable they may prove to be, there is at present no better guide to the real nature of our own time and space. They are also fascinating in their own right. The advances in physics made in the past century are such that one thing seems almost certain: time and space are stranger than Newton ever imagined.

That we can know this is in large part due to Einstein's special theory of relativity (STR), which we will be looking at in this chapter and the next. Among the counterintuitive consequences of STR, according to its standard interpretation at least, are the following:

1. Common sense says that time flows at the same rate for everyone and everything, irrespective of whether they are in motion relative to one another. Consider two twins, Alice and Bob; if Alice embarks on a long journey and returns, she will still be precisely the same age as Bob. According to STR this is wrong: Alice will be younger than stay-at-home Bob – time for her will have slowed with respect to Bob.

2. Common sense says that temporal and spatial intervals are independent of one another. If two events are separated by a spatial distance d and a temporal distance t, then this is the just a fact about how things are. According to STR this is wrong: these events will be separated by different spatial and temporal intervals in different frames of reference. Time and space thus lose their independence from each other, and we are left with a notion of *Einsteinian* (or Minkowski) *spacetime*, a unitary entity that decomposes into different spatial and temporal intervals for different reference frames.

3. Common sense says that there is a fact of the matter about which events happen at the same time as other events, irrespective of their spatial separation, and hence that simultaneity is an objective feature of the universe. This belief also falls victim to STR: as a consequence of 2, whether or not two events are simultaneous depends on the reference frame from which they are considered; observers in different reference frames will find different events simultaneous, and there is no sense in the idea that one observer is right and the other wrong.

There is no denying the strangeness of these ideas, so it is important to understand the grounds for accepting the theory that yields them, and I will provide some of the necessary background. But I will also be considering the metaphysical implications of the theory. The debate between substantivalists and relationists takes a new turn within the context of STR, but since I will be focusing on this debate when discussing GTR (in Chapters 18 and 19) I will postpone dealing with it until then. More intriguing is the impact of STR on the debate about the nature of *time*, which is commonly thought to be dramatic: influential philosophers have argued that STR's relativization of simultaneity (which is preserved in GTR) establishes beyond any reasonable doubt the static block view of time. If they are right, STR provides the solution to one of our two fundamental problems; I will be examining this issue in Chapter 17. But before entering these murky waters, the basics of STR itself must be laid bare.

16.2 Lightspeed

Einstein's special theory of relativity may have many bizarre consequences, but they all flow from just two simple principles:

- The *Relativity Postulate*: the laws of nature are the same in all inertial frames of reference, so there is no experiment that can reveal whether one is at rest or in uniform motion.
- The *Light Postulate*: the speed of light (in a vacuum) is a constant, c, which is independent of its source and the same in all inertial frames of reference.

Quite why these seemingly innocuous principles have such momentous conse-quences may not be immediately obvious, but the reasoning involved is not particularly complex, as we shall see. The difficult step, which Einstein alone took, was to take seriously the possibility that these principles *might* be true, for at the time (STR first appeared in print in 1905), this was far from obvious.

The Relativity Postulate is a strong formulation of our old friend the principle of Galilean equivalence. As Leibniz enjoyed pointing out, although Newtonian physics postulates an absolute space, there is no empirical method of detecting absolute velocity, and this fact had continued to trouble physicists over the intervening period. But as the nineteenth century drew to a close it seemed that this problem might finally be solved: a way of measuring absolute velocity had finally emerged. The key lay in the behaviour of light.

In the 1860s James Clerk Maxwell published a work on electrical and magnetic phenomena that synthesized the sizeable quantity of experimental work done over the previous centuries (and especially that done by Faraday earlier in the same century). His theory is embodied in four equations, which accounted for all the (then) observable behaviour of electrical and magnetic phenomena, and showed how electrical and magnetic forces are intimately interrelated. In particular, he demonstrated (theoretically) how electrical and magnetic fields could combine to form self-propagating *electromagnetic waves*. He could predict the velocity of these waves from his equations, which turned out to be the speed of light (which had already been discovered experimentally). It seemed improbable that this was just a coincidence, so he proposed that light itself was an electromagnetic wave, which duly turned out to be the case. Since waves can vary in frequency and wavelength, Maxwell also predicted that there would be other kinds of electromagnetic wave, and was again proved right: at the low frequency end of the spectrum are radio waves, whereas at the "higher" end X-rays and gamma radiation are to be found.

Physicists had been studying waves prior to the 1860s, and since all known waves were perturbations on or in something – water, air, the surface of drums, and so on – everyone assumed that light waves also consisted of perturbations in a medium. But there was a problem. Light can pass through a jar from which all trace of gas has been removed, so the required medium was clearly of a very special sort: it completely permeates seemingly empty space – even the space between the stars. Moreover, since light also passes through the solid glass of the jar itself, its medium extends through solid objects, as well as seemingly empty space. This invisible but all-pervasive medium was given a name: the *aether*.

Having posited this remarkable medium, questions soon arose about its properties: "aether-physics" was born. Since planets and other bodies are not noticeably retarded by anything when they move in a vacuum, the aether doesn't create any frictional drag; since light seems to travel in straight lines, even in the close proximity of large moving bodies, these bodies cannot produce any distortion in the aether, which must therefore be very rigid. One question was of particular significance: *what is the aether's state of motion?* After finding reasons for ruling out various possibilities it was generally agreed that the aether must be in a state of absolute rest (some took a further step and argued that the aether and Newtonian absolute space were one and the same entity). This consensus was to prove fruitful, for, as was soon realized, an experimental test for distinguishing absolute motion from absolute rest was now possible.

At this point some basic properties of waves (as understood in Newtonian physics) become relevant. The velocity of a projectile depends on the velocity of the object that expels it. Suppose that shells fired from a jet fighter's wing cannon move at 2000 km h^{-1} when the plane is stationary. If the same shells were to be fired when the plane is travelling at 1000 km h^{-1}, their velocity would be 3000 km h^{-1}. Waves do not behave like this; their velocities vary depending on the medium (e.g. light is slowed by glass), but they are not affected by how fast their sources are moving. The sound waves created by a plane that flies overhead move at the same speed of 1200 km h^{-1} in all directions; they do not move faster in the direction the plane is flying in. So if the plane is flying at 1000 km h^{-1}, and the pilot has an onboard device that detects and displays the sound waves radiating from the plane, he will see the waves directly in front of him moving forward at only 200 km h^{-1}, whereas in the opposite direction the waves are speeding away at 2200 km h^{-1}. Someone with a similar device on the Earth, directly below the pilot, will see the waves travelling at the same speed in each direction.

In a precisely similar fashion, Maxwell's equations predict that light rays will travel from their point of origin in all directions at the same speed: $c = 300\,000$ km s^{-1}. If we set about measuring the speed of the light waves emitted by a particular source, we will only measure them as travelling at that speed if we are at rest relative to the aether when we conduct the tests; if we are moving through the aether, we will find that the waves moving in the same direction as ourselves have a slower speed than those moving in the opposite direction. In 1881 a young American physicist, Albert Michelson, set about testing this prediction using very precise equipment that he had invented himself. Assuming that the Earth moves through the aether as it orbits the Sun, it must have different states of motion relative to the aether at different times of the year. Michelson's equipment measured the time it took for light to travel along two paths of equal distance in a laboratory; one of these paths was aligned in the direction of the Earth's motion around the Sun, the other at right angles to this direction. If, when the experiment is conducted, the Earth is moving relative to the aether, the time taken for the two trips would be different, or so theory predicted (just as the pilot in the plane gets a different speed reading for sound waves moving in the same direction as him than for waves moving upwards, downwards or rearwards). Michelson repeated the experiment at different times of the year but found no evidence whatsoever for any variations in light speed. Light, it seemed, travels at exactly the same speed in all directions, irrespective of the direction of the Earth's motion through the aether. This was an astonishing result, and a worrying one since it threatened accepted theory, so in 1887 Michelson, now working with Edward Morley, repeated the experiments to a much higher degree of precision. Again, no variations were detected, even though the equipment was sensitive enough to detect variations a hundred times smaller than those predicted to occur. All subsequent tests have confirmed the Michelson-Morley result.

The finding that light always travels at the same speed irrespective of who or what is observing it, and how fast they are moving relative to its source, is as remarkable now as it was unexpected then, for it conflicts with some of our most basic intuitions about motion. If a car is moving towards you on collision course, you can reduce the speed at which it will hit you by running away from it, or increase the speed by running towards it. The Michelson-Morley experiments show that light doesn't conform to this principle.

16.3 Compensation or revolution?

This result could not be explained by classical physics, but suggestions as to how to accommodate it were quickly forthcoming, the most notable of which was the *compensatory theory* proposed by Fitzgerald and developed in detail by Lorentz. The general idea is that when you are moving in the same direction as a ray of light, the light actually does overtake you at a slower speed than it would if you were motionless or moving in the opposite direction, but the decrease is not detected because of changes caused by your motion through the aether: your measuring instruments contract in length and your clocks slow down. Lorentz provided equations that show just how much contraction and time dilation (slowing) is needed to conceal the changes in the speed of light in different circumstances.

The solution that Einstein proposed also involves length contractions and time dilation – indeed, he made use of Lorentz's own equations – but in one crucial respect it was far simpler. The compensatory theory says that a given ray of light has different measurable speeds for different observers (in motion relative to it), but nature conspires to conceal these differences. In a bold stroke Einstein did away with the need for a conspiracy. He reasoned that, if nature makes it impossible for us to discover our state of motion with respect to the aether, perhaps we are wrong in supposing that the aether exists. Why not take nature to be as it presents itself? In rejecting the aether he also rejected the notions of absolute motion and rest, and took all inertial frames to be equivalent. He then accepted the consequence: a ray of light *really does* travel at the same speed relative to all observers, irrespective of their state of motion relative to it.

To get some impression of what this can mean in practice, suppose that Alice and Bob are in outer space. Alice points her torch in Bob's direction and switches it on. She sees the light ray moving at $300\,000\,\mathrm{km\,s^{-1}}$ towards Bob. Now Bob, as it happens, is moving away from Alice at the very high speed of $150\,000\,\mathrm{km\,s^{-1}}$, but despite this, as he measures the speed at which the light passes him using a measuring rod and a clock, he finds that it is moving at $300\,000\,\mathrm{km\,s^{-1}}$. How can this possibly be? Einstein's explanation runs thus. When Alice observes Bob she will see that his measuring rod has contracted in length and his clock is running slow (compared with her clock). Since speed is distance travelled divided by time taken, Bob's readings are now comprehensible: given his deviant equipment, it is to be expected that his readings will differ from Alice's. But this isn't quite the end of the story. Given the Relativity Postulate and the abandonment of absolute motion, Bob is perfectly entitled to regard *himself* as at rest and Alice to be moving at a speed of $150\,000\,\mathrm{km\,s^{-1}}$. From Bob's standpoint it is *his* readings that are correct and Alice's that are deviant, since when he observers Alice he will see that *her* measuring rod has contracted and *her* clock is running slow. Since (according to Einstein) both perspectives are equally valid, there is no objective fact of the matter as to who is right. And of course, there are many other perspectives on the same facts – those of potential observers in different inertial reference frames, all of whom are entitled to regard themselves as at rest, and all of whom will find Alice and Bob to have different states of motion relative to themselves.

The distinctive innovation of STR is the replacement of one absolute for another. Newton's absolute space and time are both rejected, and a new absolute or invariant quantity is introduced; the velocity of light. Since velocity involves both distance and time, the only way for a given light ray to have the same velocity in frames of reference that are in motion relative to one another is for distances and times to vary in systematic ways in the relevant frames, and the Lorentz equations allow us to calculate the precise differences. Hence the recognition of the new invariant *required* the rejection of the old: according to Newton the distances and times between objects are both absolute quantities, and so invariant across frames of reference.

Time dilation and length contraction initially strike us as utterly bizarre phenomena because they are not observed to occur in ordinary life. When travelling on a moving train, we can make the phone call we agreed to make at 2 p.m. without using the Lorentz equations to work out how much "train-time" differs from "land-time". A fast moving train does not look contracted or "scrunched up" to passers-by, but this is because the relativistic effects only become significant at enormously high speeds. The time dilation and length contraction of a train moving at $150\,km\,h^{-1}$ will be minute – about one part per hundred trillion. Even a jet travelling at twice the speed of sound will only contract by less than the width of an atom. Compared with the speed of light these speeds are snail-like. If a jet could be photographed moving at 85 per cent of light speed, it would have half its normal length, and time on board would pass at half speed compared with time on the ground. Beings who normally travelled at near-light speeds would be all too familiar with the relativistic effects we find bizarre.

The fact that we live in slow motion (comparatively speaking) does not mean that it is impossible for us to find confirmation of the predicted relativistic effects. Cosmic radiation striking our atmosphere produces showers of muons – a species of sub-atomic particle – which speed towards the Earth's surface. Even though the muons are travelling very quickly (at about 95 per cent of c), they are so short-lived (typically a millionth of a second) that they should perish long before completing their journey to the surface, but they don't. In virtue of their great speed, the length of their journey is shorter and their internal clock, as it were, runs slower, so the trip only occupies a small portion of their lifespan. On a more mundane level, very accurate atomic clocks can measure the time dilation produced by transatlantic flights. The experiment has been carried out repeatedly, and the results match the predictions of STR.[1]

When Einstein worked out some of the implications of STR for other physical quantities he discovered that as the velocity of an object increases so does its inertial mass, with the consequence that as the velocity of an object approaches that of light the amount of energy needed to move it tends to infinity, which means that no moving material body can ever be accelerated up to the speed of light.

Surprisingly, the implications of this for long-distance space travel are not as disastrous as one might think: it does not mean that stars hundreds of light-years away cannot be reached within the lifetime of a normal (unfrozen) human being. For a ship moving at near-light speed, spatial distances would shrink and time would slow to such an extent that a trip of a million light-years could take only a decade or so of ship-time.

You might think: "If it is possible for a spaceship to traverse a distance it would take light a million years to cross in only ten years, doesn't this mean the ship must be travelling faster than light, which STR tells us is impossible?" But don't forget that there are two perspectives on the light's voyage. From our point of view on Earth, the light crosses vast tracts of space in a million years, but if *per impossibile* the light ray could carry a clock and a ruler, it would be found that from *its* point of view the trip takes no time whatsoever, and involves crossing no distance whatsoever; for a photon the entire universe shrinks to zero size and time stops. Hence, from the perspectives of the ship and the light ray, the ship takes longer to make the trip than the light. From the perspective of the Earth's frame of reference, if the light takes a million years to complete the journey, the ship will take millions of years more, and the return journey will be just as long.

Unfortunately, while STR does open up the possibility of travelling to even the most distant parts of the universe within a single human lifetime (as measured by ship-time), significant obstacles of a practical sort remain. Perhaps the most intractable of these is the problem of attaining speeds approaching c. This requires acceleration, and acceleration produces inertial forces; the acceleration needed to boost a spaceship up to a significant proportion of c within a human lifespan would be far in excess of those that are survivable.

16.4 Simultaneity

So far I have said little about the aspect of STR that has the greatest (potential) significance for our understanding of time: the relativity of simultaneity. Because this topic is of especial importance I will deal with it more carefully, and rehearse the argument Einstein himself used to undermine the Newtonian (and neo-Newtonian) assumption that there is an absolute fact of the matter as to which events occur at the same time.[2]

Einstein begins by asking how we measure the time interval between events. For events that happen together at the same time at the same place, there is no problem. But what about events that are separated by some spatial distance, say at two points X and Y several miles apart? To make matters more concrete, let us suppose that X and Y are each struck by a bolt of lightning; how do we go about finding out if they are hit at the same time?

If we are somewhere between X and Y, we will have to rely on signals coming from X and Y, e.g. the observable flashes produced when the bolts hit the ground. Provided we know the speed of light (which we do), and have previously measured out the distance between ourselves, X and Y, we can easily calculate when the bolts struck. For example, if we are situated at the mid-point between X and Y, and the light reaches us simultaneously, then we know that the bolts struck at the same time. This procedure may seem rather involved and roundabout, but Einstein points out that this method (or a variant of it) is unavoidable. Since we do not have direct *instantaneous* access to what is going on in other parts of space, we have no option but to use signals that take some finite time to cross the intervening space, but as soon as we do, if we want to know *exactly* when some distant event occurred, calculations are unavoidable.

To see this it suffices to consider an alternative scenario: suppose that we have stationed observers at X and Y, and equipped them with walkie-talkie radios and clocks. We wait until after the bolts have struck and give them a call. Both observers say the same: the bolts struck at exactly 12 noon. Does this mean the bolts hit simultaneously? Only if the observers' clocks are *synchronized*. How can we go about checking this? One obvious solution is to call them both up on their radios, and say something along the lines of "Set your clocks at 12.00 *now*". But of course this is our original scenario all over again: given that radio waves travel at the same speed in all directions, they will only receive the message at the same time if we send it from midway between them.

Einstein's method does work, and can be used establish a common time system for a group of people who are stationary with respect to one another; everyone in this group will agree on which events are simultaneous and which are not. But as Einstein then went on to show, the situation is otherwise for people who are *moving* with respect to one another: it is impossible for such people to agree on a common time system. He used an example with trains to make the point.

We are to consider a train running along a straight track of indefinite length at very high speed, say $0.6c$, or 60 per cent of the speed of light. You are by the side of the track, and just as the mid-point of the train passes you lightning strikes both ends of the train, producing two flashes, which you see simultaneously. You reason as follows:

> I was standing opposite the mid-point of the train when I saw the flashes. The light from each end of the train had the same distance to travel. The speed of light is absolute – it travels at the same speed in all directions no matter what your state of motion – so the lightning must have struck both ends of the train at the same time.

However, an observer, O, sitting on the roof at the centre of the train witnesses a quite different sequence of events. Since O is moving towards the light travelling from the front of the train, he decreases the distance it has to travel, and so this light ray reaches him *before* the ray travelling from the rear of the train, which has a longer distance to cover. O reasons thus:

> I saw two flashes occur one after the other, the earlier at the front of the train, the later at the rear. These flashes could not have occurred simultaneously. Since the speed of a light ray is the same irrespective of how you are moving, and the light from both ends had the same distance to cover (since I was sitting at the mid-point of the train), if the flashes had occurred at the same time I would have seen them at the same time, but I didn't. Clearly, the lightning hit the front of the train before it hit the rear.

O's reasoning seems impeccable, and so (from his perspective) he is quite right to conclude that the two bolts did *not* strike simultaneously.

Now suppose you see a third lightning bolt strike the front of the train shortly after the first two strikes. Again, O sees something quite different: the light from the second bolt to strike the front of the train reaches him at the same time as the light from the bolt that struck the *rear* of the train. So he experiences *these* two bolts as simultaneous.

What we have here are three events, 1, 2 and 3, whose time ordering is different from different perspectives. Employing Einstein's method for determining simultaneity, you find that 1 and 2 are simultaneous, whereas 3 occurs later; applying precisely the same method, O finds that 1 occurs first, and is then followed by 2 and 3, which are simultaneous. (It is important to note that the presence of human observers is irrelevant; the same results would be obtained if there were no people involved, just automatic measuring devices.) Given the Relativity Postulate the two perspectives are equally valid; you might take yourself to be motionless, but so might O (who would thus take the ground to be moving beneath the stationary train), and there is no fact of the matter as to who is right. As a consequence, there is no fact of the matter as to whether the events in question occurred at the same time or not.

In appreciating the symmetrical character of the situation it is useful to consider it from the perspective of the person on the train; thus far we have been describing things from the ground-based point of view. So imagine *you* are on the train, and are conducting a simple experiment. You are sitting at the mid-point of your carriage, holding a switch in your hand. Connected to the switch (by equal lengths of wire) are two lamps at either end of the carriage. You flick the switch and see light arrive from each lamp at the same time. You naturally conclude that the lights came on simultaneously: after all, the light from each had the same distance to travel, and the speed of light is absolute. Now suppose that at the moment you press the switch you are directly opposite O who is sitting at the trackside. What do you think O would observe and conclude? It is easy to work it out. At the moment you flicked the switch O was racing away from you at $0.6c$ in a rearwards direction, and so O was racing *towards* the light emitted by the rear bulb and *away from* the light coming from the forward end of the carriage. So he would see the rear light first, then the forward light at a later time. Since O knows that the distance between you and the two bulbs is exactly the same, and given the absoluteness of c, O must conclude that the bulbs didn't come on at the same time.

The role of the Light Postulate is no less important; even if we assume – in line with the Relativity Postulate – that both perspectives are equally valid, the reasoning that leads to the conclusion that simultaneity is relative relies on the assumption that light rays travel at the same speed with respect to all observers, no matter how fast they are moving relative to its source. Return to the original case. O is justified in thinking that the front bolt hits before the rear bolt because (i) he is situated right in the middle of the train, so the light has equal distances to travel, (ii) the light coming from the front and rear of the train is travelling at the same speed. If O found that the light coming from the front of the train was travelling *faster* than the light coming from the rear, then he wouldn't be justified in thinking the front bolt hit first just because the light coming from it arrived first: this is to be expected if the light coming from the front is moving faster. But this isn't the case. If O measures the speed of the light when it reaches him, he will find it has the same speed in both directions.

We have been looking at just one example, but the lesson generalizes. In the context of STR, simultaneity is not absolute but frame-relative. Spatially separated events that are simultaneous from the perspective of one inertial frame are not simultaneous from the perspective of other inertial frames, and since the perspectives of all inertial frames are equally valid, there is no sense in the idea that the events in question are "really" simultaneous or not.[3]

16.5 Minkowski spacetime

The new conception of space and time was clarified by Minkowski in 1908, in a lecture whose opening words were to become famous:

> The views of space and time which I wish to lay before you have sprung from the soil of experimental physics, and therein lies their strength. The are radical. Henceforth, space by itself, and time by itself, are doomed to fade away into mere shadows, and only a kind of union of the two will preserve an independent reality.

Minkowski showed that although STR renders some quantities relative and frame-dependent, there are other quantities that are as absolute as any in Newtonian dynamics, it is just that these quantities are neither spatial nor temporal as we ordinarily understand the terms. The new clarity resulted from formulating STR as a *spacetime* theory. We have already encountered the general idea of spacetime, and two different types of spacetime structure: Newtonian and neo-Newtonian. The spacetime structure of STR is interestingly different from both.

Newtonian spacetime is rich in structure. Any two points, no matter when or where they are located, are separated by a definite spatial and temporal interval, and these intervals are invariant: they are the same when viewed from any inertial frame. Newtonian spacetime also has an affine structure: there is a distinction between curved and straight paths through the spacetime, which is also invariant. The difference between absolute rest and absolute motion is well defined; likewise the distinction between uniform motion and acceleraton. In neo-Newtonian spacetime some of this structure is absent. Affine structure is retained – the distinction between straight and curved paths remains intact – and hence there is a real difference between accelerated and non-accelerated motion. There is an invariant spatial distance between points that occur at the same time, and non-simultaneous points are separated by an invariant temporal interval. What is lost is spatial distance between points at different times. There is consequently no distinction between absolute rest and absolute motion, and the notion of absolute velocity has no sense. It is, of course, for this reason that Newtonian dynamics (in which absolute velocity plays no role) is best formulated in terms of neo-Newtonian spacetime. However, although in STR neither spatial nor temporal distances are invariant, the speed of light is. Since there are *no* invariant velocities in neo-Newtonian spacetime, it is clear that we need a different sort of spacetime to make sense of STR.

Minkowski's solution was to build a spacetime around the paths that light rays can take in a vacuum. The idea that luminal trajectories are built into the very fabric of spacetime is a natural response to the abolition of the aether, for if light is not a disturbance within any material medium, it seems there is only one candidate for the structure that constrains or determines the paths light rays can take: spacetime itself. Since all known space is flooded with electromagnetic radiation, the luminal structure of spacetime extends through the otherwise empty regions between stars and galaxies. For reasons that I will come to shortly, this invariant feature of Minkowski spacetime is often called the "light-cone" structure.

The speed of light is not the only invariant quantity in Minkowski spacetime; the so-called "interval" between points and events is the same in all inertial frames. The spacetime interval is neither spatial nor temporal, but a mixture of both. A purely spatial analogy is useful in grasping the general principle involved. By Pythagoras' theorem, the square of the hypotenuse (the longest side) of a right-angled triangle is equal to the sum of the squares of the other two sides: $z^2 = x^2 + y^2$. As Figure 16.1 illustrates, there are many ways of drawing a right-angled triangle with a hypoteneuse of a given length L.

In a similar fashion, a given spacetime interval is related in a systematic way to two other distances, but here the similarities end: first, only one of these distances is spatial, the other is temporal; secondly, to calculate the interval we do not add these distances, we subtract them. The relevant formula for interval I is:

$$I^2 = d^2 - c^2 t^2$$

where d is the spatial distance between events (given by $x^2 + y^2 + z^2$), t is the time separation and c (as usual) stands for the speed of light. Observers in different inertial frames who use this formula to calculate the interval between a given pair of events will all arrive at the same answer, even though they will disagree about the spatial and temporal distances between the two. Since all inertial frames are equivalent, there is no fact of the matter as to the correct decomposition of the interval into spatial and temporal components, and so the only objective (frame independent) fact about the events is the magnitude of the spacetime interval that separates them.

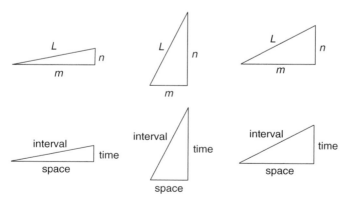

Figure 16.1 Spacetime "interval". Although observers in different frames of reference will disagree about the spatial and temporal distances between events, they will all agree on the spacetime interval between them – interval is thus an invariant quantity in Minkowski spacetime. Calculating interval is similar to calculating the length of the hypoteneuse of a right-angled triangle using Pythagoras' theorem: the square of the interval between two events is systematically related to the squares of the spatial and temporal distances between them.

I will say a little more about the standard way of construing the luminal structure of Minkowski spacetime, a device that provides a useful aid in arriving at a proper understanding of the feature of STR that has the greatest philosophical significance – the relativity of simultaneity. The key notion is that of a *light-cone*, a choice of terminology that is apt in one sense but misleading in another. Imagine a light source O in outer space, which is turned on for a brief time. Light spreads from the source in all directions, and so forms a spherical surface or wave-front. As the seconds tick away, the spherical surface expands. Depicting the expansion of this spherical front in a single two-dimensional diagram is impossible (it can be approximated with a sequence of pictures, each showing a slightly larger sphere). However, if we use only two dimensions to represent space, the expanding sphere becomes a succession of increasingly large circles, which together form the surface of a cone. We can represent all the light arriving at the source O by the surface of a second cone, which expands backwards into time. When this is combined with the future-directed cone (representing light emitted by the source) they together yield the characteristic egg-timer shape shown in Figure 16.2.

Note the following.

- The points on the surfaces of the cones are those that can be connected by light rays travelling in a vacuum, and are said to be "light-like" connected. The spacetime interval between such events is zero. Zero interval is not like spatial distance, since events separated by very large spatial distances can be connected by an interval of zero.
- The region of spacetime that lies outside the cones is known as "absolute elsewhere"; the spatiotemporal separation between points in this region is such that only a signal travelling faster than light could connect them; such points are said to be "space-like" separated. Events that are very close together in time but spatially at a considerable distance fall into this category. The interval between space-like separated points is positive.
- The points lying inside the cones constitute the regions known as "absolute past" and "absolute future"; their spatiotemporal separation is such that they can be connected by signals travelling slower than light, and they are said to be "time-like" connected. The interval between time-like separated points is negative.

Figure 16.2 The past and future light-cones of spacetime point O. A four-dimensional structure, shown here with one spatial dimension suppressed. Region F is O's "absolute future", region P is O's "absolute past". The points on the surfaces of the cones are light-like connected to O; the regions within the cones can be connected to O by processes travelling at sub-light speeds. For a signal originating at any point outside the cones to reach O it would have to travel faster than light – this region is known as "absolute elsewhere".

- For any space-like separated events there is an inertial frame in which they are simultaneous, and so differ only in their spatial coordinates.
- For any time-like separated events there is an inertial frame in which they occur at the same spatial coordinates, and so differ only in their time of occurrence.
- Assuming that no physical influence can travel faster than light, events that are time-like connected can be causally related, but no causal relation can hold between space-like events. It is for this reason that Minkowski spacetime is sometimes said to embody the causal structure of the universe.

As shown in Figure 16.3, it is possible for two planets X and Y to causally interact because they are linked by paths through spacetime that are always time-like; X and Z, on the other hand, cannot causally interact, because all the spacetime paths between them are space-like at least in part. For obvious reasons, causal influence is sometimes said to "thread the light-cones".

For our purposes, the most significant facts to draw from the above is that although STR relativizes simultaneity, it does so within very strict limits: the relativization applies only to space-like separated events. Events that are time-like separated are non-simultaneous in all frames of reference. Figure 16.4 shows the upshot of this for a single spacetime point O (which lies where the tips of the two cones meet).

The temporal ordering of all the events lying within the cones is the same in all frames of reference, although the temporal distances between them are not. Three "simultaneity hyperplanes" are shown slicing through the region of absolute elsewhere; each hyperplane is an extended three-dimensional region of four-dimensional spacetime, and consists of a collection of space-like separated events (or spacetime locations) that are simultaneous in different frames of reference centred on O. The middle (horizontal) hyperplane represents the events that are simultaneous in O's frame of reference; the other two hyperplanes represent events that are simultaneous in the frames of reference that are moving relative to O – e.g. that belong to two spaceships close to O that have different states of

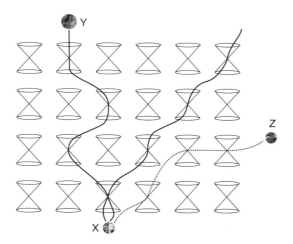

Figure 16.3 Threading light-cones. The solid lines represent time-like trajectories – the sort that objects travelling slower than light have. The dotted line shows a space-like trajectory – only something travelling faster than light can have one of these. Since no time-like path links X and Z, these two planets are causally isolated from each other.

motion. As can be seen, each of these alternative hyperplanes contains events that are *not* simultaneous in O's frame of reference. Indeed, they both contain events that, from O's perspective, lie in the past and the future (but *not* O's *absolute* past or future, i.e. the regions lying within the upper and lower cones). Only a few of the many possible alternative simultaneity hyperplanes that are centred on O are shown: *all* the events that are space-like related with O – all the events in O's "absolute elsewhere" – are simultaneous with O from *some* inertial frame. A similar state of affairs obtains at each spacetime point.

Figure 16.4 Simultaneity hyperplanes falling within the space-like connected region of "absolute elsewhere".

Looking at pictures like these it easy to fall into thinking that from the perspective of each spacetime point the temporal ordering of *half* the entire universe is frame-relative according to STR, but this is not necessarily so: it depends on the age, shape and size of the spacetime in question, and the location of the spacetime point. If current estimates for the age, size and rate of expansion of our universe are approximately correct, the bulk of spacetime may well lie in our absolute past and future. Light-cone diagrams conceal these complications because they conform to the useful (but potentially misleading) convention of indicating light-like connections with lines at forty-five degrees from the horizontal (given its role as an absolute standard, light is taken to move one unit through space per one unit of time).

In the context of the comparatively small distances we are mostly concerned with in everyday life, STR does not altogether abolish the present, it merely renders it slightly fuzzy. For events occurring in an ordinary sized room the region of indeterminacy is minuscule – a few millionths of a second. For events a couple of hundred miles apart, it is slightly larger, of the order of a thousandth of a second. It should also be borne in mind that, although the temporal ordering of events within these regions is frame-relative, there is no question of this relativity extending to causal ordering: the relativity of simultaneity applies only to events that are so *nearly* simultaneous (by everyday standards) that nothing travelling slower than light – and this includes subluminal causal influences – can connect them.

Maudlin suggests that "A test of how well one has internalized the spirit of Relativity is how one is tempted to answer the question: what would it be like to travel at 99.99% the speed of light?" (1994: 57). "I would be flattened like a pancake and feel time passing very slowly" is a bad answer. Only slightly better is "Although I would be flattened and slowed down, I wouldn't notice because my brain processes would be slowed down too, and since my rulers and measuring tapes would be contracted to the same extent as my body, nothing would measurably change in size." But, as Maudlin points out, the truly relativistic answer is:

> right now you *are* travelling at 99.99% the speed of light – in some perfectly legitimate inertial frame. It is no more correct to say that you are now at rest than that you are infinitesimally close to moving at light speed. Indeed, except for things travelling at light speed, it makes no sense to attribute any absolute velocity to any object. Not being invariant, sub-light velocities are not real properties of objects. There can be no dynamical effect of "travelling near the speed of light" because there is no such objective states as travelling near the speed of light. Things don't shrink or slow down. Rather, there are always infinitely many ways of expressing space-time intervals in terms of distance in space and elapsed time. None of these ways, represented by the various inertial frames, is any more valid, or less valid, than any other.
>
> (1994: 57)

Reading this you may be tempted to think "So, the curious effects predicted by STR are nothing but illusions, of the sort to be found in a hall of distorting mirrors!" and to an extent you would be right. Length contraction and time dilation are certainly linked to the behaviour of light, and there is a sense in which they are always in the eye of the beholder. But the analogy extends no further. The image seen in a curved mirror is truly a distortion, in that someone standing next to the mirror and looking in your direction would not see you flattened or stretched. However, if you and I pass one another travelling at near-light speed, and I look across and see you flattened and slowed, I am not seeing a distorted image of you. I am seeing you as you truly are, just as someone who shares your velocity and who doesn't see you flattened and slowed *also* sees you as you truly are. We naturally think of shape and duration as invariant intrinsic properties; STR renders them perspectival; objects and processes still have determinate shapes and durations, but these are now frame-relative properties (only interval is frame-invariant). We can no longer speak of *the* shape of an object. We can speak only of its shape in such and such inertial frame, and likewise for events and their durations. Moreover, it would also be a mistake to suppose that the frame-relativity of shape is nothing more than an optical effect, a matter of how a thing *looks*, for in this instance appearances do not deceive: an object's apparent, frame-relative shape is an accurate guide to the shape of a hole that the object can successfully pass through (and the shape of the hole is itself frame-relative). Similarly, an external clock on a fast-travelling spaceship that is seen to run slow when viewed through telescopes on Earth is an accurate guide to the rate at which processes within the ship are unfolding; on their return the astronauts will have aged less than those who remained behind. All this may take some getting used to, but it is by no means obviously incoherent.

17 Relativity and reality

17.1 Reality unconfined

We turn now from the physics of STR to its metaphysical interpretation. I want to focus on some influential lines of argument that attempt to draw very significant conclusions about the nature of time from the distinctive spacetime geometry of STR. These arguments takes as their target all those dynamic models of time according to which there is an ontological asymmetry between past, present and future. As we have seen, there are several such models. Presentists of all persuasions are agreed that the past and the future are unreal; the growing block theory of Broad and Tooley is more liberal, and withholds reality only from the ever-shrinking future. In both cases, the present has special ontological significance, constituting as it does the frontier between what is real and what is not.[1] In outlining these theories I tacitly assumed that the present extends throughout the universe and is absolute (i.e. non-perspectival, non-frame-relative). Let us call models of time that posit an absolute and universal present, and hence an absolute and universal tide of becoming (or annihilation) "classical dynamic models", or CDMs. Given what we have seen so far, it is obvious that there are certain tensions, to say the least, between CDMs and STR. One of the questions we will be addressing in this chapter is whether these tensions are fatal to CDMs. We will also be exploring the options for *non-classical* dynamic models of time in the context of STR.

The mere fact that STR is a *spacetime* theory does not in itself mean it is a hostile environment for CDMs. Minkowski spacetime diagrams depict worldlines representing all stages of an object's history stretching across the page, and future light-cones are represented in precisely the same way as past light-cones. But while this may well give the impression that a Minkowskian world is a static block of events, this interpretation is by no means obligatory. Newtonian physics can be formulated as a spacetime theory. Although there is no distinction drawn between past, present and future in Newtonian spacetime diagrams, it does not follow that there *is* no such distinction in a Newtonian world; spacetime diagrams of the standard form may not include complete information about what they depict. More threatening is what STR has to say about the present itself. The most obvious challenge posed by STR to CDMs runs thus:

> For all CDMs the division between what is present and non-present has ontological significance. Under STR, however, the distinction between what

is present and non-present is dependent on an arbitrary choice of reference frame. Isn't it absurd to suppose that the division between what is real and unreal can be similarly dependent?

Consider an event E on the recently discovered planet that orbits Epsilon Eridani (10.5 light-years from Earth). According to the Earth's frame of reference E lies in the present, but according to Jupiter's frame of reference (let us suppose) it lies in the future. If – like CDM theorists – we hold that whether an event is real or not depends on where it stands in relation to the present, what is real relative to one state of motion is not real relative to another. Relative to the Earth, E is real. Relative to Jupiter it lies in the future, and so is unreal. Do we really want to relativize reality in this way? If not, then the only obvious option is to subscribe to the static block conception, and hold that all events are real, irrespective of their spacetime location.

The suspicion that CDM theories and STR are irreconcilable can be put on a firmer footing. Rietdijk (1966) and Putnam (1967) argue that, provided we make certain plausible assumptions, it can be proved that the *only* model of time compatible with STR is the static block.

Let R be the relation of "being real with respect to", so Rxy means that x is real with respect to y. Let us further suppose that R is reflexive (a thing is real with respect to itself), symmetrical (if Rxy then Ryx) and transitive (if Rxy and Ryz, then Rxz). As already noted, in a Newtonian world it is natural to suppose that each successive present is a three-dimensional universe-wide reality-slice. Or, in the jargon of spacetime theories, the present is a single three-dimensional simultaneity hyperplane of momentary (or very brief) duration.

Now consider Figure 17.1. This may look a little crowded, but to start with we need concern ourselves with just three elements: E_1, E_2 and E_3, each of which is a spacetime location. E_1 is the spacetime point at which two spaceships S_1 and S_2 that are in motion relative to one another make a fleeting contact; two of their radio antennae brush against one another as they pass. The pointed arrows represent E_1's light-cones, and P_1 is the simultaneity hyperplane of S_1.

Now look at E_2 and E_3, each of which is space-like related to E_1. Since E_2 lies on P_1, it is simultaneous with E_1 relative to S_1, whereas E_3 lies in the future with

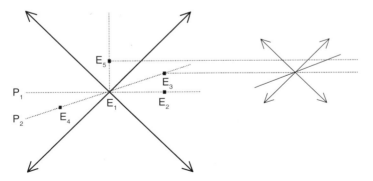

Figure 17.1 STR and the status of distant times. Two spaceships, S_1 and S_2, make fleeting contact. P_1 is the simultaneity hyperplane of S_1 and P_2 is the simultaneity hyperplane of S_2. An event (E_3), which lies in S_1's future, is present for S_2; an event (E_2), which is present for S_1, is past for S_2.

respect to the latter. If we suppose that the growing block model applies, then E_1 is R-related to E_2 but not E_3. More generally, everything up to and including P_1 is real, but nothing later than P_1 is yet real.

But this is far from the end of the story. P_2 is the simultaneity hyperplane of S_2, the *other* ship at E_1, and since the ships are moving relative to one another the contents of this hyperplane (the events that those on board S_2 can justifiably regard as being simultaneous at the time in question) will be different from those in P_1 – recall Einstein's train. In fact, as can easily be shown, P_2 will be "tilted" with respect to P_1 in the way depicted in Figure 17.1. As can be seen, from the perspective of P_2 it is E_3 that is simultaneous with E_1, whereas E_2 occurs *earlier* than either E_1 or E_3.

This yields a significant result. Although from S_1's perspective E_3 has yet to occur, it no longer seems to make sense to suppose that E_3 is unreal. Since S_2 is real with respect to S_1 (recall that the two ships are touching), and E_3 is real with respect to S_2, how can E_3 *not* be real with respect to S_1? More formally, given the transitivity of the "is real with respect to" relation R, since E_3 is R-related to E_1 (via P_2), and E_1 is R-related to E_2 (via P_1) it follows that E_3 is R-related to E_2.

We have focused on just two of the alternative simultaneity planes running through E_1, but since the same reasoning extends to all other locations that are space-like related to E_1, all spacetime locations within this region are real with respect to the latter. Furthermore, it can be shown that the same considerations extend to locations that are *within* E_1's forward light-cone, and so in its absolute future, such as E_5. For any such location, there will be some frame of reference (such as the one depicted to the right of Figure 17.1) containing a location that is R-related to both E_1 and E_5. Given this result, reality cannot be confined; it is a contagion that spreads to all points future, and the growing block model is thus shown to be false.

Presentism is similarly afflicted, and not just because reality inexorably extends forwards in time; it also extends backwards, and for exactly the same reasons. Suppose we say that only what is present is real, and take this to mean that only what is simultaneous with E_1 exists. From the perspective of S_1 the location E_4 lies in the past, but E_4 also lies in the simultaneity plane of S_2, and so is real relative to S_2. Since S_2 is real relative to S_1 (they are in physical contact at E_1), then E_4 must be too.

If this reasoning is valid, dynamic models of time are in serious trouble. Putnam goes further:

> I conclude that the problem of the reality and determinateness of future events is now solved. Moreover, it is solved by physics and not philosophy. We have learned that we live in a four-dimensional and not a three-dimensional world, and that space and time – or, better, space-like separations and time-like separations – are just two aspects of a single four-dimensional continuum with a peculiar metric . . . Indeed, I do not believe that there are any longer any *philosophical* problems about Time; there is only the physical problem of determining the exact physical geometry of the four-dimensional continuum that we inhabit.
> (1967: 247, original emphasis)

This is a momentous conclusion, but is the reasoning that supports it valid? And even if it is valid, are there ways in which it can be resisted? There are, and they fall into two camps. Some dynamists respond to the Rietdijk/Putnam argument by rejecting STR, at least in its standard form; I will be looking at this strategy in §17.4. Others adopt a compatibilist approach, and try to find a way of accommodating dynamism *without* rejecting any of the key principles of STR. We shall start by looking at the options open to compatibilists.

17.2 Compatibilism

A very simple and direct way to bring about the reconciliation is noted by Sklar. The dynamist faces the problem that under STR different events are simultaneous in different frames of reference, so if we suppose that (i) simultaneous events are R-related to one another (where Rxy means "x is real with respect to y"), and (ii) R is a transitive relation, then we cannot avoid the result that all events are equally real. But, Sklar asks, why assume that R *is* transitive? Since anyone who embraces STR will accept that simultaneity is both non-transitive and frame relative, what is to prevent them from taking a step further and holding that R is non-transitive and frame relative as well? It could even be argued that anyone who refuses to take this extra step is failing fully to enter the relativistic spirit. To be sure, it is odd to suppose that if x and y are real with respect to one another, and likewise y and z, it is possible that x and z are *not* real relative to one another, but so is the idea that simultaneity can fail to be transitive. Is relativizing existence any less odd than relativizing simultaneity? Those who dismiss any connection between temporality and existence may be inclined to think it is, but exponents of dynamic models of time do not share this view; anyone who believes that time and existence are intimately connected might be very tempted to suppose that if simultaneity is frame-relative, existence must be too. Sklar concludes, "The science can change the philosophy and put the dispute in a new perspective, but it cannot resolve the dispute in any ultimate sense" (1977: 275).

While this relativization shows that STR does not in itself entail the static block conception, but does so only with the benefit of a questionable metaphysical assumption, it does nothing to assist us in choosing between dynamisms: it seems equally compatible with the growing block model and the different versions of presentism. In a later essay (1981) Sklar argues that some presentists may find themselves in an uncomfortable position. The presentists in question are those who reject the reality of both past and future on the grounds of "epistemic remoteness": unlike what is happening here and now, our knowledge of the past and future is based on more or less unreliable inferences of one kind or another. Since the here and now is causally isolated from the events to which it is space-like related, the epistemically motivated presentist should surely place these events in the same category as past and future events, and hold them all to be unreal. Presentism thus collapses into solipsism of the most extreme sort: other than *this* spacetime point nothing is real. However, while these considerations may lead epistemically motivated presentists to reconsider their position, those who reject the past and future on purely metaphysical grounds – e.g. the conviction that when an event is past it is utterly *gone* – will not feel the same inclination.

Sklar suggests that relativizing the relationship of "determinate reality" is "by no means inconsistent or patently absurd" (1975: 275), but since relativizing here means rendering nontransitive, some may disagree. Of course, compound present-ists willingly embrace the nontransitivity of coexistence, for they see no incoher-ence in the idea that events may undergo absolute annihilation as well as absolute becoming, and I suspect they are right. But any dynamist who takes a different view on this issue, and who also inclines towards compatibilism, will have to look elsewhere. The version of compatibilism proposed by Stein (1968, 1991) may well fit the bill.

17.3 Time fragmented

Stein suggests that anyone who wishes to reconcile STR with "real becoming" must answer two questions:

- What is the spatiotemporal nature of the "stages" in which reality comes into being?
- What is the spatiotemporal criterion that, at any given stage, distinguishes the "definite" from the "not yet settled"?

He cites the following as uncontroversial general principles to which answers to these questions should comply:

(i) The fundamental entity, relative to which the distinction of the "already definite" from the "still unsettled" is to be made, is the *here and now*; that is, the space-time point . . .

(ii) If the state at point b is "already definite" as of point a, then whatever is definite as of b is (*a fortiori*) definite as of a. (Thus the notion "is already definite as of" is to be a *transitive relation between points*.)

(iii) The state at any point a is already definite as of a itself.

(iv) For any point a, there are points whose state is unsettled as of a.

(1991: 148)

He further assumes that there is a definite time-orientation throughout Minkowski spacetime, an assumption needed to make sense of the claim that what lies in the *future* (rather than the past) is not yet definite or settled. With these assumptions in place he goes on to establish, in rigorous fashion, that for any given point x, the only points that are R-related to x (and so those that are already definite as of x) are those that lie in x's causal past: the points on or in the past light-cone of x, and, of course, x itself. Points that are outside x's past light-cone, those that are space-like related to x, have the same status as the points in x's future light-cone; they are *not* real or definite with respect to x. No other assignment of a domain for R is compatible with the stated assumptions.[2]

The relativization of becoming to individual points has the effect of fragmenting reality. No longer is there a universe-wide division between what is real and determinate and what is not; the division is different for every point, since every point has a different past light-cone, and on Stein's model, what is real and definite for a given point does not extend beyond the past light-cone. As a consequence, no two observers will fully agree on what is real and definite unless they are coincident, for it is only then that their past light-cones will entirely coincide. If these observers are in motion relative to one another, they will still agree about what is real and definite, but they will disagree about what is present. But since for Stein everything in the absolute-elsewhere has not become – and so is not yet determinate – this disagreement does not concern anything that is factual as of the relevant point.

Stein does not take himself as having proved that objective becoming actually occurs in Minkowski spacetimes; he remains neutral on the issue. His aim is simply to show that the Rietdijk/Putnam considerations do not rule out the *possibility* that becoming occurs, and many commentators agree that he has succeeded.[3] Stein's contention that assumptions (i)–(iv) are uncontroversial is debatable – as he notes himself, not all writers on this topic share them – but the important point is that they are both reasonable and defensible. Rietdijk and Putnam's objection succeeds against CDMs, models of time that posit a single universe-wide ontological division between what is real (or definite) and what is not, but, as Stein notes, this choice of target seems singularly misconceived. The notions of "temporal becoming" and "spatially extended now" seem to be conceptually independent, so there is no obvious or compelling reason to think that the former requires the latter, and this is especially true in the context of the special theory. STR explicitly denies the objectivity (or frame-independence) of a spatially extended present, so given the assumption that real becoming *is* an objective process, it is clearly unfair to inflict on the dynamist a spatially extended division between what is as of now determinate and what is not: "to insist (without supporting argument) upon a notion of 'present [spatially distant] actualities', in assessing special relativity, is simply to beg the question, since it is fundamental to that theory that it rejects any such notion" (Stein 1991: 152). Sklar is guilty of the same error, for he too assumes that the dynamist will want to transplant a concept of extended becoming that, while perfectly appropriate in a Newtonian setting, is quite alien to the environment of STR.

Sklar's approach also requires the dynamist to adopt a nontransitive notion of "real or determinate with respect to". Stein's approach does not. Provided we stipulate, as Stein does, that the R-relation can only hold between a pair of points if one is in the causal (or absolute) past of the other, R can be transitive without risk of becoming universal (i.e. spreading from the past to the future). If any point can be R-related only to points in its causal past, although it follows from Rab and Rbc that Rac, it also follows that c is in the absolute past with respect to a, and the same applies generally. The key factor is that while R is transitive and reflexive, it is not *symmetrical*: if b is definite as of a, by virtue of lying in a's causal past, it does not follow that a is definite as of b. If it did, then we would again be in the situation in which all points are equally definite with respect to one another. The relation ". . . is real *as of* . . ." in the Broad–Tooley version of the growing block theory is also asymmetrical: 1900 is real as of 2000, but as of 1900, the year 2000 is not real.

Stein's compatibilism may be coherent, but in one respect it is ambiguous. The avowed task is to reconcile a metaphysic of "becoming" with STR, but when it comes to specifying what this metaphysic actually involves, he is less than fully clear. He talks of "a notion of temporal evolution as (in some sense) *a becoming real, or becoming determinate, of what is not yet real or determinate*" (1968: 14, original emphasis), and of a distinction between a part of the world's history that "had already become" or is "ontologically fixed and definite" and a part that "is not yet settled" (1991: 148). Is the "not yet settled" part of reality wholly non-existent? Or does it exist, but only in a sketchy, partially complete form? It may not matter. Perhaps the notion of a real but not fully determinate event can be made coherent; if so there are dynamic models of time other than those we explored previously. What does seem clear is that Stein's account of becoming *can* be applied to models of time that posit a clear distinction between what is real and what is not, where "becoming" is *absolute* becoming, the *ex nihilo* creation of time and events – or rather, spacetime *points* and events.

An understandable reaction to what has been said so far would be:

> I think I have grasped what Stein's account of becoming amounts to; we cannot ask what is real as of a given moment of *time*, if by this is meant a temporal cross-section of the universe as a whole, but we can ask what is real as of a given *place* at a time, a spacetime point. And what is real as of one point is different from what is real as of every other point. What I can't do is form an intuitively clear *picture* of what this point-relativized dynamism amounts to.

A universe that becomes à la Stein *is* hard to visualize, but since this can be put down to contingent (and understandable) limitations on our imaginative abilities – we can only easily picture planes of becoming that are spatial – this is not necessarily a problem for Stein's model. Dorato provides some suggestive descriptions: the universal tide of becoming of the CDMs is replaced by water flowing through "an uncorrelated, non-denumerable set of narrow creeks" (1995: 184) and shattered "into small disconnected fragments, that is, convex infinitesimally small regions in the future lobe of the light-cone at any point p within [the Minkowski manifold] **M**" (1995: 185).

Perhaps the easiest way of putting some pictorial flesh on the model is to consider things from the perspective of individual worldlines, Dorato's "narrow creeks" of becoming. Consider a collection of point-particles scattered through a region of space; from one moment to the next the worldlines of these particles *grow*, and as they do so they drag their past light-cones with them. To generate an image, imagine a large cloth sheet lying on the ground that is gradually pushed into a shape resembling a mountain range by several rigid rods slowly rising up from the ground beneath. Looking down from above, not only do we see the entire cloth, but we see the peaks associated with each growing worldline all rising together, in synchrony. But these are artifacts of the mode of representation to which nothing in the envisaged reality corresponds. There is no one privileged perspective within this world to which the world as a whole is revealed, or to which the world as a whole is factual; there are just the different perspectives of each point, and what is real for each point is different (although of course, for neighbouring points the difference is infinitesimally small). Consequently, there are no temporal relations between the different peaks. All that is real from the perspective of each peak is what lies on and beneath the downward pointing part of the sheet (i.e. its causal past). In Figure 17.2 the arrows point towards what is *not* real as of each peak.

There is one further respect in which pictures like Figure 17.2 might mislead: the undulations are not purely *spatial*, they are *spatiotemporal*. Figure 17.3 depicts the past light-cones at two neighbouring locations on the worldline of a particle, P_1

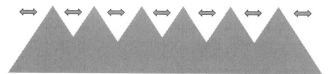

Figure 17.2 Stein's model: becoming relativized to spacetime points. As worldlines grow, their past light-cones expand. Despite appearances, the portions of the cones indicated by the arrows are not real relative to one another.

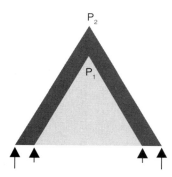

Figure 17.3 The coming-into-being of the distant past.

and P_2; the smaller triangle exists in the causal past of the larger, and the difference between the two represents the regions of spacetime that are real relative to P_2 but not P_1.

The regions of spacetime indicated by the arrows at the bottom are *very* remote in the causal pasts of the points at the apex of the light-cones: billions of light-years away. Consequently, some of the events that are real for P_2 but not for P_1 are not merely a long way off in space, they are a long way off in *time*: billions of years in the past. It is not just worldlines that grow in Stein's model, it is *light-cones* and all they contain; so as our worldlines grow, so too do the regions of the very distant past that are real relative to us. Indeed, as your worldlines grow, by far the greater proportion of what becomes real for you lies in the spatially and temporally remote past, whereas what becomes present is always strictly confined to the here and now. Light, in Stein's model, does not merely *reveal*; it *makes real*.

Stein provides a way of retaining a form of the growing block metaphysic in the context of STR, but his model does not provide a congenial home for presentism. However, a simple modification does the trick. We retain the idea that what is real is always relative to a single spacetime point; but rather than regarding the whole past light-cone of this point as real and determinate, we exclude the interior region, and hold that only the *surface* of the cone is real; the final step is to regard the entire past light-cone of a point as the *present* of that point. For understandable reasons, Hinchliff calls this position "cone presentism".[4]

From the standpoint of common sense, cone presentism has a highly appealing feature. You may recall how surprised you were when you were taught that everything you see is actually in the past, due to the time-lag generated by the finite speed of light; astronomers observing distant stars and galaxies see them as they were millions or billions of years ago. If cone presentism is true, the naïve view is vindicated: everything you see is in the present (or at most only a fraction of a second in the past – the time it takes the brain to process the retinal image). The doctrine has a second appealing feature. By restricting the present to the here and now (a single point), Stein's model has the consequence that (strictly speaking) we are always alone: no one else shares our present. This is no longer true under cone presentism, since everyone and everything on our past light-cone shares our present.

As might be expected, cone presentism also has some counterintuitive features. Symmetry and transitivity must be denied. If event E is in my present (by virtue of being on the surface of my past light-cone), I am not in E's present (since I am not on the surface of E's past light-cone, but rather on the surface of E's *future* light-cone). If an event D lying within my past light-cone occurs on the surface of E's past light-cone, then although D is present for E, and E is present for me, D is *not* present for me (it is past). But no doubt the most counterintuitive feature of all is the way in which temporally distant events are rendered present, for example, a supernova that occurred a billion years ago, but is currently being observed by an astronomer. While this does conform with the phenomenology of perception – as we have just noted – it also seems very odd, to say the least. In saying that an event that occurred a billion years in the *past* can also be *present*, are we not stretching the notion of "the present" beyond breaking point?

However, given the way STR renders temporal distances frame-dependent, this objection does not have the force it otherwise would (e.g. in Newtonian spacetimes). Since the temporal (and spatial) distances between two distinct events can vary enormously from one reference frame to another, there is no objective fact as to the *true* spatiotemporal distance between the events in question. Consequently, the cone-presentist does not equate "the present" with a collection of events (and points) that occur *simultaneously*, where "simultaneously" means "at the same time", or "zero temporal distance apart"; rather, the present is composed of events (and points) that are at zero spacetime *interval* from one another. Unlike its pre-relativistic counterpart, cone presentism is not a doctrine about *time* at all, it is a doctrine about spacetime, of the Minkowski variety, and in this context "present" takes on a new meaning.[5]

17.4 Absolute simultaneity: the quantum connection

Dynamism can be rendered compatible with STR, but the cost to our common-sense understanding of reality is more than some dynamists are prepared to pay. For those falling into this camp an approach that may appeal is the *refusnik strategy*:

> There *is* a single, universe-wide division between what is real and what is not, a division that coincides with the present. Since STR denies this, STR must be false, and the same applies to theories such as GTR that also relativize the present.

This stance has the merit of exemplifying a healthy scepticism towards current scientific theorizing, for it may well be that Einstein's physics will one day meet the same fate as Newton's, and be displaced by a quite different theory with superior explanatory and predictive power. However, the strategy suffers from two defects.

First, if there is an absolute and universe-wide present there must be an absolute simultaneity relation. Since no such relation is required or possible in STR, an opponent could reasonably argue thus:

> I agree that we *could* have found evidence for absolute simultaneity. For example, we might have discovered that some signals or influences travel instantaneously – they are received at the same moment they are sent, irrespective of the distance they have to travel; in which case we could have defined an absolute simultaneity relationship via these signals. But no such evidence has been found. The fact that your pet metaphysical model of time requires such a relation does not, in itself, give us any reason to believe that your model applies to our world.

And this has some force. In the absence of any reason for believing that events in our world *are* absolutely simultaneous, compatibilist dynamisms seem less costly than the CDMs, for at least they do not require us to posit facts for which there is no empirical evidence.

Secondly, STR may indeed one day be replaced by another theory, but since the relativistic empirical effects – such as measurable time dilation and length contraction – are well confirmed, the successor theory will have to explain them in its own terms. And so must anyone seeking to defend CDMs.

Tooley is one dynamist who is alert to these problems and also committed to absolute simultaneity and a universe-wide division between what is real and what is not real. Rejecting the refusnik strategy he sets himself the task of devising an alternative to STR that:

- is closely related to STR, and compatible with all the observational data that are usually taken to confirm it;
- entails that events stand in relations of absolute simultaneity; and
- is superior to STR in other respects, and so preferable independently of its conforming to the requirements of CDMs.

I will conclude by briefly summarizing Tooley's position (elaborated in Tooley 1997: Ch. 11), which although bold is not without its rationale.[6]

Tooley starts by showing that absolute simultaneity can in fact be *defined* within the framework of STR provided we make certain metaphysical assumptions about spacetime. The assumptions Tooley proposes are these:

- Spacetime points are substantival entities, and likewise spacetime itself is the sum of these entities.
- Spacetime points can be causally related to one another.
- Every spacetime point causes at least one other spacetime point to exist, and was itself caused to exist by a spacetime point.
- Causally related spacetime points occur at the same place (i.e. on the same line through spacetime).
- If P, Q, R and S are four spacetime points, such that the existence of P causes the existence of Q, and the existence of R causes the existence of S, then the lines defined by PQ and RS are parallel.

The picture, then, is of an absolute substantival spacetime that is uniform and causally self-propagating over time. We now assume that spacetime (rather than any aether) is the medium through which light propagates, and that the speed of light is constant relative to absolute space (which is what we would expect if natural laws are uniform). As for simultaneity, Tooley accepts Einstein's operational definition: two events E_1 and E_2 occur simultaneously if light from each would arrive at an object O which is both equidistant and at rest relative to them (i.e. in the same inertial frame) at the same time. Starting with this definition we only reach the conclusion that simultaneity is relative if, like Einstein, we assume that all inertial frames are on an equal footing, and this Tooley rejects. If absolute space exists, then some inertial frame is motionless with respect to it, and so we can say that "Two events E and F are *absolutely simultaneous*" means the same as "E and F are simultaneous relative to some frame of reference that is at rest with respect to absolute space" (1997: 344).

According to STR in its standard form, light has the same speed in *all* directions in *all* inertial frames. In Tooley's modified theory, light has different speeds in different inertial frames. To focus on the simplest case, a light ray will have a speed of $(c - v)$ in a frame moving at velocity v in the same direction as the ray, and a speed of $(c + v)$ in a frame moving at v in the opposite direction to the ray. Since empirical evidence supports the "same speed in all inertial frames" prediction, doesn't Tooley's theory fall at the first hurdle: compatibility with known empirical data? Not in the least. What the empirical data actually supports is the weaker hypothesis that, for any two locations L_1 and L_2, observers in all inertial frames will agree on the *round-trip* speed of a light ray travelling from L_1 to L_2 and back again – the speed that results from dividing the total distance travelled by the time taken. But as Einstein himself recognized in his original 1905 paper, there is no empirical evidence that the *one-way* speed of light is constant in all inertial frames; he simply regarded it as a reasonable assumption, and made it true by definition. There have been many subsequent attempts to devise experiments that would demonstrate that the one-way speed of light is constant, but thus far none has been successful. In the absence of any countervailing empirical evidence, Tooley feels free to abandon Einstein's assumption, and hold that light travels at different speeds in inertial frames that are in motion relative to absolute space. As for why this is not detected, he relies on a Lorentz-style compensatory theory: physical processes in moving frames are affected in ways that systematically conceal the variations in light speed that actually obtain. Since the original Lorentz transformations presuppose that the one-way speed of light is the same in all inertial frames, they will not perform the required job, but there is an alternative that will: the so-called "ε-Lorentz transformations" (the terminology is due to Reichenbach), which specify the necessary compensations corresponding to different assumptions concerning the one-way speed of light. Winnie (1970) showed that a suitably modified version of STR entails the ε-Lorentz transformations, and that the modified theory is inconsistent only if STR is itself inconsistent.

That there should be an empirically equivalent competitor to STR is not surprising – as we saw during our discussion of conventionalism, there is always room for such a competitor – but is there any reason for preferring Tooley's version of STR over the standard form? Discounting any purely metaphysical preferences we might have, in the absence of any evidence for an absolute simultaneity relation, defenders of the status quo (standard STR) will object that theories such as Tooley's are guilty of positing a real *relationship* – absolute simultaneity – which we have no reason to believe exists. It is at this point that Tooley turns to quantum theory.

Defending the possibility that absolute simultaneity *could* exist in our universe, Lucas remarks:

> if some superluminal velocity of transmission of causal influence were discovered, we should be able to distinguish frames of reference, and say which were at rest absolutely and which were moving. If for instance, we were able to communicate telepathically with extraterrestrial beings in some distant galaxy, or if God were to tell us what was going on in Betelgeuse now, then we should have no hesitation in restricting the principle of relativity to electromagnetism only: hence it cannot be an absolute principle foreclosing absolutely any possibility of absolute time. (1999: 9–10)

And in this he is surely right. Were we to discover that a truly instantaneous connection exists between objects at different places in space, then assuming the connection has some detectable effects, not only would the notion of absolute simultaneity have a real application, but we would have a way of determining which frames of reference are at absolute rest and which are not: it is only with respect to frames truly at rest that the relevant changes would occur at precisely the same time. The idea that there are – *contra* Einstein – instantaneous connections in nature has gained favour in recent years as the significance of a now famous theorem due to John S. Bell has gradually sunk in.

In 1935 Einstein, Podolsky and Rosen (EPR) came up with an ingenious argument designed to show that quantum mechanics, at least in its orthodox form, cannot provide a complete description of nature. According to quantum orthodoxy, particles only acquire definite properties when these same properties are measured: a photon only acquires a definite polarization when its polarization is measured; an electron only acquires a definite momentum when its momentum is measured; and so on. EPR devised a thought-experiment that (as they thought) cast doubt on this doctrine. Quantum theory predicts that it is possible (in principle at least) to create pairs of particles that are *correlated* with one another, in the sense that there is some property that both particles possess, which is such that if the property has a certain value for one of the particles it will have a certain value for the other, *irrespective of the spatial separation of the particles*. To take an artificially simple example, suppose that the property is P, and the value of P for both particles must add up to 1. Then if particle A is measured to have a P of 0.25, particle B will have a P of 0.75, whereas if P for A is 0.5, it will be 0.5 for B, and so on. (Actual instances of P include quantum properties such as spin and polarization, whose numerical values take different forms.) Suppose that we create a pair of correlated particles that speed off in opposite directions, and measure the value of P for A at time t, and find that it is 0.6. We know at least this much: from t onwards the value of P for B is 0.4. But when precisely did B acquire this property?

One possibility is that neither A nor B possesses a determinate value of P until time t; at the very same moment that A is measured as possessing a P of 0.6, B acquires a P of 0.4. It is necessary that the particles acquire determinate values *absolutely simultaneously.* If this does not occur, if the P-value of B remains truly indeterminate for some period after *t* (for some frame of reference), there is nothing to prevent our measuring the value of P for B and finding a value other than 0.4, which we know is nomologically impossible. But this can only be nomologically impossible if the value of P for B becomes fixed at precisely the same time as P for A is fixed. However, we are now confronted with a situation in which what is done to A instantaneously affects B, irrespective of how far the two particles are spatially separated from one another: it is only because we conduct a P-measurement on A at *t* that B acquires a determinate P-value at *t*. How can B "know" what is being done to A this quickly?

If B's state *is* instantaneously affected by the measurement performed on A, one thing seems clear: STR cannot provide a complete account of the spatiotemporal relations that actually exist, since superluminal causal connections between spatially separated material things are explicitly ruled out by the theory. For obvious reasons, this is Tooley's preferred interpretation.

But there is a second possibility, which for understandable reasons Einstein found more palatable, namely that both particles *always* had determinate values for P: at the moment they were created A had a P-value of 0.6 and B a value of 0.4. The measurement of A's P-value at *t* merely reveals a property that A always possessed, and this is why if we were to measure the P-value of B at *t* we would find it to be 0.4 – this was the value B had all along. Hostile to the idea of superluminal "spooky" action-at-a-distance, this is the option that EPR preferred. And hence their conclusion: quantum mechanics cannot provide a *complete* description of nature. Since quantum theory does not ascribe a P-value to A or B until a P measurement is conducted on one of the particles, if we suppose that their P-values are fixed when they are created, there are properties that exist but quantum theory does not recognize.

The situation changed radically in 1964 when Bell showed that EPR's preferred interpretation was not in fact an option. Bell proved that the quantitative predictions made by quantum mechanics are incompatible with the possibility that A and B possess determinate P-values before any measurements are made. As Tooley puts it:

> the thrust of the Einstein-Podolsky-Rosen thought-experiment is no longer merely that the Special Theory of Relativity, or else quantum mechanics, is incomplete. It is rather that either the Special Theory of Relativity is incomplete, or quantum mechanics is *false.*   (1997: 359)

The Aspect experiments on quantum-correlated particles, carried out in the early 1980s (and successfully replicated on many subsequent occasions) revealed that the quantitative predictions of quantum theory for this type of situation are in fact correct, and so, given Bell's result, the EPR interpretation is falsified: the particles cannot possess determinate P-values prior to measurements being conducted. If B's P-value becomes fixed only when A's P-value is measured, it seems that some events at least occur absolutely simultaneously.

Or so Tooley concludes. And he is not alone. Although Tooley does not mention the fact, there is a well-known and respected interpretation of quantum mechanics – due to David Bohm – which is highly non-local and requires (and posits) a preferred reference frame with respect to which non-local interactions are instantaneous.[7]

But these matters are complex; the correct interpretation of quantum theory remains controversial, likewise the relationship between quantum theory and relativity. Nonetheless, the possibility of physics recognizing a preferred reference frame (and so abandoning Lorentz invariance) remains real. After a thorough investigation of recent debates Maudlin (1994: 240) concludes that the space-like connection established by the Aspect experiments:

- does *not* require superluminal matter or energy transport;
- does *not* entail the possibility of superluminal signalling;
- *does* require superluminal causal connections; and
- can be accomplished *only* if there is superluminal information transmission.

There are ways of accommodating the quantum results within STR but, as Maudlin notes, the cost is high:

> Embedding quantum theory into the Minkowski space-time is not an impossible task, but all the available options demand some rather severe sacrifices. Bohm's theory and orthodox collapse theories seem to require the postulation of some preferred set of hyperplanes, ruining Lorentz invariance and jettisoning the fundamental ontological claims of the Special Theory. Lorentz invariant theories might be developed using explicit backwards causation, so that the birth of every pair of correlated particles is already affected by the sorts of measurements to which they will be subject, perhaps hundreds of years later . . . One can achieve Lorentz invariance in a theory with instantaneous collapses by embracing a hyperplane dependent formalism, but the exact ontological implications of such a view are rather hard to grasp . . . Or finally, one can avoid collapses and retain locality by embracing the Many Minds ontology, exacting a rather high price from common sense.
>
> . . . the common thread that runs through all these proposals is that no results are to be had at a low price. Indeed, the costs exacted by those theories which retain Lorentz invariance is so high that one might rationally prefer to reject Relativity as the ultimate account of space-time structure.
>
> (1994: 220)

It is with considerations such as these in mind that Lucas writes:

> [although] the Special theory made time seem spacelike . . . quantum mechanics is kind to time . . . [it] redresses the balance. We no longer need feel obliged to construe time in a non-temporal way in order to be truly scientific and philosophically respectable.   (1999: 11)

But while quantum mechanics *alone* may indeed be kind to our ordinary understanding of time, we have already seen in §6.9 that quantum mechanics may yield a very different result when integrated with GTR. Attempts to introduce gravity into quantum theory give rise to the quantum *problem of time*: if time exists in the equations of quantum gravity theory, it is not obvious where.

What can we conclude? Given the manner in which STR relativizes simultaneity, it is easy to suppose that Minkowski spacetime can only be a static block, and hence agree with Putnam's claim that the problem of the reality and determinateness of the future is now solved, and "solved by physics and not philosophy". But this is wrong, for there are compatibilist and non-compatibilist ways of reconciling dynamic models of time with the empirical evidence that supports STR. Stein's point-relativized dynamism preserves the spirit of STR while retaining an ontological division between what is as yet real and what is not; following on from this lead, cone presentism goes a step further, and restricts reality to what is present (relativistically construed). Tooley's modified version of STR goes still further, and by rejecting the one-way constancy of the speed of light, retains a universe-wide plane of becoming, without abandoning empirical equivalence with orthodox STR. Indeed, we have seen that there may well be support from physics for an absolute simultaneity relation. But as this support is rooted in the still murky and much disputed arena where quantum theory and relativity come together, it is too early to know precisely what impact physics will eventually have on our understanding of time in our universe. However, even if STR has not provided the definitive solution to the static–dynamic controversy, it has certainly cast new light on it, and forced us to entertain previously unforeseen possibilities. The debate has moved on, even if where it will lead as yet remains unclear.

18 General relativity

18.1 The limits of STR

In this chapter several threads of our inquiry thus far come together. Our topic is Einstein's *general* theory of relativity (GTR), the product of ten years' arduous labour, which followed the completion of STR in 1905. During this intervening period Einstein wrote "never in my life have I tormented myself anything like this . . . Compared to this problem the original relativity theory [STR] is child's play".[1] The metaphysical implications of this elegant and intriguing theory are of interest in their own right, for it remains our best account of our universe's large-scale spatiotemporal structure. But it is of interest for two further reasons.

Since GTR posits a variably curved spacetime manifold, it constitutes a challenge to Poincaré's conventionalism: the claim that, although non-Euclidean spaces are mathematically intelligible, we could never, in practice, have reason to accept a theory that ascribes a non-Euclidean geometry to our world. We will be looking into the considerations that led Einstein (and most other physicists) to accept a theory that does precisely this. GTR is also of interest in the light of the discussion in §§14.2–3, where we arrived at the conclusion that in worlds where material objects behave *as if* they inhabit a variably curved space, there is a strong case for supposing that the space in question is substantival rather than relational. We shall see that, although there are strong grounds for construing GTR in a substantivalist manner, there are also difficulties of a hitherto unsuspected variety.

As for why Einstein found himself obliged to develop GTR in the first place, it was not because he discovered a flaw in STR, but rather because the latter is limited in its aspirations: it was designed to explain some basic physical phenomena, but not all (hence the label *special* theory of relativity as it only applies to certain special cases). STR provides an account of the difference between accelerated and inertial motions (Minkowski spacetime has a well-defined affine structure), and a harmonious integration of the dynamics of moving bodies and Maxwell's electromagnetic theory, but it was silent on one crucial phenomenon: gravitation, the force that keeps the planets in orbit around their stars, that prevents us from floating off into outer space, and which provided Newton with his greatest triumph.

There are two main reasons why Newton's theory of gravity cannot be smoothly slotted into the framework of STR. First, Newton's theory of gravity presupposed that time and space were separate and invariable quantities; this presupposition was abandoned by Einstein with STR, so he needed to develop a theory of gravity that was compatible with length and time dilation and the relativity of simultaneity. It is easy to see how relativistic effects lead to anomalies when combined with Newtonian gravitation. The strength of Newton's gravitational force between a pair of bodies depends on their distance apart. The fact that under STR this distance will be different in the inertial frames of the relevant bodies (if they are in relative motion) means that observers on the two bodies will arrive at different results when they calculate the strength of the gravitational attraction between them. Secondly, Newtonian gravity is a force that acts between all material bodies *instantaneously*, irrespective of their distance. Since STR puts a finite limit on causal propagation, Einstein had to find an alternative explanation of gravitational effects.

The explanation he came up with was remarkable. Whereas Newton regarded gravity as a *force*, which acted at a distance to attract material bodies towards each other, Einstein explained gravitational effects without appealing to such a force, and did so in a way that dispensed with the need for instantaneous "action-at-a-distance". Einstein replaced Newton's attractive force with mass-induced space-time curvature. Quite why this suffices to keep us rooted to the Earth's surface will emerge in due course.

18.2 Equivalence

Einstein started serious work on the problem of gravity in 1907, and his first breakthrough occurred in November of that year: "I was sitting in a chair in the patent office at Bern, when all of a sudden a thought occurred to me: 'if a person falls freely, he will not feel his own weight'." He was later to call this "the happiest thought of my life".

The fact Einstein grasped is no doubt a good deal more obvious these days than it was in 1907. Suppose you are in a lift, going up, and the cable snaps. The lift cabin begins to fall freely. What happens? Does the roof of the cabin rush down and bang into your head? No. As Galileo discovered, all objects, irrespective of their size, weight and composition, fall at exactly the same rate when released – gravity is not a discriminating force – and so the cabin and all its occupants will fall at the same speed. In fact, you will find yourself weightlessly drifting around the cabin – just as you would if you were in outer space. We have all seen films of astronauts floating around in weightless conditions, and this is just how it would be for you in the lift cabin. The analogy is closer still since astronauts in orbit around the Earth are – in effect – *falling* towards the Earth all the time (they remain in orbit only because they also have a considerable forward directed velocity). In short, if you "go with the flow" near a gravitating body, you don't feel any force; to avoid feeling the pull of gravitational forces, you need only to be allowed to continue with your free-fall path. You only feel a force inside a gravitational field if something *prevents* you from continuing along your free-fall trajectory, if something else applies a force to you, such as a floor, or the Earth's surface.

Thanks to this insight, Einstein realized that the relativity principle he used in STR could be extended. Roughly speaking, the relativity principle of STR says that the laws of physics are the same in all inertial frames, that is, those that are at rest or moving in straight lines (and not under the influence of external forces). But if the effects of gravity disappear in free-fall, then we can also say that the laws of physics take the same form in frames that are freely falling in gravitational fields as they do in inertial frames. Suppose you are in the cabin of a windowless spaceship, enjoying weightless conditions. You wonder about your state of motion. Are you drifting aimlessly in interstellar space, or are you accelerating towards a nearby planet under the influence of its powerful gravitational field? Surprisingly, alarmingly, there is nothing you can do to find out. All experiments that you can perform within the cabin will yield exactly the same results if you are accelerating under gravity as they would if you are not. Nature does not distinguish between inertial frames and frames that are freely falling in gravitational fields. This became known as the "equivalence principle".

Einstein recognized another respect in which the presence of gravity is undetectable. Suppose you wake up in your windowless space cabin to find that you are no longer weightless; in fact, it feels just as it would if the cabin were lying motionless on the Earth's surface. Nothing is floating in mid-air, you can walk about as normal, you can feel the difference between up and down (a disturbing absence in weightless conditions), objects fall when dropped, and so forth. Can you conclude that you are in fact on Earth (or at least on some massive body)? After a few moments thinking back to the (Newtonian) physics you learned in school, you realize that the answer is no. Newtonian physics can explain all the relevant phenomena in two different ways:

- you are being held motionless near a planet or star, and feeling the effects of gravity; or
- you are not subject to any gravitational forces, but are being uniformly *accelerated* (by a silent rocket motor, say); we know that acceleration gives rise to forces (when a plane takes off you are pulled back into your seat).

Once again, no test you could carry out inside your windowless laboratory would tell you if you were inside a gravitational field, or being accelerated by a rocket. (Of course, both could be happening, but to keep things simple we will focus on the simple either/or case.) This was a further key point that Einstein grasped: so far as their physical effects on a body are concerned, the "pulls" generated by gravitational attraction and acceleration are indistinguishable.

The equivalence principle in this form is particularly significant. Why? Because whereas acceleration is a well understood and purely local phenomenon, gravity – at least as construed by Newton – is a mysterious force linking all bodies in the universe, which acts instantaneously. Here was a promising hint as to how the problem posed by gravity might be solved: given that there was no way of telling, locally (over small distances), between gravitational effects and acceleration effects, perhaps there is *no difference between the two*; perhaps gravity and acceleration are essentially the same phenomenon. It was several more years, which involved numerous detours and false starts, before Einstein finally worked out how this might be.

18.3 Spacetime curvature

Starting from the equivalence principle, there are several lines of reasoning that lead to the idea that gravity is not a force acting *in* or *over* spacetime, but rather a modification of the very structure of spacetime itself.

The equivalence principle yielded a novel prediction: *light* should be affected by gravity (this was generally denied at the time, since it was widely believed that light was massless). Einstein reasoned thus. If you were in a sealed laboratory being accelerated by a rocket motor and aimed a light ray across the room, it would hit the opposite wall at a lower position than if the laboratory were not moving – the velocity (and so direction) of light is independent of the source, and the laboratory changes position during the time the light takes to cross the room. If accelerated motion and gravitational attraction are equivalent, then if the laboratory is held stationary in a gravitational field, the light ray should also hit the wall lower down. The diagrams in Figure 18.1 depict the apparent path of the light ray from the perspective of someone within the laboratory.

As we saw in the previous chapter, Minkowski spacetime is structured around the trajectories of light rays. Not only do they define the absolute past and future of every point, but they are the paths of shortest possible distance: the interval between light-like connected points is zero (and so there is, from the perspective of light, no spatial or temporal distance between them at all). Consequently, if gravity affects the path light takes, and one is already assuming that we live in a Minkowski spacetime, it is not a huge leap to the idea that gravity affects the structure of spacetime itself.

There were other considerations, of a more concrete sort, that convinced Einstein that acceleration would produce warpings of both time and space. Extending the reasoning he employed in STR, he imagined a group of people on a

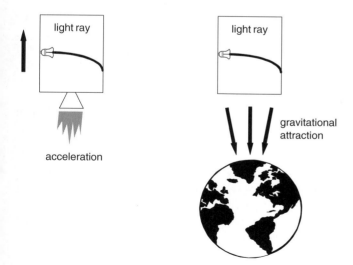

Figure 18.1 The equivalence principle and light. Light travelling across the cabin of an accelerating spacecraft follows a curved path. If gravity has the same (local) effects as acceleration, light crossing the cabin of a stationary spacecraft within a gravitational field should follow a similar trajectory.

rotating disc who perform various measurements, of both distance and time, and worked out exactly what they would find. Suppose you have the role of external observer, and have measured the circumference and radius of the disc before it starts moving, using a string and ruler. Once the disc is in motion, you pass the ruler to someone on the disc, Sam, who uses it to make similar measurements. As he sets about measuring the circumference, you observe that Sam's ruler is *shortened* (due to the Lorentz contraction), and realize that the number he will arrive at for the circumference will be larger than yours. When he measures the radius, on the other hand, the ruler is not pointing in the direction of rotation, but at ninety degrees to it, so it will not be shortened (length contraction occurs in the direction of motion), and his measurement will agree with yours. When Sam puts the results of his measurements together, he finds that the ratio of the circumference to the radius is greater than 2π, which as we saw earlier is a result that is indicative of a non-Euclidean space. If we now imagine that there are some people at the edge of the disc who, at regular intervals, compare their watches with that of someone at the centre, we know that these watches will not keep the same time. Since the people at the edge are moving faster than the person in the middle, and STR predicts that the faster you move the slower time passes for you (relative to others who don't share your motion), time will pass more slowly at the edge of the disc.

What is the relevance of these disc-based results to gravity? Once again the equivalence principle comes into play. Rotation is a form of acceleration, and if acceleration is associated with time dilations and changes in spatial geometry – in relativistic terms, warpings of the *chronogeometry* of spacetime – then, given the equivalence principle, gravity must be too.

What Einstein actually concluded was this: gravity is *nothing but* the warping of space and time. Material bodies do not exert a gravitational pull on one another, rather material bodies warp space and time, and these warpings produce the effects we associate with gravity – the effects Newton explained in terms of an attractive force. It was mysterious how anything could act-at-a-distance in the required way. Einstein does away with action-at-a-distance, since gravity for him doesn't exist as a force operating between bodies. Instead, matter bends the spacetime in its vicinity; the more matter, the greater the distortion, and the matter-induced curvature is transmitted through spacetime at the speed of light, not instantaneously.

By way of an analogy, think of walking on the icy surface of a frozen lake. As you get to the middle, the ice suddenly cracks beneath you. The cracks *spread* across the surface, from the centre outwards, at a finite speed. Matter-induced spacetime curvature works rather like this. A large concentration of matter induces strong curvature just where it is, and the spacetime in the immediate vicinity is very curved; the spacetime in the next region isn't directly affected in any way by the matter, but *is* affected by the adjacent spacetime, so the curvature is transmitted through spacetime, from region to region, like cracks spreading through ice. But it gradually weakens. As you get further away from the mass, the curvature of each successive spherical region of spacetime diminishes. The important point is that all gravitationally generated causal relations are *local* in GTR, they are directly transmitted from one region of spacetime to the neighbouring regions of spacetime, they don't work magically across the distances separating material bodies. Since all

Figure 18.2 Mass-induced spacetime curvature. The greater the concentration of mass, the greater the curvature. Inertial paths cease to be Euclidean straight lines in the vicinity of mass-concentrations. When the concentrations are large, this becomes noticeable.

the material bodies in the universe induce curvature in spacetime, the overall shape of spacetime is the product of their combined influence, and as the bodies move about the overall shape of spacetime alters; spacetime structure is thus dynamic rather than static. There is a standard two-dimensional analogue. Think of a big flat sheet of rubber stretched out like a trampoline. Now drop some heavy metal balls on to it. The rubber will be bent, the deflection from flat being greatest near each ball. Now move the balls about. The overall shape of the surface will change; it will be differently curved.

At this point the concepts introduced during our earlier discussion of curved space become relevant. In Newtonian (and STR) physics the natural path for a body to take when it isn't subject to any forces is a straight line, and a body moving along its natural path will continue to do so for ever, without slowing or speeding up. These inertial paths are those that *spacetime forces on a body*, and a body only deviates from an inertial path if some external force acts on it. We have seen that when a body falls under the influence of gravity, it does not experience any forces provided it "goes with the flow" and continues along its natural free-fall trajectory, and this suggests that free-fall paths in gravitational fields are themselves inertial paths. The path of a spaceship that quickly flies past a planet will be deflected; it will be *as if* the planet is pulling at the ship, causing it to move in a curve rather than a straight line. But if the spaceship is not being accelerated by its engines, but simply continuing on its natural free-fall path, as it passes the planet it won't feel any *force* acting on it (just as you didn't feel any force pulling on you in the falling lift). The ship's path bends inwards because the planet's mass is *altering the inertial path of the spaceship*: the planet is not pulling the ship towards it like a magnet, it is altering the shape of spacetime. In a curved spacetime the equivalents of Euclidean straight lines are the *geodesics* of the spacetime (the paths of shortest length), and mass affects the curvature of these geodesics: the larger the mass, the more the local geodesics deviate from Euclidean straight lines.

18.4 Feeling the grip of spacetime

To see how all this applies to some simple cases of motion and explains the gravitational effects we are familiar with in everyday life, we will start by ignoring time, and focus on purely spatial curvature.

On the face of it, Newton's account of gravity as an attractive force has a significant advantage over Einstein's: it explains why objects *fall towards* the Earth when dropped, and why astronauts who have spent months in space – and have become unaccustomed to gravity – *feel the Earth dragging them down* when they return. How can these seemingly real instances of attraction be explained by alterations in the geometry of space? As it happens, quite easily. In response to the question "Why do we feel pulled downwards?" Einstein answers, "We don't, we feel *pushed up*". We are prevented from following our natural free-fall trajectory (which is determined by our prior motion and the matter-controlled shape of space) by the surface of the Earth, and so we feel a force on us, as the Earth keeps preventing us from continuing on our free-fall path. Why is our free-fall motion downwards? It isn't. It is a greatly elongated corkscrew or spiral motion, since the Earth is moving forwards (around the Sun) and rotating; given the distortion in space produced by the Earth's mass, our free-fall slightly inclines us into the Earth's centre, and the Earth's surface pushes against us, preventing us from following our natural path. So *the only force applied to us is due to the Earth's surface preventing us from following our inertial path*. There is no gravitational force pulling us, just an alteration in the curvature of our local space.

If we accept the Einsteinian picture, it is not difficult to see why Newton thought that gravity was a force. Suppose that Bob and Sue set off to roller-skate around an artificial metal planet, using magnetic roller skates, which stick to the planet's smooth metal surface.[2] They set off along parallel paths, four metres north and south of the equator, each of which circumscribes a circumference of the planet, and they never deviate from these paths. Let us suppose that there is no gravitational pull exerted by the planet on Bob and Sue; they just coast along, on its perfectly smooth surface.

Bob and Sue don't realize that the planet they are on is spherical; they think it is a great big flat plane, which is just how it looks viewed from the surface.

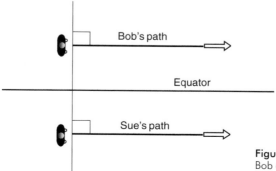

Figure 18.3 Roller-skating at the equator.
Bob and Sue, viewed from above, just prior to setting off on their around-the-world trip.

Consequently, they assume that, since they set off along parallel paths, the distance separating them will remain constant, unless some force pushes them off their original paths. Hence their surprise when they notice that they are slowly drifting nearer to each other. In an attempt to stave off collision they each take hold of the opposite ends of one of Bob's ski-poles, which they keep extended between them. But while this prevents their drifting closer together, they now feel a constant inward pressure. In effect, *they have to be pushed apart* to keep from coming together. Not surprisingly they conclude that some force is pulling them together: a mysterious force of attraction.

But if we look at them from above, from an orbiting spacecraft, we see that there is no force involved; nothing is pulling them together. Given the way they started off, they are merely following the "great circle" paths shown in Figure 18.4, and these paths gradually converge and intersect. These are their natural (inertial) paths given the intrinsic geometry of the Earth's surface, and the pressure on the stick is due to the force needed to deflect them from these paths. When they hold the stick between them, they are forcing themselves to follow paths that are not inertial given the Earth's geometry, and this is why they feel a force.

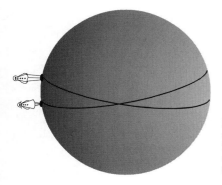

Figure 18.4 A meeting of ways. Bob and Sue continue to move straight ahead, without deviating to the right or the left; but since each is moving along a different great circle, their paths converge, and eventually cross.

Suppose now that a fat uncle follows them. Sue finds that he set off just behind her, and follows exactly her path. Since she is attributing her movement towards the equator to the action of a force, and since the fat uncle moves like her, she concludes that the force acts more strongly on greater masses, since it manages to make her uncle move just like she does. Finally, when they find that all bodies move in the same way, no matter what their masses, they conclude that the attractive force is proportional to the mass of each object: the more massive, the more force, and the same resulting path.

So Sue and Bob are led to postulate a force that looks like Newtonian gravitation: a universal force that is proportional to mass. But they are making a mistake. There is no mysterious force of attraction, only a non-Euclidean intrinsic geometry. Force-free objects are doing just what they should do: following straight-line paths unless acted on by a deflecting force.

The examples we have looked at so far involved *space*, and might seem compatible with a broadly Newtonian view of the universe. But of course, what is really being curved, according to Einstein, is the spacetime manifold. This means that there are time-dilation effects produced by gravity, as well as space-bending effects: clocks tick more slowly in the vicinity of large material objects; the stronger the spacetime curvature, the slower the clocks tick. Could Newton have explained gravity in space-bending terms, if he had given up the absolutist conception of space as something whose shape is fixed and unalterable, and independent of the bodies in space? At first sight it might seem so, but in fact the empirical data wouldn't have permitted it.

If gravity is a curvature in *space alone*, then we would expect all bodies starting from the same place and time to move along the same curved path under the influence of gravity. For example, suppose you are at the other side of the room from me, and I throw a ball towards you at the same time as I fire a bullet at you, aiming in the same upwards direction (so the bullet should miss your head). Both ball and bullet set off in the same direction, but one is moving much faster than the other. They won't travel in a straight line: due to the influence of gravity their paths will curve downwards. But will the curvature of both paths be the same? No, clearly not. The slow-moving ball's path will be strongly curved and the bullet will move in almost a straight line. This difference means we cannot explain the effects of gravity is terms of the curvature of space. It is quite different if we move to considering the event in spacetime, rather than space alone.

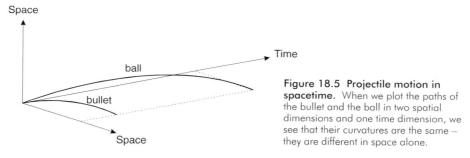

Figure 18.5 Projectile motion in spacetime. When we plot the paths of the bullet and the ball in two spatial dimensions and one time dimension, we see that their curvatures are the same — they are different in space alone.

When time is explicitly introduced, we can see that, although both ball and bullet travel the same distance, the ball is in flight for longer, and when this is taken into account, the curvature of its trajectory (in spacetime) is precisely the same as that of the bullet, which is what we would expect if both are in inertial motion through a medium with the same intrinsic curvature.

Hence we see why it must be spacetime, not space alone, that is the medium for the motion of masses, and so Newton could not have explained gravity in terms of purely spatial curvature.[3]

18.5 Evidence

It is one thing to show, in a qualitative way, by appealing to images and analogies, that gravitational effects can be explained by spacetime curvature, but another to show that this account is superior to alternative accounts. It is here that empirical evidence and precise quantitative predictions enter the fray. Despite the enormous difference in basic concepts between GTR and (orthodox) Newtonian theory, when it comes to predicting how material bodies should behave there is much common ground. For almost all the situations we can create in terrestrial laboratories the two theories generate virtually identical predictions, and for the most part the same applies to the behaviour of the planets. When planning interplanetary probes, for example, scientists still use Newton's equations rather than Einstein's: they are easier to work with, and the extent to which their predictions differ from Einstein's are too small to have any significant effect. But there are differences.

In the 1900s there were only two known discrepancies between Newton's law of gravity and the observed motions of the planets: one involved the Moon (and eventually proved to be due to erroneous observations) and the other Mercury. The orbit of Mercury is not a perfect ellipse. It completes each trip around the Sun at a slightly different position from where it set off. Newtonian theory could account for much of this "perihelion shift" but not quite all of it. There was a residual advance of 43 arc-seconds per century; a small difference, but a significant one, since astronomers at the time reckoned that their measurements were accurate to within 0.01 arc-seconds (equivalent to a hair's breadth at a distance of 10 km). The observed perihelion shift is precisely predicted by Einstein's equations. "I was beside myself with ecstasy for days", he wrote after making this discovery in 1916. An even more dramatic confirmation occurred in 1919. Einstein worked out that starlight passing very close to the Sun on its way to the Earth should be deflected by 1.75 arc-seconds. Dyson and Eddington (Director of the Greenwich observatory and Secretary of the Royal Astronomical Society respectively) mounted an expedition to Principe (off the coast of West Africa), where a total eclipse was due (only then are the stars adjacent to the Sun visible). The confirmation of Einstein's prediction made headlines around the world. In more recent years, using the more sensitive measuring equipment that has become available, there have been many further confirmations.

GTR does more than improve on the accuracy of Newton's theory. It has enabled physicists to postulate, and in some cases confirm the existence of, entirely new phenomena, such as spinning black holes, gravitational lensing and gravity waves. One of the theory's earliest and most striking predictions – that given certain plausible assumptions the universe as a whole must be either contracting or expanding – was confirmed by Hubble's discovery of the red-shift in 1928. We now know the universe to be expanding; what we don't know is whether it will contract again or not. More on this shortly.

These successes do not mean that GTR is true; Newton had many successes to his credit as well. But anyone who favours a generally realist stance with regard to scientific theories certainly has reason to take the theory seriously, especially in the absence of serious alternatives. Even if the theory in its current form will one day be supplanted, its core insight – that gravitational effects are manifestations of spacetime geometry – may well be retained.[4]

18.6 Equations

Not surprisingly, given that they have to accommodate all the different ways a four-dimensional continuum can bend and twist, the equations Einstein used to make these predictions are complex, and can only be fully understood by those with a thorough grounding in differential geometry; Einstein himself found the mathematics far from easy. They consist of a set of ten coupled, non-linear, partial differential tensor equations. The tensor formulation (see §13.4) means the various quantities represented are independent of any particular coordinate system; the tensor formalism also allows the equations to be expressed in very concise ways. One such is:

$$\mathbf{G} = 8\pi\mathbf{T}$$

On the left-hand side of the equation, the "Einstein tensor", **G**, describes the curvature of spacetime in a metrical form. On the right the "stress-energy tensor", **T**, describes the distribution, density and pressure of mass-energy and momentum at the same region of spacetime.[5] We know from STR that matter and energy are interconvertible, and various non-gravitational fields (such as the electromagnetic) also contribute to the energy-density – but to keep things simple I will often overlook these complications and refer simply to "matter distributions". So in slightly less schematic form we have:

[Spacetime curvature] = 8π [Mass-energy and momentum density]

The equations apply to every spacetime point – they are field equations – and tells us that, for any point P, the curvature in P's immediate neighbourhood is related to the mass-energy and momentum density around P. In effect, the equations establish a law-like connection between the intrinsic geometry of spacetime and the distribution of mass-energy through spacetime. Or, as Wheeler succinctly puts it, "matter tells space-time how to bend, and space-time tells matter how to move". Needless to say, given their complexity (concealed by the schematic forms used here) finding exact solutions for the equations is a formidable task. In practice, physicists usually make drastic simplifying assumptions (e.g. that mass is uniformly distributed) in order to render them manageable.[6]

Wheeler's formulation could mislead. It should not be taken to mean that mass *causes* spacetime to curve. Einstein's equations do no more than posit a systematic relationship between spacetime curvature and distributions of matter: certain distributions of matter can only exist with certain spacetime curvatures. There is no "one causing the other" here (or at least, the theory is agnostic on this issue); there is simply a constraint on the way physically possible worlds that conform to GTR might be.

18.7 Relativistic cosmology

When it comes to the *actual* geometry of the universe, Einstein's equations tell us less than we would like to know, for there is no one–one correspondence between global (or cosmic) matter-distributions and spacetime structures. The same distribution of matter can exist in spacetimes of different curvatures and topologies. To solve the equations at all, we need to assume certain boundary conditions (such as whether the universe is finite); the equations themselves do not tell us which boundary conditions apply in the real world, and different boundary conditions yield different geometries for the same matter distributions. More generally, in exploring the ramifications of GTR it is sometimes useful to consider universes very different from our own (e.g. entirely filled with a uniform dust and rapidly rotating); any possible universe whose matter distribution and spacetime geometry conforms to Einstein's equations is said to be a *model* of GTR. There are an infinite number of these "cosmological models", and at present we do not know which model best corresponds to our universe. Although recent astronomical results have allowed us to rule out certain classes of models, many possibilities remain open.

Before turning to the implications of GTR for the general shape and structure of the universe, mention should be made of the theory's most notorious by-product. In 1916 Schwarzschild used an exact solution of Einstein's equations to model the warping of space and time around a spherical star. The Schwarzschild solution clearly implied that if a concentration of mass exceeds a certain "critical value", the spacetime distortion will be such that nothing, not even light, will ever be able to escape from it.[7] Anyone unfortunate enough to cross the "event horizon" surrounding one of these concentrations would never be able to return. It was a while before the physics community settled on a name for objects in this condition – "compressed stars", "dark stars" and "frozen stars" were among the early candidates – but the term "black hole", coined by John Wheeler in 1967, was the one that stuck. Stephen Hawking has called them "possibly the most mysterious objects in the universe" (Thorne 1994: 11), despite the fact that a good deal has subsequently been discovered about their properties, with Hawking himself being a notable contributor. Interest in the phenomenon started to increase after 1930, when Chandrasekhar showed that stars with a mass above what is now called "the Chandrasekhar limit" should collapse into black holes when they run out of nuclear fuel to burn. Black holes were no longer a merely theoretical possibility. Evidence for their existence has accumulated rapidly in recent years, and it may well be that "super-massive" black holes – with billions of times the mass of the average star – can be found at the heart of most galaxies.

Of all the conundrums that black holes pose, perhaps the most interesting is the fate of the matter inside them. Does it reach a state, finite in volume, beyond which it cannot be compressed any further, or does it keep on collapsing until a "singularity" – a dimensionless point of infinite density – is formed? Since current physical laws "break down" (do not apply) to circumstances where infinite energies come into play, most physicists would have preferred the former answer, but it was not to be: in 1964, Penrose published a "singularity theorem" to the effect that all black holes must contain singularities. Subsequent work over the next decade convinced most physicists that this is indeed the case. Since a singularity is not a point *in* spacetime (at least in the usual sense), matter falling into a black hole vanishes from the universe altogether.[8]

Moving swiftly from the realm of the very small to the very large, I will outline the cosmological implications of GTR. The most significant assumption made in relativistic cosmology is that the universe is homogeneous and isotropic.[9] This so-called "cosmological principle" – a latter-day formulation of the Copernican principle that our location is in no way special – initially strikes one as highly unrealistic. How can a universe in which matter is concentrated into stars, galaxies and clusters of galaxies (not to mention black holes), all separated by huge tracts of empty space, be considered homogeneous? Size is the solution: at scales measured in hundreds or thousands of light-years, the distribution of matter is statistically uniform. And although a typical star contains a considerable quantity of matter compared with a typical human body, given the scale of cosmological distances (driving to the nearest star at normal motorway speeds would take millions of years) individual stellar matter concentrations have insignificant effects on the large-scale curvature of spacetime. (Think of the effect of a grain of sand on a taut trampoline.) The Cosmic Background Explorer (COBE) satellite data on the microwave radiation that pervades the universe – usually supposed to be a remnant of the big bang – supports the isotropy assumption; the temperature of the universe seems very much the same (to one part in a hundred thousand) in every direction.

The cosmological principle has an important geometrical consequence. If the distribution of matter is uniform, then spacetime (and so the universe) must have a constant average curvature. This constant curvature can be positive, negative or zero. No other options exist. While this result simplifies things considerably, many possibilities remain open. Quite how many possibilities was not at all obvious when Einstein formulated GTR – much of the relevant mathematics had not been done – and is still a topic of some controversy today.

When Einstein was formulating his field equations it was widely assumed the universe was static. The available astronomical data supported this assumption: the relative motions of the stars appeared to be small (the existence of other galaxies had still to be established). But a static universe posed a problem for Einstein. His equations told him that a universe would not remain static for long, since gravitational effects would lead to an overall contraction; portions of matter would move towards one another, due to the curvature in spacetime that they themselves create. Determined to save appearances, Einstein introduced an additional factor into his equations, the "cosmological constant", λ, which was used to represent a repulsive force whose value is fixed so as to precisely counterbalance the "attractive" effects of gravity. The resulting theory of the overall shape of the universe – which appeared in Einstein's 1917 paper "Cosmological considerations on the general theory of relativity" – had an appealing simplicity. The universe Einstein described was of uniform positive curvature and *closed*: finite in volume but without boundaries or edges. Recall the example of Flatlanders confined to the surface of a sphere; the spatial component of Einstein's universe is analogous: it is the three-dimensional surface of a four-dimensional hyper-sphere. (Remember, too, that, as in the Flatland case, there is no interior to the sphere.) Despite its appeal – not the least of which was the solution to the long-standing problem of how space could be finite without possessing a boundary that someone could easily cross – this cosmological model was not entirely satisfactory, as Einstein was well aware. Not only does the cosmological constant have the air of an *ad hoc* face-saving device, but it has a peculiar feature in its own right. Other known forces (such as magnetism) become weaker with distance, but the λ-force becomes stronger with distance. A global force of this kind, acting across distances, also conflicted with the Machian aspirations Einstein still held dear.

Despite these difficulties, for a brief time Einstein hoped that GTR, when combined with the homogeneity and isotropy assumptions, would have just one cosmological solution. These hopes were soon dashed. Other solutions – all compatible with the constraints deriving from the cosmological principle – soon appeared. In the 1920s the Belgian cleric-mathematician Georges Lemaître, and the Russian geo-physicist Aleksandr Friedmann, discovered a variety of *non-static* solutions to the GTR equations; some of these models were of expanding universes, others of contracting universes, some employed a cosmological constant, others did not. Moreover, a serious flaw in Einstein's initial model was discovered by Lemaître in 1927. It turned out that Einstein's static model, using a λ-force that precisely cancels gravitational attraction, was radically unstable: the slightest irregularity in matter distribution would result in expansion or contraction. So even if Einstein's model did apply, our universe would be unlikely to have remained static for long; it would be expanding or contracting. But the most influential discovery came from an astronomer. In 1929 Hubble published convincing evidence, based on the "red-shift" of distant galaxies, that the universe is actually expanding. The rate of expansion is fast (at least by terrestrial standards), and increases with distance; the more distant a galaxy, the faster it is receding. Hubble's discovery led to renewed interest in the non-static models of Friedmann and Lemaître, not to mention the latter's deduction that an expanding universe must have originated from a far smaller "cosmic egg": what we now call the "big bang singularity". Einstein embraced the expanding model of universe, and repudiated the cosmological constant – calling it his "greatest mistake" – not surprisingly regretting having missed the opportunity of going down as discoverer of the dynamic cosmos.[10]

By the end of the 1930s relativistic cosmology had stabilized somewhat. Most physicists (with the notable exception of Lemaître) followed Einstein in rejecting the cosmological constant; there were still those who resisted the idea that the universe

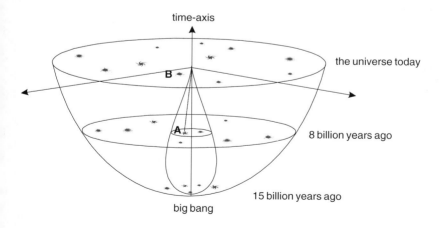

Figure 18.6 The visible universe. The "teardrop" shape shows the region of the expanding universe that is visible from present day Earth (located at the intersection of the two spatial axes). Light emitted by objects outside this region has not had time to reach us yet. Light from galaxy A has taken 8 billion years to reach us, so we see it as it was 8 billion years ago. Galaxy B will not be visible from Earth for another 8 billion years or so.

originated in a big bang, but it was widely assumed that it did, and so attention turned to other questions.[11] How old is the universe? What is its rate of expansion? Is its curvature positive, negative or zero? Is it open or closed? The answers to these (still unresolved) questions depend on a number of interrelated factors, not least of which is the average mass-density of the universe, which has proved difficult to estimate with any precision. If the density is above the "critical value", W, the universe will eventually slow to a halt and then collapse back in on itself to a "big crunch"; if the density is less than W it will keep on expanding forever.[12] Recent evidence suggests that the mass-density falls a good way short of W.

Figure 18.7 shows some of the more influential cosmological models of the period in question. A and B are the closed and open Friedmann (or Friedmann–Robertson–Walker) models, of positive and negative curvature respectively. C depicts the 1932 Einstein–De Sitter model, which was a popular model for some sixty years but currently out of favour (it cannot be reconciled with current estimates of the age of the universe). This model posits a flat and open spacetime, with a steadily decelerating rate of expansion. D shows Lemaître's (1931) "hesitating universe" model: after an initial fast expansion, the universe (whose curvature is positive) slows for a while before accelerating again. This model – which alone of the four includes a cosmological constant – has recently found favour with some cosmologists, who argue that a repulsive force of some kind is needed to explain the observed rate of expansion.

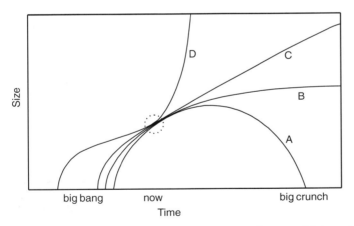

Figure 18.7 The standard big bang models. The different models have different implications for both the distant future and the distant past (Luminet 2001: 64).

Despite their various differences, these models share a common feature: all are topologically timid. The spatial components of these spacetimes are all *simply-connected* (in the manner of a plane); none are *multi-connected* (in the manner of a torus).[13] Friedmann pointed out the compatibility of GTR with multi-connected topological manifolds in 1924, but the suggestion fell by the wayside, despite the fact that Einstein's field equations are entirely neutral with respect to this issue. However, the prejudice in favour of simply-connected models lifted in the 1980s and 1990s, and the possibility that our universe is multi-connected began to be taken seriously.[14] The number of possible multi-connected spaces of constant curvature (flat, positive and negative) is vast, and many are highly complex. But even the simpler cases (such as the hypertorus) have the potential for creating mischief, in the form of topologically generated "ghost images" of nearby galaxies.

The shaded circle in Figure 18.8 represents the region of the universe that is observable; objects outside this region are so distant that the light they emit has yet to reach us, so we cannot detect them (see Figure 18.6). The central square with a bold boundary represents the entirety of the actual universe (only a few stars and a single galaxy are depicted); the remainder of the squares are filled with ghost images produced by light from the same few stars and galaxies weaving endlessly back and forth through the multi-connected spacetime manifold before being detected by telescopes on Earth.

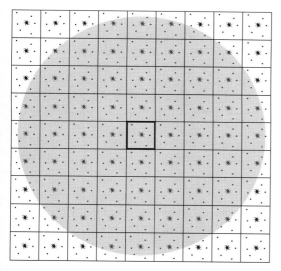

Figure 18.8 Topological ghost images on a cosmic scale, creating an illusion of infinity. The night sky of this small toroidal universe is filled with repeated images of the same few stars and galaxies (Luminet 2001: 116).

In one respect this depiction is misleading. Since a galaxy emits light in all directions, although we would always be observing the same astronomical objects, we would be observing them from many different angles: we would see each galaxy from behind, below, the right, the left (relative to the Earth), and all angles in between. Also, since light from a given object can make many "passes" through a multi-connected space before arriving in our neighbourhood, the heavens will be filled with images of the same galaxies at different stages of their career, and in different configurations. In principle, this opens up the possibility of seeing our own sun and solar system as they were at different points in the past. All we need do is detect a ghost image of our own region of the galaxy, which originated in the desired historical period, and point a sufficiently powerful telescope in the appropriate direction. Unfortunately, if Gott is correct (2001: 80) in his calculations (based on assuming a toroidal topology), we can expect the closest image of our galaxy to be 5 billion light-years away, and hence 5 billion years old – a billion years before the Earth itself formed.

In combination, these factors make the "ghosts" very difficult to detect. It would be far from obvious to Earth-bound astronomers that they are observing

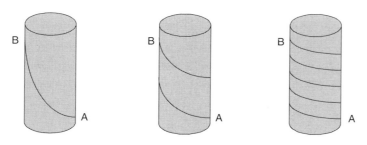

Figure 18.9 Winding ways. The length of a path linking two points on the surface of a cylinder depends on how many times it winds around *en route*. The same holds in multi-connected spaces; in many instances there is an additional complication: there may be more than one route between any two points (in a toroidal space, for example, the "windings" can go in two opposite directions). Since the speed of light is finite, ghost images will often be considerably out of date.

repeated images of the same galaxies. Nonetheless, in principle it should be possible to ascertain whether this is in fact the case. Various methods of detecting the distinctive signatures imposed by multi-connected topologies on galactic distributions have been proposed, and the search for these patterns is being pursued by a number of cosmologists. Although these enquiries are still in their preliminary stages, there is no decisive evidence so far that our universe is *not* multi-connected (Luminet 2001).

Interestingly, the possibility of a multi-connected universe was noted by Schwarzschild (who, as we have seen, would later make significant contributions to GTR) as early as 1900.

> One could imagine that as a result of an enormously extended astronomical experience, the entire universe consists of countless identical copies of our Milky Way, that the infinite space can be partitioned into cubes each containing an exactly identical copy of our Milky Way. Would we really cling on to the assumption of infinitely many repetitions of the same world? ... We would be much happier with the view that these repetitions are illusory, that in reality space has peculiar connection properties so that if we leave any cube through a side, then we immediately reenter it through the opposite side. (Luminet and Roukema 1999: 2–3)

Schwarzschild is certainly correct about the way we would be inclined to interpret such findings; the "strange geometry" hypothesis would have much the greater appeal. So we have here a further kind of case that undermines Poincaré's claim that we would always opt for a cosmological hypothesis that posits the simplest (i.e. Euclidean) geometry.

19 Spacetime metaphysics

19.1 Substantival spacetime

Newtonian space is both substantival and absolute: able to exist in the absence of matter, possessed of an immutable Euclidean geometry, it is entirely unaffected by the presence and distribution of matter within it. The spacetime of GTR is very different; its geometry is variable, and affected by the presence and activities of material things. GTR's spacetime is certainly not *absolute*, but is it *substantival*? Einstein certainly came to think it was. In his 1920 Leyden lecture "Ether and the theory of relativity" he remarked that the theory "has, I think, finally disposed of the view that space is physically empty" (1922: 18). It is not hard to see why he adopted this view.

We saw in §14.2 and §14.3 that in variably curved spaces – for example, worlds where non-Euclidean "holes" occur at irregular intervals – the relationist faces a near-impossible task in seeking to explain the behaviour of bodies solely in terms of object–object forces and relations.[1] Since the internal tensions suffered by a body moving through such a hole cannot be explained by alterations in the shape of space, they must be explained in terms of relations between the body in question and other bodies; since the other bodies may be in any number of different configurations, the laws involved – if it proves possible to formulate any – will be horrendously complex.

Since, according to GTR, variations in the geometry of spacetime are related to the movements of matter, areas of strong localized curvature sitting motionless in empty tracts of spacetime are unlikely to be encountered. However, GTR does posit dynamic equivalents of the non-Euclidean hole, which prove just as problematic for the relationist, the most dramatic example of which is the phenomenon of *gravity waves*. Large rotating concentrations of mass (e.g. a pair of black holes) generate outwards spiralling ripples in the fabric of spacetime, and these mobile geometrical disturbances of varying curvature affect the shape and trajectories of any material objects they encounter (although the waves themselves are *un*affected by any matter they pass through). A gravity wave traversing the Earth stretches and squeezes it in directions at right angles to its own. Moreover, the precise way a gravity wave affects a material body will depend on the geometry of the local spacetime, which varies from place to place depending on local matter distributions. Is it possible to explain the subtle and complex motions in material systems that would be produced by a gravitational wave (or worse, many such waves acting simultaneously) solely in terms of forces and laws that relate material bodies themselves? Perhaps, although it would not be easy. But even if it could be done, the resulting theory would bear no resemblance to GTR. It would be an altogether different type of theory.

The way in which GTR makes use of variations in the geometry of *empty* spacetime regions in explaining the motions of bodies renders relationist reinterpretation highly problematic. The spacetime manifold is not just a convenient heuristic device; it is an agent in its own right.[2] We would be led to the same conclusion if cosmologists were to discover evidence that our universe is multi-connected: what manner of laws governing objects and their spatial relations could explain the behaviour of objects in such worlds? It is easier and more plausible by far to find in favour of a strange but substantival spatiotemporal manifold.

There is a further fact, specific to the spacetime of GTR, that seems decisively to confirm its substantival character: gravitational *energy*. The following description of the fate of a binary system comprising two black holes provides an illustration of the way energy can be transmitted back and forth between objects and spacetime:

> As they depart for outer space, the gravitational waves push back on the [black] holes in much the same way as a bullet kicks back on the gun that fires it. The waves' push drives the holes closer together and up to higher speeds; that is, it makes them slowly spiral inward towards each other. The inspiral gradually releases gravitational energy, with half of the released energy going into the waves and the other half into increasing the hole's orbital speeds . . . the closer the holes draw to each other, the faster they move, the more strongly they radiate their ripples of curvature, and the more rapidly they lose energy and spiral inward.   (Thorne 1994: 358–9)

Here we see objects transferring part of their energy into spacetime, which then transmits it at light speed through empty space in the form of gravitational waves. How can something that can behave like this be anything other than substantival in nature?

19.2 Mach's Principle

The question of whether the spacetime of GTR is substantival or not is sometimes addressed by focusing on a different but related issue: does GTR confirm Mach's Principle? It will be recalled that, in the context of Newtonian dynamics, Mach asserted that there was no need to posit an absolute space to explain inertial phenomena, for the latter could all be explained solely in terms of relationships between material things (e.g. the water rises in Newton's rotating bucket because it is moving relative to the centre of gravity of the universe, not absolute space). Transferred into the context of GTR, Mach's Principle can be restated in a weak (WMP) or strong (SMP) form, depending on whether or not we include only inertial structure (which Mach himself was primarily concerned with), or all geometrical features.

- WMP: The inertial structure of spacetime is uniquely determined by the distribution of mass-energy through all of spacetime.

- SMP: All geometrical features of spacetime are uniquely determined by the distribution of mass-energy through all of spacetime. ⟩

As the stronger principle, SMP entails WMP.[3] It may well be that SMP must be satisfied if relationism is to be viable, but the converse does not follow. Even if GTR were Machian in this strong way, the properties it ascribes to spacetime are such that it is hard to see how it could fail to be *substantival*. The fact that an entity's structure is entirely dependent on how some other entities are distributed within it does not *in itself* entail that only the latter entities exist. As it happens, and despite Einstein's own hopes, GTR does not satisfy Mach's Principle, in either form, for, as already noted, the same distributions of matter are compatible with very different spacetime structures. It is worth exploring this point a little further, for the light it casts on the extent to which spacetime structure and mass distribution are interwoven in GTR.[4]

The environment of STR is extremely hostile to the Machian: the distinction between inertial and non-inertial paths is built into the fabric of Minkowski spacetime, and is entirely independent of the presence or absence of material bodies. In a universe consisting of just one particle, the distinction between inertial and non-inertial (straight versus curved) paths and worldlines remains valid. It is not hard to see why the innovations introduced by GTR raised hopes among Machians. The theory states that the geometry of spacetime is dependent on the distribution of material bodies; furthermore, the theory predicts that a stationary object that is surrounded by a large quantity of rotating matter will experience inertial effects similar to those that would be produced if the object itself were rotating – which supports Mach's contention that inertial effects are entirely due to the relative motions of bodies. But deeper explorations soon reveal respects in which GTR is profoundly non-Machian.

SMP fails because the same mass distributions can occur in spacetimes of different topologies. The spacetime that houses a finite matter system may itself be finite (closed) or infinite (open) – the matter distribution itself does not determine which. So far as we can tell from observations, our universe is expanding and (at least approximately) isotropic (similar in all directions), and homogeneous (similar at all points). Ellis (1978) has shown that there is a model of GTR that is compatible with these observations but where spacetime is inhomogeneous and anisotropic. The matter in this universe pours out of one singularity and is sucked into another; globally this universe is not expanding, but it seems to be in the region around the source singularity.

If WMP were true, two consequences would follow. First, universes that contained no matter would be devoid of inertial structure; if inertial structure is determined by matter distribution, then no matter implies no structure. Secondly, it would be absurd to suppose that all the matter in the universe could be rotating; or at least, if the matter-field *were* rotating it would have no empirical consequences (since the relations between material things remain the same as in the non-rotating case), and so the distinction between a stationary and rotating matter-field would have no physical content. Unfortunately for the Machian, there are models of GTR that refute both predictions.

In most open GTR universes, regions of spacetime that are a long way from any matter have the structure of flat Minkowski spacetimes, so it is reasonable to suppose that in an *entirely* matter-free GTR universe the spacetime would take the same form. But since in Minkowski spacetime the distinction between inertial and non-inertial worldlines is well defined in the absence of matter, it would be here too. And as Sklar notes, GTR allows for stranger possibilities:

> Spacetime curvature has its own gravitational self-energy. So it is possible to have non-zero curvature in an empty universe, or for there to be regions of curved spacetime whose deviation from flatness is supported by no matter at all but simply by the self-energy of the curved spacetime region.   (1992: 78)

In empty worlds such as these the inertial structure is quite definite, but quite different from that of flat Minkowski spacetime.

There are several solutions to GTR's field equations that posit global rotation. The Oszváth and Schücking solution characterizes a globally rotating, spatially closed, homogeneous dust-filled universe; Kerr's solution is for an isolated rotating mass; in Gödel's solution matter is represented as a perfect fluid present throughout spacetime, which is rotationally symmetric about all points. Although all the matter in a Gödel universe is rotating, there are experiments that would reveal that for each spacetime point there is a certain distinctive plane, and light (or any other particle) sent out along this plane will trace out a spiral path (just like a marble propelled from the centre of a rotating turntable). Consequently, it would seem to *all* observers that they themselves are at the centre of the universal rotation. This is a bizarre state of affairs, but, as we shall see, the Gödel universe has still stranger properties.

There are, then, models of GTR that are radically non-Machian in character, in both the strong and weak sense, and so GTR is not a Machian theory. But there are also models that *are* Machian. Closed Friedmann models, which characterize finite isotropic homogeneous dust-filled universes, are not globally rotating, and their inertial and metrical structures are uniquely determined by their matter distributions. They thus satisfy WMP. (Since matter-fields of the same type may also occur in *open* spacetimes, they do not uniquely determine topology, and so fail to conform to SMP.) Since Friedmann models provide good approximations for our universe – they are the standard way of representing big bang cosmologies – the Machian cause is not entirely lost. But it should not be forgotten that, like all GTR models, these are only approximations. The stress-energy tensor specifies only the gross, large-scale patterns within the matter-field; local variations are simply not registered. So given the non-Machian character of GTR itself, there may well be regions within the observable parts of our universe where WMP fails. And, of course, it could be that the parts of the universe we can observe are not representative of the larger picture; if so the models of GTR that correspond most closely to the global structure of our spacetime may be very different from the Friedmann models that satisfy the strictures of WMP.

19.3 The hole argument

While it is generally conceded that GTR renders substantivalism (of some description) hard to avoid, the mathematical formalism in which the theory is couched renders this commitment problematic. In the 1980s the significance of "the hole argument" – a version of which dates back to Einstein (in about 1913) – was recognized by Stachel, and deployed against orthodox substantivalist interpretations of GTR by Earman and Norton. Reading the already considerable literature on the hole argument brings some clarity to how the issue of substantivalism fares in GTR, as well as new puzzles and perplexities.

As we have already seen, modern spacetime theories start by positing a manifold of spacetime points and then proceed to add additional structures. In the case of GTR, the manifold, M, is a collection of points possessing certain topological properties – it is four-dimensional and continuous – and the additional structures consist of a metric field and a matter-field, defined by the metric tensor, g, which specifies the distances and geometrical relations between points, and the stress-energy tensor T, which provides an approximate representation of the distribution of matter and energy. Given that all models of GTR have the three components [M, g, T] the substantivalist is faced with a question: if spacetime is a substance, which of these components *constitute* this substance? Is it the manifold, or manifold + metric, or is it manifold + metric + matter-field? Since the last answer can be ruled out – it obliterates the distinction between spacetime and matter – the real issue is whether we take spacetime to be M alone or [$M + g$]. Earman and Norton argue that the substantivalist should opt for M alone:

> The advent of general relativity has made most compelling the identification of the bare manifold with spacetime. For in that theory geometric structures, such as the metric tensor, are clearly physical fields in spacetime. The metric tensor now incorporates the gravitational field and thus, like other physical fields, carries energy and momentum . . .
> If we do not classify such energy bearing structures as the [gravitational] wave as contained within spacetime, then we do not see how we can consistently divide between container and contained. (1987: 518–19)

Traditionally substantivalists have viewed space as a sort of container: it is that extensive medium that contains the material world. Earman and Norton argue that if we wish to retain this fundamental idea in the spacetime framework of GTR we must identify the "container" (space) with the bare manifold M, for if we include the metric we no longer have a sharp distinction between spacetime and matter, between container and contained. According to GTR the metric field contains energy, and hence has a "material" aspect since matter and energy are interconvertible.

So, let us suppose that the substantivalist adopts this view: spacetime is identical with the manifold of points. A problem now arises.

GTR is a *generally covariant* theory, which (roughly) means that the mathematical form of the laws of physics are the same no matter which system of coordinates are used to map spacetime points. A coordinate system is a way of assigning numbers to the manifold's points; GTR requires that we can smoothly assign four numbers to each point in the manifold – thus guaranteeing that spacetime is four-dimensional and continuous (there are no gaps between points), and hence differentiable. There are any number of different coordinate systems that one can adopt (different ways of assigning numbers). General covariance is a desirable property for a spacetime theory to have. Something would be wrong if the mathematical form of basic laws depended on an arbitrary choice of coordinates, so all spacetime theories (not just GTR) are nowadays presented in a generally covariant form. But while desirable, general covariance also brings a potentially problematic freedom.

If we take one model of GTR [M, g, T] we can apply a *diffeomorphism* – a permutation (or switching about) – of points in M satisfying certain restrictions; for example, the deformation of the original must be smooth (no topology-destroying breaks or tears). This transformation – call it *h* – yields a different model [M, *h**g, *h**T]. Although this new model describes a universe that is observationally equivalent to the original, there is a significant divergence: it spreads the matter and metrical fields over the points of the manifold in a different way, and so the same physical processes are differently located in spacetime.[5] Since there are any number of diffeormophisms that can be applied to the original model, there are any number of ways the material contents of the universe can be located with respect to the manifold.

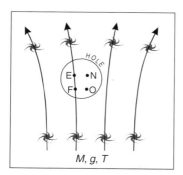
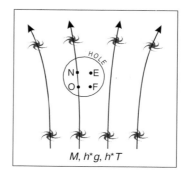

Figure 19.1 The hole diffeomorphism. The hole transformation affects what happens where. The diagrams show the same four galaxies in free-fall motion, in an expanding universe. In the original GTR-model, the galaxy that traverses the hole passes over a sequence of points that includes E and F; in the second GTR-model, E and F are unoccupied, and the galaxy passes over a sequence of points that includes N and O (which are themselves unoccupied in the first model). The locations of processes outside the hole-region are unaltered.

To illustrate the problem this freedom poses for the substantivalist Earman and Norton focus on a particular kind of transformation: a *hole diffeomorphism*. The "hole" is a particular region of spacetime; it is not a hole *in* spacetime (i.e. an absence of points) nor need it be empty of matter. The diffeomorphism leaves all assignments of matter to points *outside* the hole region unchanged, but alters the assignments *within* the hole; hence the transformation assigns different spatiotemporal locations to the material events occurring within the hole. So, for example, the original model might say that the worldline bundle of a given galaxy passes through a spacetime region E, whereas the new model (after the hole transformation) says that the galaxy doesn't pass through E: it occupies a new collection of spacetime locations. This is problematic because it can be viewed as a failure of determinism. As Norton writes:

the physical theory of relativistic cosmology is unable to pick between the two cases. This is manifested as an indeterminism of the theory. We can specify the distribution of metric and matter fields throughout the manifold of events, excepting within the region designated as The Hole. Both the original and the transformed distribution are legitimate extensions of the metric and matter fields outside the Hole into the Hole, since each satisfies all the laws of the theory of relativistic cosmology. The theory has no resources which allow us to insist that only one is admissible.   (1999: 5)

With reference to our earlier example, given complete data about what occurs outside the hole, GTR cannot predict whether or not the galaxy in question will pass through region E. More generally, since we can select *any* region as the hole, GTR cannot predict which events will happen where, even given maximal data concerning what happens outside the chosen hole. As Hoefer puts it:

If A is a spacetime point that will (in fact) exist around here sometime tomorrow, the past plus the field equations do not determine whether A will underlie me, or you, or some part of a star in a distant galaxy.   (1996: 9)

The hole transformations do not render anything that is observable indeterminate; they leave all predictions concerning the relative distances between objects intact. The indeterminism only extends to what happens where: the spread of material processes over the manifold of points. An indeterminism of this sort will not be a problem for a relationist who regards spacetime as a fiction, but substantivalists cannot dismiss it lightly. For any model of GTR that accurately represents the observable universe, there are many other models, observationally equivalent to the first, that situate any given material process on a different collection of spacetime points.

Earman and Norton recognize that we cannot reject a theory just because it is indeterministic in some respect, or else we would have to reject quantum theory and much classical physics. However, they argue that the *way* in which determinism fails in the case of GTR is deeply suspect: it doesn't fail for any reason *of physics*, but rather because of a commitment to a particular *metaphysic*.

The argument . . . does not rest on the assumption that determinism is true, much less on the assumption that it is true *a priori*, but only on the assumption that it be given a fighting chance. To put the matter slightly differently, the demand is that if determinism fails, it should fail for a reason of physics.
(Earman 1989: 180)

As for why the failure of determinism is due to reasons of metaphysics rather than physics, recall Leibniz's argument against Newton. We called worlds where all inter-body relations are the same "Leibniz equivalents". Leibniz is committed to the doctrine of Leibniz equivalence: the claim that there is no physical difference between universes that do not differ in any observable way, but do differ in the way the bodies are related to a hypothetical underlying space. Since Newton is committed to the view that the material bodies *could* be differently located with respect to absolute space, and that this difference is a genuine physical difference, Newton rejects the principle of Leibniz equivalence. Earman and Norton argue that in the context of GTR there is only a failure of determinism if we reject Leibniz equivalence: for if we accept Leibniz equivalence then *there is no longer any physical difference* between the models of GTR that are generated by hole transformations. It is only if we insist that the spacetime points in M are real physical individuals that the transformations represent physically different states of affairs. If we reject this metaphysical commitment to substantivalism, we are free to accept Leibniz equivalence, and determinacy is restored. As Norton puts it:

> If we deny manifold substantivalism and accept Leibniz equivalence, then the indeterminism induced by a hole transformation is eradicated. While there are uncountably many mathematically distinct developments of the fields into the hole, under Leibniz Equivalence, they are all physically the same. That is, there is a unique development of the physical fields into the hole after all. Thus the indeterminism is a direct product of the substantivalist viewpoint. Similarly, if we accept Leibniz equivalence, then we are no longer troubled that the two distributions cannot be distinguished by any possible observation. They are merely different mathematical descriptions of the same physical reality and so should agree on all observables . . . the manifold substantivalist advocates an unwarranted bloating of our physical ontology and the doctrine should be discarded.   (1999: 8)

19.4 Metrical essentialism

Since the hole argument (at least in its original form) applies to any spacetime theory that is formulated in a generally covariant way, not just GTR, those inclined (for general metaphysical reasons) towards substantivalism may well be tempted to dismiss it as no more than an artifact of a particular mode of mathematical representation. But while there is something to be said for this response – after all, we have already seen that there are very strong grounds indeed for supposing that the spacetime of GTR is substantival – it carries little weight unless accompanied by a viable alternative formalism, one that renders GTR immune to the hole argument. Needless to say, such alternatives are not easy to come by. But there are less radical responses available, responses that meet the threat posed by the hole diffeo-morphism while retaining GTR in its standard form. Indeed, there are several such responses, and a full survey cannot be supplied here.[6] But to provide some indication of the way explicitly metaphysical considerations lie at the heart of the debate I will explore the rationale of an approach that several writers have found appealing.

Earman and Norton argue that if one is to embrace both GTR and substantivalism, then one should endorse *manifold substantivalism*; that is, of the three components of GTR models $[M, g, T]$ we take M alone to represent substantival spacetime. If we do, then it makes sense to think of the very same spacetime points being differently related to g and T, and so the very same spacetime points can have different metrical properties and be differently related to the spread of mass-energy. This particular conception of substantivalism is crucial to the success of the hole argument. For the hole transformations to be *possible* we must be able to make sense of the very same spacetime points being differently related to the metrical fields and matter-fields. But suppose we adopt a different version of spacetime substantivalism: *metrical essentialism*. On this view, we take substantival spacetime to be identical with a particular metrical field; that is, spacetime points that are structured by a metric. The distance relations between points (and their geometrical properties) are now taken to be essential to them: it no longer makes sense to suppose that a given point could exist in different metrical relations to other points.

If the substantivalist adopts this position, the hole argument cannot get off the ground: the hole alternative involves numerically the same points bearing different metrical relations to one another, so if we adopt metrical essentialism, the hole alternative is metaphysically impossible. Consequently, the empirically equivalent alternatives generated by the hole transformations do not exist, and the associated failure of determinism is averted.

An example may make things clearer. Suppose our original model of GTR situates the physical events in France at a given time on a collection of points situated where we normally take France to be – between Germany and the UK. Call this collection of points P(F), and call the collection of points that underlie the events in Germany at this time P(G). The proponent of the hole argument maintains that there is an observationally equivalent model of GTR in which P(G) underlies the goings-on in France, and P(F) underlies the goings-on in Germany, so the very same points that in the original model were where we normally take Germany to be are now where we normally take France to be, and the corresponding "French" points have been shifted eastwards. Since the equations of GTR together with occurrences outside the Franco–German region do not allow us to choose between these two models, we have a failure of determinism. The metrical essentialist will argue thus:

> The competing model represents the same spacetime points existing in different relations to both material processes *and* other spacetime points. This makes perfect sense if we hold that spacetime is a manifold of points that can be rearranged in this sort of way without losing their identities. But this conception of point identity is not obligatory. If we adopt the view that the identity of a point is fixed by its spatiotemporal relationships with *other points*, the competing model cannot exist. A spacetime is a collection of points with a *specific spatiotemporal organization*.

Since GTR invites a substantivalist interpretation, the fact that metrical essentialism immunizes it against the hole argument – which threatens such an interpretation – is one reason for adopting the doctrine. Another is the suspect character of the main alternative: manifold substantivalism.

Recall what *M* consists of: a collection of points that form a smooth four-dimensional continuum. In the absence of a metric there are no spatial or temporal distances, no affine structure; the light-cone structure is absent and so past and future are not distinguished. Our spacetime is four-dimensional (at least on the macro-scale), but it seems very plausible to suppose that something more is needed, beyond mere four-dimensionality, for a collection of points to be a *spacetime*. How could a collection of points for which the distinction between *space and time* does not exist constitute a spacetime? Yet this distinction does not exist prior to the imposition of the metrical field on *M*. The manifold substantivalist might object:

"Look, in moving beyond Newton and adopting relativity, you've had to take on board some pretty strange ideas about space and time; common sense has largely gone out of the window. Might it not be that the truth about space and time is stranger still?" This might seem to be a reasonable response, but it is still problematic. In urging us to recognize that spacetime is a real entity in its own right, the substantivalist points to the useful work that spacetime does in unifying and explaining physical phenomena such as acceleration, inertial motion and light propagation. But it is the *metrical* field of a GTR spacetime – which brings with it inertial structure, spatiotemporal distances and light-cone structure – that does this work, *not* the bare manifold, which in itself explains very little (cf. Hoefer 1996: 12).

It seems, then, that substantivalists have a good reason for rejecting manifold substantivalism in favour of metrical substantivalism, of some form or other. What of the charge that only manifold substantivalism saves the distinction between "container" and "contained" from obliteration? Earman and Norton argue that, since the metric field carries energy and momentum (as shown by the phenomena of gravity waves), if we allow spacetime to include the metric field, we lose the sharp distinction between spacetime and its contents. But the metrical essentialist can ask:

> It is true that Newton assumed space was devoid of *material* characteristics such as energy and momentum, but why must the contemporary substantivalist seek to retain the distinction between container and contained in its traditional form? In accepting GTR we have accepted that Newton was wrong about a good deal, why not accept that he was wrong about this too? Since the spacetime of the metrical substantivalist contains energy and momentum, the case for recognizing spacetime as a *substance* is all the stronger: spacetime is now a physical entity in its own right, one which has causal powers just like any other physical entity – spacetime becomes all the more substantival!

On the face of it, this seems a viable position (we shall be considering it from a different vantage point in the next section).

Earman and Norton have a further argument for excluding the metric from spacetime, an argument that appeals to the very nature of GTR. According to GTR the metric is *dynamic*: the geometry of spacetime is (in part) a function of the distribution of matter – g and T are interdependent – so different matter-fields yield different metrics. But surely, they continue, an *essential* feature of relativistic spacetime must be a feature that is *invariant*, a feature that occurs in all models of GTR. Since metrical structure is dynamic, varying from model to model, it is surely right to classify it as an accidental, rather than essential, feature of spacetime. Hence we are led to the view that spacetime *essentially* consists of the manifold (viewed as a four-dimensional continuum of points) and nothing but.

It is hard to see that this argument has any real force. For one thing, if we were to follow the logic of Earman and Norton's argument all the way, and hold that the essential features of spacetime must be invariants, then topological properties become non-essential too, since there are models of GTR with different topologies. We would thus be led to the *simple point substantivalism*: spacetime is a collection of points that need not be related to one another spatiotemporally at all. Needless to say, this is a singularly unappealing doctrine. But more importantly, there is no compelling rationale for the doctrine that essential properties must be invariant across models. If we were still working in a Newtonian framework, where space endures through time, then a dynamic metric would clearly be incoherent: since the *very same* spatial points would have different metrical properties at different times, we could scarcely claim that metrical properties are essential to points. But in the four-dimensional framework of GTR we have abandoned enduring points; a spacetime is composed of momentary points, ephemeral entities that do not endure. There is surely no obstacle to maintaining that the metrical properties of *these* entities are essential to them.

Metrical essentialism comes in different forms. Hoefer's preferred version leads him not only to reject primitive identities for points, but to hold that points are to be individuated by reference to both their relations to other points and material processes. Consequently, he maintains, substantivalists should embrace Leibniz equivalence and hold that all "qualitatively" (observationally) identical worlds are physically identical. If we take this step, it no longer makes sense to suppose that all the material bodies in the universe could have existed three feet to the east of where they actually are.

While this position may have a certain appeal from a purely scientific point of view – observationally equivalent worlds are physically identical – it is less appealing at the metaphysical level. If spacetime is a genuine substance, in worlds where physical law makes spacetime structure independent of material bodies, why couldn't God (or chance) locate the material system elsewhere? As Nerlich stresses, given the symmetries of Newtonian spacetime, the substantivalist has a perfectly respectable explanation of why Leibniz equivalents are observationally equivalent. But they are not "qualitatively identical" if we take (as a substantivalist should) the domain of real properties to include spacetime location: worlds in which the same objects are differently located are *not* qualitatively identical from the standpoint of a full-blooded substantivalism. Hoefer construes the doctrine of primitive identity as entailing that two actual spacetime points could have *all* their properties systematically interchanged, and (seemingly) assumes that if we reject this doctrine we should hold that all of a point's actual properties are essential to it. But the metrical essentialist need not go so far. The option of holding that points have their metrical relations to other points essentially is all that is required, and this limited essentialism allows the substantivalist to accept the logical possibility of the same spacetime existing without material objects, or with material objects differently distributed (as in Newtonian worlds). Indeed, it would be quite in the spirit of substantivalism to maintain that the identities of *material objects* are essentially linked to their locations in spacetime, a doctrine that precludes the possibility of there being two distinct worlds in which numerically different objects occupy the same locations in the same spacetime.

But these are issues of comparative detail. The important point is that metrical essentialism, however the fine print is formulated, seems a clear and well-motivated form of substantivalism. There is a cost. Spacetime points lose the ontological autonomy they have in manifold substantivalism and, to this extent, the substantivalist's position might be thought weakened: an attribute of traditional "substances" is their ability to exist independently of all other entities. It is not clear, however, that the loss is to be regretted. The doctrine that spacetime is a real entity does not in itself entail the view that this entity has component parts that are capable of independent existence. It may be that not all real substances can be broken into parts in the way of macroscopic material things. Last but not least, it is worth noting that Newton himself seemed to subscribe to a form of metrical essentialism. To quote once again from *De Gravitatione*: "The parts of duration and space are only understood to be the same as they really are because of their mutual order and position, nor do they have any hint of individuality apart from that order and position which consequently cannot be altered" (Huggett 1999: 112).

If GTR makes metrical essentialism an appealing position for the substantivalist, it is also a position that is vulnerable to Foster's argument. Foster argues that the same underlying spacetime could (metaphysically speaking) sustain a number of different physical geometries. This possibility poses no problem for anyone who takes spacetime points to have primitive identities and who takes a particular spacetime to consist of a particular collection of points. Put together, these doctrines permit the same spacetime to have different intrinsic and physical geometries. But adopting metrical essentialism rules out this response to Foster's scenarios. If the "underlying space" and "physical space" have different geometries, they must be composed of different points, and hence be numerically distinct. Of course, as we have seen, this fact only yields an anti-realist conclusion if we subscribe to geometrical monism. By accepting that a single spacetime can possess two distinct geometries at the same time, one physical and one intrinsic, anti-realism is averted. Hence the substantivalist who adopts metrical essentialism should also embrace geometrical pluralism.

19.5 An outmoded debate?

Earman and Norton argue that unless substantivalists equate spacetime with M the sharp distinction between "container" and "contained" is lost; construing spacetime as $[M + g]$ or $[M + g + T]$ infects spacetime with material attributes, since the metric field g possesses paradigmatic physical properties such as energy and momentum. I suggested above that the substantivalist might welcome this result, since it confirms the status of spacetime as a real entity. However, in a recent paper Rynasiewicz has argued that a quite different conclusion should be drawn: spacetime does indeed take on distinctively physical attributes in GTR, but, rather than vindicating the substantivalist position, this shows that the issue between substantivalists and relationists no longer has any clear sense; it is a debate that has been rendered pointless and irrelevant by developments in science.

> The fundamental fallacy here is the supposition that the alleged issue can be stably formulated in terms that transcend any particular historical or conceptual context . . . The distinction between "space" and "matter", or alternatively, between "container" and "contents", necessary for the debate, though immediate to common sense and sustainable to the microlevel for the earlier participants of the debate, ultimately evaporates in the course of development

of the physics of the aether. Relativity, instead of restoring the original contrast, has made it only more problematic. The current attempts to continue the debate do so by misguided attempts to legislate how that distinction should be projected onto a framework that has left it behind.   (1996: 306)

Summarizing rather brutally, Rynasiewicz's argument involves the following claims:

- Lorentz's aether can equally well be viewed as a successor to Newton's absolute space or Descartes's subtle fluid. Unlike Newtonian space it is the bearer of physical properties, since at every point within it the electromagnetic field has a definite value. But since the aether is at rest, it is natural to view it as playing the same role as Newton's space, and hence take the distinction between inertial and non-inertial motion to be grounded in the aether's structures. Is the aether *spatial* or *material* in character, or *something neither spatial not material*? "The concepts of space and matter are not themselves sufficiently determinate to settle the issue. The question is not one of fact . . ." (1996: 288).
- The spacetime of GTR is simply the Lorentzian aether in relativistic guise. It can no longer be regarded as being at absolute rest, but it continues to ground the distinction between inertial and non-inertial movement and possesses physical properties. As such, it is neither clearly spatial nor material in character, and there is no way of projecting on to it the distinction between space and matter that has lost the clear rationale it possessed in the days of Newton and Descartes.

It must be conceded that the concepts of space and matter have evolved considerably, and so transposing the classical substantivalist–relationist controversy into the contemporary context does involve extending concepts beyond their original sphere of application. But as Hoefer forcefully argues, "sometimes – indeed quite often – there is one natural and compelling way to make such extensions, all other decisions looking forced and awkward at best. To identify Lorentz's ether with absolute space is natural; to equate it with a Cartesian subtle matter, forced at best" (1998: 457). Unlike the aether, Descartes's fluids could not be at rest; indeed, they were meant to explain inertia and gravity, as these apply to solid bodies, through their motions. Furthermore, unlike both absolute space and Lorentz's aether, they do not pervade the interiors of solid material things. As for GTR, we have already seen that there is a strong case for identifying substantival space with the metrical field. Since the latter possesses its own energy, the question arises as to whether it is legitimate to regard *this* field in particular, rather than the others now recognized by physicists, as the latter-day counterpart of Newton's absolute space. Einstein thought it was. In his Leyden lecture he noted "a remarkable difference" between the gravitational and electromagnetic fields. Whereas it is quite conceivable that a part of space could exist without its electromagnetic field, there can be no space, nor any part of space, without a gravitational (i.e. metric) field, "for these confer upon space its metrical qualities, without which it cannot be imagined at all" (Einstein 1922: 21). Another writer makes the point thus:

Granting certain similarities of the metric field to other physical fields, still the metric plays a unique role in any relativistic theory. For the metric is indispensable. There is a theory of what the universe would be like without electromagnetism or without strong gauge fields. One can describe universes that are complete vacua, with stress-energy tensors identically zero. But there is no theory of space-time with electromagnetism but with no metrical structure. Without affine structure and light-cone structure and covariant differentiation the laws governing other fields could not be written. It is the physics itself, not just philosophical fancy, that singles out the metric as peculiarly intertwined with space-time. (Maudlin 1989: 87)

There is a further and quite different reason for supposing that the terms of the classical debate remain relevant. GTR and quantum theory remain unreconciled. Some of the more radical proposals for reconciliation hope to show that the macroscopic spatiotemporal features of the universe are the product of inter-relations between pre-spatial entities. Were this approach to prove successful, there is a clear sense in which classical *relationism* would stand vindicated: the universe would consist of objects that, far from requiring spacetime to exist, generate spacetime – or rather, a world that manifests itself to us in spatiotemporal form – by their modes of interrelation. Of course it may be that the successor to quantum mechanics and GTR will take a different form entirely, one that truly does render the substantivalist–relationist debate redundant. But equally, it may not.[7]

19.6 GTR and time

What of the implication of GTR for the nature of time? The impact of GTR on the static–dynamic debate is potentially enormous, but also intriguingly ambiguous. The field equations do not decisively point one way or the other, but there are particular models of the theory that do, and others that might.

Nothing in GTR is obviously incompatible with the static block conception. As in STR, simultaneity – when defined by Einstein's light-signalling method – is relative to inertial frames, albeit with a difference. The inertial frames of STR are valid throughout the whole of Minkowski spacetime; given the possibility of variable curvature there are no *global* inertial frames in typical GTR worlds (even though the distinction between inertial and non-inertial motion is at each place well defined). The fact that simultaneity is localized as well as relativized in GTR might lead one to expect that the latter theory is even more hostile to dynamic models of time than its predecessor. But this is not so.

A certain class of GTR models are surprisingly friendly to those versions of dynamism that posit a universe-wide tide of becoming. The manifolds in the models in question can be exhaustively partitioned into "foliations" of non-intersecting global (three-dimensional) hyperplanes (or "leaves") that are orthogonal to the time-like geodesics. By regarding all the points on each of these hyperplanes as simultaneous, we can assign a global or *cosmic* time to all events, despite the fact that the events in question are not causally connectable by a light signal, and so fall into the region of space-like connected events that evade Einstein's operational definition of simultaneity. The models in question only apply to universes that are isotropic and homogeneous, but, as we saw in §18.7, the Friedmann models used by

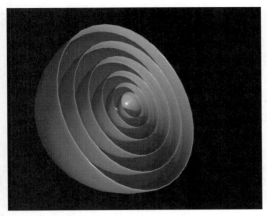

Figure 19.2 Cosmic time.
The Friedmann–Robertson–Walker models of
GTR assume an even distribution of mass-
energy throughout the universe; the resulting
spacetimes can be regarded as consisting of
nested non-intersecting global hyperplanes,
shown here in cross-section, with one spatial
dimension suppressed. Each "shell"
represents the entirety of the universe at a
given moment of time. This distinctive
foliation generates a consistent universe-wide
time-ordering for events. However, there is a
sense in which cosmic time is just one
reference frame among others: the effects of
relative motion on measurements of
temporal intervals (and simultaneity) remain.

cosmologists for modelling our universe fall into this category (as do the
Schwarzschild models used for modelling the solar system), so it is at least possible
that they apply to our world, even if the hyperplanes in question will be somewhat
uneven due to local variations in matter densities.[8]

If our universe is of this form, it may be possible for the dynamist to embrace
relativity without resorting to Stein's stratagem and relativizing "becoming" to
spacetime points. Dynamic (or compound) presentists will regard a universe
possessing a cosmic time as a succession of thin hyperplanes, created and annihilated in turn. Rather than a hollow shell, growing block theorists will view the
universe as akin to an expanding solid sphere (assuming the big bang), growing as
hyperplane is added to hyperplane;[9] hence Lucas:

> In many of the models that cosmologists use . . . there is a worldwide cosmic
> time that flows, if not evenly and uniformly, at least generally and universally.
> There are thus also preferred hyperplanes (not necessarily flat) of simultane-
> ity constituting a worldwide present and separating a real unalterable past
> from a possible future not yet actualized.   (1999: 10)

But further assumptions are at work here, which require additional justification.
Even if our universe allows the postulation of worldwide simultaneity planes, it
does not necessarily follow that there is an objective difference between past and
future of the sort Lucas here describes. Since the cosmic time of GTR is perfectly
compatible with the static block model, the dynamist cannot simply assume that
the events we call "future" have a different status to those we call "past". Lucas
himself appeals to quantum considerations: he construes the collapse of the wave-
packet as a real physical occurrence in which a definite actuality emerges from a
spread of mere probabilities, but he concedes that this interpretation is controversial and rejected by many. Others will attempt to locate a cosmic asymmetry in
causal or entropic considerations; others may be content with a *directionless*
dynamism.

However, as Torretti has noted "this remarkable revival of absolute time is more
than compensated by the complete destruction of universal time order in other
conceivable G[T]R worlds" (1999a: 77). Among the GTR worlds Torretti has in mind

are those corresponding to the Gödel solutions mentioned earlier. In these worlds there are no global time slices, and no globally consistent temporal ordering. But these spacetimes have a more remarkable property: they contain time-like paths (one through each point), which, when followed in a given direction, eventually lead back to nearly the same point. These paths are known as "closed time-like curves" or CTCs. If you were to travel along one of these paths, although your watch would always be moving forwards you would end up approaching the place and time from which you originally started. In this sense, a trip along a CTC is a journey into the past. A worldline that forms a CTC is no different locally than an ordinary worldline – which is why the hands on your watch would continually advance – but viewed *as a whole* the events along a CTC cannot be temporally ordered in the usual way, since each of these events occurs both earlier and later than all the others. As can be seen in Figure 19.3, your approach (A) to your original point of departure (D) occurs both later and earlier than the latter.

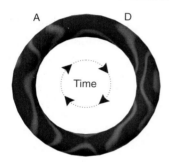

Figure 19.3 **A closed time-like curve.** Anyone who completes a trip around this spacetime loop finds themselves back where they started, in time as well as space. The time of departure (D) occurs earlier and later than the time of arrival (A) – as do all the other times on the curve.

Although CTCs have been found in many other solutions to the GTR equations, the relevant models represent matter distributions quite unlike those that obtain in our universe. Can we conclude that our universe is free of CTCs? Not necessarily. There may well be no naturally occurring CTCs in our universe, but this does not rule out the possibility of creating them by artificial means. Tipler (1974) showed that a quickly rotating cylinder of very dense matter would make backward time travel possible; by flying your spaceship around the cylinder a few times you could return to Earth before you set off. Since it is not clear that a cylinder of finite length would generate the kind of spacetime warping required for this, it may well be that a Tipler time machine could never be constructed. More recently Thorne and others have worked on the possibility of *wormhole* time machines of the sort mentioned in §8.9.[10] Again, it is as yet far from clear that constructing such a device is possible. Leaving aside the formidable technological difficulties – for example, the creation of sufficient "negative energy" to keep the wormholes open – the question of whether GTR and quantum theory will allow such things to exist is unresolved.

Gödel himself inclined to the view that whether CTCs actually exist is irrelevant; the mere fact that they are physically *possible* (assuming GTR is true) suffices to establish the block view. This is the upshot of his "modal argument":

The mere compatibility with the laws of nature of worlds in which there is no distinguished absolute time, and [in which], therefore, no objective lapse of time can exist, throws some light on the meaning of time also in those worlds in which an absolute time *can* be defined. For, if someone asserts that this absolute time is lapsing, he accepts as a consequence that whether or not an objective lapse of time exists (i.e., whether nor not time in the ordinary sense of the word exists) depends on the particular way in which matter and motion are arranged in the world. This is not a straightforward contradiction; nevertheless, a philosophical view leading to such consequences can hardly be considered satisfactory. (1949: 561–2)

In talking of "lapses" of time, Gödel's is talking about absolute becoming: "The existence of an objective lapse of time . . . [means] that reality consists of an infinity of layers of "now" which come into existence successively" (1949: 558). It is not clear whether Gödel was committed to the growing block view or presentism, but his contention that *real* time, time as we ordinarily understand it, is dynamic may well be correct. The more intriguing and more controversial suggestion is that time in GTR worlds *cannot* be dynamic because in some physically possible GTR worlds time is static. Dynamists might be tempted simply to dismiss Gödel's claim:

Fine, GTR worlds with CTCs are block worlds where time is static, but mass-energy in our world is differently organized, and the actual universe may well possess a global time function whose simultaneity planes are perfect candidates for becoming; the fact that *some* GTR worlds are static does not mean that *all* are.

But this reply, as it stands, is insufficient to counter Gödel's point.

Suppose it were the case that by simply rearranging mass-energy in *our* world wormholes allowing fast and unhindered access to past and future times could be opened up. Producing the required conditions might not be practically possible for civilizations with technological capabilities similar to our own, but this limitation is of no consequence. If in *our* universe CTCs are physically possible, if they could be produced simply by rearranging matter, then the past and future times to which these CTCs would provide access *must* be real. Unless they are real, it would not be physically possible to travel there, but we are assuming that it is.

It is clear what dynamists must do to counter Gödel's argument: they must deny that the relevant kind of CTCs *are* physically possible in our universe. This might seem problematic, given the fact that there are solutions to the GTR equations that allow CTCs, but as we have seen on previous occasions, the GTR equations need supplying with boundary conditions, certain assumptions concerning the large-scale structure of the universe. The boundary conditions applied by the dynamist are of a distinctive sort; the presentist will maintain that times earlier and later than the present do not exist; the growing block theorist will maintain that times later than the present do not exist. With these stipulations in place, the potentially problematic paths through spacetime are *not* physically possible in our universe. These stipulations, it is true, do not derive from GTR itself, but then few dynamists claim that the laws responsible for the systematic creations and/or annihilations that render time dynamic can be derived from current scientific theories. If future developments in physics provide an underpinning for dynamic time, so be it; if not, then dynamists will have to rely on other considerations to support their view.

In the same paper Gödel makes another suggestion aimed at undermining the position of dynamists. He is talking about beings living in the "Gödel world" corresponding to his own solution to the GTR equations:

> the decisive point is this: that for *every* possible definition of a world time one could travel into regions of the universe which are past according to that definition. This again shows that to assume an objective lapse of time would lose every justification in these worlds. For in whatever way one may assume time to be lapsing, there will always exist possible observers to whose experienced lapse of time no objective lapse corresponds (in particular also possible observers whose whole existence objectively would be simultaneous). But, if the experience of the lapse of time can exist without an objective lapse of time, no reason can be given why an objective lapse of time should be assumed at all.   (1949: 561, original emphasis)

Remarkably, in the Gödel world, scenarios similar to the one described by Williams (see §3.8), where an entire life is lived at a single time, are a real possibility.[11] But more important in the current context is Gödel's final point. Savitt (1994: 468) expands it into an argument, which (in slightly abbreviated form) runs like this:

1. It is physically possible that some galaxy in a physically possible universe is inhabited by creatures just like us. Let us suppose this universe is the Gödel world.
2. The direct experience of time of the creatures just like us in the Gödel world will be just like our own direct experience of time.
3. It is possible to have direct experience of time just like ours in a universe in which there is no objective lapse of time.
4. Our direct experience of time provides no reason to suppose that there is an objective lapse of time in our universe.
5. Our direct experience of time provides the only reason to suppose that there is an objective lapse of time in our universe.
6. Since there is no objective lapse of time in the Gödel world, there is no reason to suppose that there is an objective lapse of time in our universe either.

As we have seen, there are dynamists who would not accept 5. Some growing block theorists appeal to content-asymmetries (such as the fact that we have no detailed knowledge of the future), and Tooley rests his case on causal considerations. While proponents of the block view reject these arguments, we saw in Chapter 4 that, although a good deal of progress has been made on the task of explaining content-asymmetries without appealing to temporal dynamism, work remains to be done. But what of the point Gödel himself is making? Does the fact that his GTR worlds can be just like ours (locally) prove that it is possible to have direct experience of time just like ours in a block universe?

No. Once again, a question is being begged. A dynamic theorist who holds that the character of *our* direct experience of time (in particular the "immanent flow" discerned in §7.6) depends on temporal becoming would be justified in maintaining that the creatures in the Gödel world could not possibly have temporal experience exactly resembling our own for the simple reason that in their world there is no temporal becoming. This said, as noted in §7.8, it is far from clear that the "flowing" character of our experience *does* depend on time's being dynamic. If this is the case, then it may well be that "Gödelians" could have conscious lives that are phenomenologically indistinguishable from our own. But of course, if a more penetrating phenomenological analysis were to reveal that experience such as our own could only exist in a temporally dynamic universe, this verdict would have to be revised.[12]

20 Strings

We have now encountered a number of ways in which contemporary physics threatens common-sense views of space and time. With STR comes the intermingling of space and time and the abandonment of absolute simultaneity; with GTR comes curved spacetime; there are interpretations of quantum theory that posit backward causation, multiple universes and instantaneous action-at-a-distance. Mention has also briefly been made of an attempt to integrate these two theories. We saw in §3.8 that canonical quantum gravity theory has significant and potentially disturbing implications for time, even if one interpretation of what these are is empirically suspect (see §6.9 and §7.8). There is another theoretical programme aimed at integrating (or replacing) GTR and quantum theory that more than merits a mention: string theory. This too has radical implications, but they concern space rather than time. If current versions of string theory are to be believed, we live in a universe of at least nine but possibly ten (or more) spatial dimensions, not just three.

String theory is still in the early stages of development. New versions are continually being developed, and there is as yet no empirical evidence that strings themselves exist. Given their size (see below), direct detection may never be possible. But although the theory is far from universally accepted, it is also taken seriously, and in the opinion of many it is our best prospect for a theory of everything. And as we shall see, the theory not only expands the dimensions of space, but it provides space with a new role.

20.1 Higher dimensions

The idea that we live in a nine- or ten-dimensional space might seem quite absurd. Why is our freedom of movement so restricted if these extra dimensions exist? Why don't objects disappear into them? Before considering how string theorists reply to these questions, it is worth considering how evidence of concealed dimensions could emerge in a simple context provided by Flatland.

Imagine a universe containing a Flatland in which certain (non-organic) physical objects are *three*-dimensional. These objects are homogeneous, stationary, immovable and indestructible solids of fixed but differing orientations with respect to the two-dimensional plane in which they are embedded.

If we suppose that the Flatlanders, and their senses, are wholly confined to their two-dimensional plane, the three-dimensional nature of these objects would be wholly concealed from them. All they can detect using their senses are their two-dimensional cross-sections. If the thought were to ever to cross their minds (there is absolutely no reason why it should) that these immobile solids *were* three-dimensional, the available evidence would not allow them to determine their shapes. The way a cube manifests itself in Flatland depends on its orientation: it might appear as a square, but if it is tilted with respect to the plane, it will be a quadrilateral with sides of unequal lengths; if it is not fully embedded in the plane, it will appear as a (not necessarily equilateral) triangle, or even a straight line.

What would have to change in this scenario for the Flatlanders to begin to suspect the truth? Suppose the objects are as before, but the Flatlanders can also push them about. They would find that objects of the same (two-dimensional) size and appearance require more effort to shift than others. But this fact could be explained in terms of the objects in question possessing different material constitutions or densities. A mathematically sophisticated Flatlander might propose an alternative explanation of these mass differences, pointing out that these differences would be what you would expect if material objects of the same constitution were in reality three-dimensional solids. But in the absence of any independent evidence of a third dimension of space, it is unlikely that anyone would take this explanation seriously. Imagine a medieval alchemist explaining the difference in weight between equal volumes of water and gold in an equivalent manner, in terms of invisible extrusions into a *fourth* dimension of space.

However, the three-dimensional hypothesis would become much more plausible if it were discovered that objects could be moved without actually touching them. For example, by pushing a square towards a rectangle from a certain direction, the rectangle starts to move before the square comes into contact with it (in the Flatland plane), and at this same moment it becomes correspondingly harder to push the square forwards. This would arise if objects were embedded in the plane at different angles, so that they could come into contact above or below the plane (see the two small blocks tilting towards one another on the right and to the rear in Figure 20.1). If a systematic dynamics explaining such phenomena could be worked out by positing a hidden third dimension, it might well be accepted on methodological grounds (not least because action-at-a-distance would be eliminated).

Figure 20.1 Hidden depths: static three-dimensional objects embedded in a plane. Flatlanders would be oblivious to the true geometrical character of these objects – if they could move them this could change.

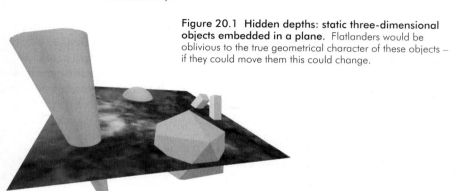

This example shows that beings whose own movement (and perception) is confined to n dimensions could come to believe that certain *objects* are of $n + 1$ dimensions. There is no reason why higher dimensional dynamics need stop here. It could turn out that to explain the observed interactions our Flatland physicists are led to hypothesize that they are dealing with four-dimensional objects: neither we nor they can visualize such things, but their mathematical properties can be worked out, and likewise for objects of even higher dimensionality. But what of *space itself*? How could there be more spatial dimensions than there appear to be?

One possibility is enforced confinement. A Flatland physicists could argue thus:

> Since we know that there are three-dimensional objects in our universe, we must be living in a three-dimensional space, but for some reason *we* are restricted to moving in a two-dimensional plane. We are in the grip of a force-field that prevents us from enjoying the full spatial possibilities our universe offers.

This is not an entirely unreasonable hypothesis, but it is vulnerable to Occam's razor. In the absence of any reason to posit the force-field, the [two-dimensional space + three-dimensional object] hypothesis is more economical; there is, after all, no obvious metaphysical reason for supposing the three-dimensional objects must be completely immersed in a surrounding spatial medium.

Considerations of a quite different kind could suggest the existence of additional spatial dimensions. Perhaps movement in the additional dimension is not only possible, but happening all the time, but this dimension is so small that the movements through it are miniscule, and so not easily detected. Suppose that our Flatland physicists have developed very precise measuring equipment. After conducting a series of experiments on colliding bodies they discover that the numbers don't add up; objects have more energy than their current theories predict they *can* have. Numerous accounts of why this should be are proposed, but only one proves viable. The proposal runs thus. The additional energy is due to the fact that objects are vibrating in a previously unsuspected way. Flatland space is extended a tiny way into the third dimension – the space has a slight *thickness* – and all objects are constantly vibrating along this extra axis, but the distances they move are so small they cannot be observed directly. The phenomenon is illustrated (in much exaggerated fashion) in Figure 20.2. Figure 20.2 shows Flatland space edge-on, and, as can be seen, it has a slight thickness. The object depicted is exploiting this additional degree of freedom by moving up and down as it slides along. In reality, the up–down movements are so small they cannot be directly detected.

The reasoning that leads string theorists to believe there are additional spatial dimensions is far more complex and subtle than anything to be found in this simple fictional scenario, but the underlying reasoning is comparable.

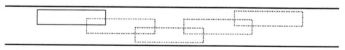

Figure 20.2 Hidden dimensions: Flatland with added depth. If the extrusion into the third dimension were sufficiently small, Flatlanders would be oblivious to its existence – any "wobbles" would be undetectable.

20.2 Kaluza–Klein theory

A mechanism not dissimilar to the one just described was first proposed as a piece of serious physics by Kaluza in 1919, and developed further by Klein in 1926. Although Einstein formulated GTR in terms of a four-dimensional (3 + 1) spacetime, the equations he used could be extended to universes with additional spatial dimensions. Kaluza derived the modified equations for a (4 + 1) universe and made an astonishing discovery: the five-dimensional equations included Maxwell's equations for electromagnetism. A theory that was designed to explain only gravity *also* explains electromagnetism when an additional spatial dimension is introduced. This unexpected unification at the mathematical level suggests that a single physical mechanism might be responsible for two very different phenomena: perhaps both gravity and electromagnetism are nothing but spatial perturbations, with gravity consisting of disturbances in the familiar three dimensions of space, while electromagnetism consists of disturbances in the additional fourth spatial dimension.

This is an exciting idea, but for it to be anything but a mathematical curiosity a plausible story of how the additional dimension is related to the four of GTR is needed. Kaluza supplied this too. He proposed that the extra dimension was curled into a small circle attached to each point of the (3 + 1) spacetime manifold. (Hence the Maxwell connection: mathematically speaking, electromagnetism is a "U(1) gauge theory", and U(1) is the group of rotations around a circle.) Figure 20.3 shows a simplified Kaluza–Klein space. Only a few points are represented; in reality there would be a circle at every point.

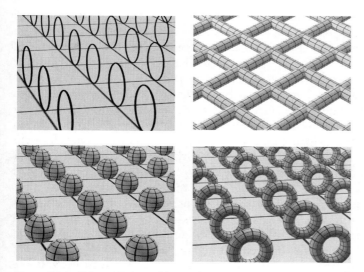

Figure 20.3 Kaluza–Klein and beyond. The two uppermost images are alternative ways of depicting Kaluza–Klein space. Ordinary three-dimensional space is shown in Flatland form, with many points omitted: only a grid-like structure remains. The fourth spatial dimension takes the form of a tiny Planck-scale circle, one for each point in three-dimensional space. In the 1970s, physicists explored the idea that the problems with the original version of Kaluza–Klein theory could be solved by positing still more spatial dimensions. The two lower images depict two variants of five-dimensional space, with the two curled up dimensions having the topology of the sphere and the torus. But these proposals soon ran into difficulties (cf. Greene 1999: 186–201).

An electron moving through spacetime is free to occupy different positions in the extra dimension. In effect, it changes its location on the circles as it moves. This motion is not detectable because the circles are so small: according to Klein's calculations, their size is of the order of the Planck length, 10^{-33} cm. Since modern particle accelerators can only probe down to distances of around 10^{-16} cm, it is not surprising that the "wobbles" of particles moving in the extra dimension go undetected.

For all its initial promise the Kaluza–Klein theory soon ran into difficulties. The theory predicted mass-charge ratios for the electron that differed greatly from the measured values. Quantum theory burst on to the scene, and its successes were such that rival approaches – such as Kaluza–Klein theory – were soon obscured. Before long many new particles had been discovered, providing new challenges: discovering the precise properties of the new particles was a difficult task, and finding a theoretical scheme that explained these properties proved harder still. The idea that additional curled up (or "compactified") dimensions might exist – and have an important explanatory role – was not entirely forgotten, but it was fifty years before its significance re-emerged.

20.3 The standard model

By the mid to late 1970s all the main planks of the "standard model" of particle physics were in place. This theory could account for all the experimental results in particle physics that could be carried out at the time with great accuracy, and was consistent with both quantum theory and special relativity. The general framework in which it was formulated is known as *quantum field theory*. In more recent decades, as more powerful accelerators have been built, it has become possible to test further predictions of the model, and these too have been vindicated. Evidence for the elusive "top quark" was discovered only in 1995.

The standard model was put together with remarkable speed. As late as the 1960s physicists were baffled on two fronts: particles and interactions. Hundreds of different types of particle had been discovered, and the existence of many more types was suspected. There was evidence that all the "strongly" interacting particles were composed of just three more basic particles – quarks – but there was much that remained mysterious. Four different and seemingly fundamental interactions had been discovered: the electromagnetic, the weak, the strong and gravity. Unlike gravity and electromagnetism, the weak and strong interactions work over very short distances, typically within atomic nuclei.[1] Physicists only possessed a fully worked out quantum theory for one of these interactions: electromagnetism. The weak and strong forces were not completely understood, and the best account of gravity – GTR – was not itself a quantum theory. All this was soon to change. By the early 1970s the Weinberg–Salam *electroweak* theory (proposed in 1968), which unified the weak force and electromagnetism, was firmly established, and in 1973 a promising account of the strong interaction – *quantum chromodynamics* – was proposed, and confirmed in 1975 with the discovery of the predicted "charmed quark".

Thereafter things rapidly fell into place. The electroweak theory and quantum chromodynamics were combined to provide a complete account of three of the four known interactions, and the standard model was born. Three basic types of particle were recognized: quarks (which combine to form protons, neutrons and other more massive particles); leptons (which include electrons and muons); and force carriers, particles which mediate the various interactions (gluons for the strong interaction, photons for electromagnetism, and weak gauge bosons for weak interactions). Before long there was talk of a *grand unified theory* (GUT) – first proposed by Georgi and Glashow in 1974 – which would show the electromagnetic, weak and strong forces to be modes of a single more basic kind of interaction. Support for this idea came with calculations that suggested that, while the three interactions are of very different strengths in (comparatively) low-energy contexts, this changes at (extremely) high-energy levels, where all three have the same strength. Why should the interaction strengths converge in this way unless they are all related? GUTs provide an answer: there is only one underlying mode of interaction.

Work on quantum field theory and GUTs continued, with mixed success, but even if success had been forthcoming a problem would have remained: the "grand" unification – like the standard model itself – deals with only three of the four interactions. No mention is made of gravity. This omission did not affect the standard model's ability to provide highly accurate predictions of particle interactions – gravity is by far the weakest of the four interactions, and has negligible effects on such minuscule and isolated masses – but it was clear that the omission would have to be rectified if a truly complete physical theory was ever to be devised. Unfortunately, all attempts to incorporate gravity into the quantum field theory framework of the standard model failed. Some of the general properties of the particle that would mediate gravitational interactions – *the graviton* – could be worked out, but when the normal rules of quantum field theory were applied they generated nonsensical results. When two gravitons approach one another the theory predicted that the force between them becomes infinite (which is impossible); other quantum equations generate infinite quantities, but none of the techniques used to eradicate them can be extended to the gravitational case. Although the spacetime of general relativity can vary in curvature, spacetime itself is always smooth (and effectively flat over very short distances). Quantum field theory delivers something quite different. The "uncertainty principle", which lies at the heart of quantum theory, entails that all quantum properties will be subject to significant fluctuations over small spatial and temporal intervals, and the

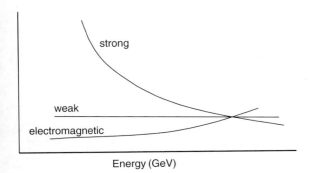

Figure 20.4 Intimations of unity.
If the strong, weak and electromagnetic forces have the same strength at very high energies, perhaps they are merely different modes of a single, more basic, type of interaction.

smaller the intervals, the larger the fluctuations. When the uncertainty principle is applied to the vacuum, the energistic eruptions it permits affect the geometry of spacetime. Over very short Planck-scale distances the fluctuations are of such violent intensity that spacetime comes to the boil. The result is quantum *foam*, a spacetime of dramatically varying geometry and topology, where tiny wormholes are continually being created and annihilated. The idea that spacetime resembles a roiling frothy liquid at the very small scale is not in itself problematic – the disruptions even out over larger areas, and in any case are far too small to have observable consequences – but it does make the incorporation of GTR into quantum field theory highly problematic.

20.4 Strings

For reasons that will shortly become clear, string theory is a descendant of the Kaluza–Klein theory, and so could be viewed as dating back to 1919. However, the theory in its current form originated in work done during the 1970s and early 1980s, work largely ignored at the time due to the successes of the standard model, with a landmark paper by Scherk and Schwarz appearing in 1984. For a detailed account of the origins of the theory see Davies and Brown (1988) and Greene (1999).

From the mathematical perspective string theory is considerably more daunting than either quantum theory or relativity, but some of its essential features are easily grasped. All the various particles of the standard model are considered to be structureless points, these point-particles have various properties – velocity, position, mass, electric charge, spin, colour (a charge-like property of quarks) – and different types of particle have different combinations of these properties. As we have already seen, there are hundreds of different particle types. String theory does away with this "particle zoo" and posits just one fundamental entity: a "string". As its name suggests, a string is *not* point-like. It is one-dimensional and so *line*-like; it has length but no breadth. But not much length: strings are roughly 10^{-33} cm long – the now-familiar Planck-length – and so resemble point-particles when viewed from afar (i.e. at all distances that can be probed by current technology). Strings can be closed, like loops, or open, like short lengths of bendable wire. As a point-particle moves through space it traces out a *worldline*. As strings move through space they trace out *world-tubes* (closed) or *world-sheets* (open). Strings interact by combining and splitting apart. Some simple closed string interactions are depicted in Figure 20.6.

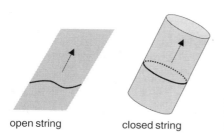

open string   closed string

Figure 20.5 A "world-sheet" and a "world-tube".
String-theory's substitutes for the "world-lines" of classical point-particle physics.

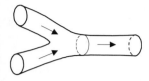
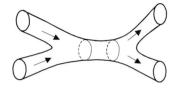

Figure 20.6 Simple string interactions: world-tubes undergoing fusion and fission.

As well as being able to join and split apart, strings can also vibrate. Most of us have probably experimented with the vibrational properties of cords in the school playground. If two people each hold one end of a (longish) skipping rope and move their arms up and down in random fashion, the rope starts to jiggle chaotically; but by timing the movements in the right way, orderly oscillations emerge. Slow rhythmic movement will produce the pattern in Figure 20.7a; speed up, and the pattern in Figure 20.7b emerges; speed up some more and the pattern in Figure 20.7c will develop. Each of these vibrations corresponds to a *resonant* frequency of the rope: the peaks and troughs are evenly spaced, and an exact whole number of waves can fit between the two endpoints. Elementary strings can vibrate – or pulse – in a precisely analogous fashion. Three possible modes are shown in Figure 20.8.

(a)   **(b)**   **(c)**

Figure 20.7 Resonant oscillations.

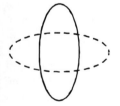

Figure 20.8 Resonant modes: some simple cases. The physical properties of a string depend on the way it vibrates. Different resonance patterns generate the properties associated with the various types of particle recognized by the standard model (e.g. electron, up-quark, photon), whereas different amplitudes and wavelengths are associated with different energies and masses.

Different vibration patterns in a guitar string produce different sounds. The different vibrational modes of an elementary string generate different physical properties. One pattern of pulses causes a string to behave like an electron, another like a photon, another like a neutrino, and another like a weak gauge boson. The way this works is most obvious in the case of mass. Quite generally, the energy carried by a wave is a product of its amplitude (the greater the distance between peaks and troughs, the greater the amplitude) and wavelength (the distance between one peak and the next). Higher amplitudes and shorter wavelengths correspond to more frenetic vibrations that possess greater energy; think of the way a gentle pluck on a guitar string creates a gentle vibration, whereas a more forceful pluck (which transfers more energy) produces a more energetic vibration. Since we know from STR that mass and energy are interconvertible, it follows that more energetic strings are also more massive. The same applies, although less obviously, to all the other properties particles can possess, such as spin, colour (in the quantum chromo-dynamics sense) and electric charge. As Greene puts it: "What appear to be different elementary particles are actually different 'notes' on a fundamental string. The universe – being composed of an enormous number of these vibrating strings – is akin to a cosmic symphony" (1999: 146).

The standard model can handle a vast variety of different particle types, and even generate predictions about the properties of particles as yet unobserved, but it doesn't explain *why* the particles have their various properties; *that* they do is taken as a brute fact. String theory goes deeper. It is no longer a mystery why certain combinations of mass, spin, charge, and so on are found: the relevant property combinations correspond with different vibrational modes of a single fundamental type of entity.

I have yet to mention the most significant advance achieved by string theory. One of the resonance patterns generates precisely the properties the standard model ascribes to the *graviton*, but whereas the standard model could not be extended so as to incorporate this particle in a consistent way, its existence is guaranteed by string theory. It is there right at the outset, as a basic mode of string oscillation. Moreover, it is possible to make sense of interactions between gravitons in string theory. There are none of the problematic and ineradicable infinities that arise in quantum field theory. In large part this is due to the extended nature of strings themselves. The equations of the standard model deliver problematic results because the interacting particles are zero-dimensional points that can become arbitrarily close to one another; in string theory interactions are not concentrated into single points, but are always spread over small regions of space (as in the world-tubes depicted in Figure 20.5). The extended character of string interactions prevents unwanted and physically unrealistic infinities from occurring.[2]

Strings may be able to explain a lot, but their remarkable properties do not stop here. For one thing, being approximately of Planck length, they are very small. The figure of 10^{-33} cm (a millionth of a billionth of a billionth of a billionth of a centimeter) is easy to write down or say, but hard to grasp. Atomic nuclei are very small – billions could fit into a pinhead – but strings are a hundred billion billion times (10^{-20}) smaller still. The known universe is very big. The star nearest to us, Alpha Centauri, is a hundred million times further from us than the Moon (a mere 240 000 miles); the diameter of our galaxy, the Milky Way, is about 20 000 times greater than the distance between here and Alpha Centauri; there are about a hundred billion other galaxies in the universe. After pondering on this for a moment, imagine a single atom boosted in size so it fills the known universe. How wide would the individual strings in this atom be after this boost? Not much wider than an average house.

The minuscule size of strings means we may never be able to detect their existence directly; we would need an accelerator a million billion times more powerful than any now available to smash matter into individual strings. The explanation for the minuscule scale of strings is as remarkable as the scale itself:

internal string tension. Calculations suggest that this is of the order of 10^{39} tons: the so-called "Planck tension". It is because strings are so taut that they find themselves pulled into so compact a size.

20.5 Calabi–Yau space

One other aspect of string theory merits a special mention, for it is the aspect that is most relevant to our primary concern: space. Investigations into the earliest versions of the theory revealed a problem. Although the infinities associated with point-interactions did not arise when particles were taken to be one dimensional rather than zero dimensional, the equations generated predictions that featured negative probabilities; another reliable sign that something has gone awry. Further work suggested a surprising solution. It was discovered that the negative probabilities were connected with the number of different directions in which strings can vibrate. If string vibrations are confined to the directions available in three-dimensional space the negative probabilities cannot be avoided; moving to four or five dimensions doesn't help. But move to *nine* spatial dimensions (plus one for time) and the problem disappears completely. Does this mean that string theory must be false? No, for as we saw earlier, Kaluza and Klein discovered a way in which our universe could have more spatial dimensions than it seems to have: if the additional dimensions are curled up enough they would be effectively invisible. Since strings are themselves very small, they do not need much room in which to vibrate: the six extra dimensions need only be of the Planck magnitude.

String theorists were thus led to resuscitate the discarded Kaluza–Klein theory, but in a more radical guise. Kaluza and Klein introduced one extra dimension to unify relativity and electromagnetism (and the attempt failed); string theory introduces six extra dimensions, but in so doing provides a unified framework that encompasses all known particles and all known interactions.

The extra six dimensions can be compactified in a multitude of different ways, each compactification corresponding to a different six-dimensional manifold. We saw earlier how in Kaluza–Klein theory the extra dimension takes the form of a circle attached to each point of the familiar four-dimensional spacetime manifold. This strategy can be extended: instead of attaching one circle we attach six – in effect a six-dimensional torus – to each point in ordinary space. But while this simple solution is often adopted (it makes calculations easier) it turns out not to deliver quite the right results. It has been demonstrated that to get the right results the extra dimensions must compactify on to a particular sort of six-dimensional space: a *Calabi–Yau* manifold. Unfortunately, Calabi–Yau manifolds themselves come in many thousands of forms, and it is not yet clear which is required for string theory (recent work has suggested that different Calabi–Yau manifolds are related through "conifold transitions", which may simplify things). Being six-dimensional, there is no way to visualize a Calabi–Yau manifold in all its complexity, so Figure 20.9 is not realistic.[3] It is of course rather odd to think that with every move we make each and every part of us is continually shifting through these contorted six-dimensional manifolds, but don't forget their size: the resulting "wobbles" are exceedingly slight.

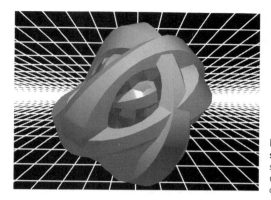

Figure 20.9 Concealed corridors of space. A single point in four-dimensional spacetime, with an impressionistic rendering of the additional six dimensions at the Planck scale.

This multiplication of dimensions is only one of the ways in which string theory requires us to revise our preconceptions concerning space. Another concerns topology. As was noted earlier, quantum theory predicts that at very small scales space resembles a turbulent frothing liquid, with the spatial fabric being repeatedly torn apart and stitched back together again, with the consequence that space does not have a stable unvarying topology. General relativity allows space to bend and flex in response to variations in mass-energy density, but it does not permit topology-destroying tears in the manifold (which is one reason why black hole singularities are thought by many to be a disastrous consequence of GTR). String theory does. Work in the 1990s proved that a certain kind of topology transformation (a "flop") could occur within a Calabi–Yau manifold without drastic physical consequences. Indeed, it now seems that such processes are commonplace, assuming that current versions of string theory are pointing us in the right direction.[4]

More intriguing still is the way in which string theory provides space with an entirely new *role*. Newton ascribed two functions to space. By virtue of being the medium in which all physical things exist, it constrains (or makes possible) the ways material bodies can move, and hence the ways in which they can be spatially related; it also grounds the distinction between inertial and non-inertial motion. With the advent of GTR, space (or spacetime) continues to fulfil these functions, but it loses its "absolute" status: not only is its structure no longer fixed, but its structure is responsive to movements of matter. In string theory all this is retained – space still determines possible movements – but it does something more: it directly controls the *properties* that material things possess and can possess.

Why does an oboe sound different from a saxophone? To a large extent it is because of the way air can vibrate in the interior spaces of the instrument; different shapes permit different patterns of vibration, which results in different tonal qualities. As we saw earlier, elementary strings take on different physical attributes depending on how they vibrate, and as with the air in a musical instrument, the way they can vibrate depends on the shape of the space they inhabit. As a string moves through space, pulsing as it goes, it travels through the twisting byways of the compactified Calabi–Yau manifolds, the complex geometry of which significantly affects the vibrational patterns available to the string, and hence its physical properties:

> Since tiny strings vibrate through all of the spatial dimensions, the precise way in which the extra dimensions are twisted up and curled back on each

other strongly influences and tightly constrains the possible resonant vibrational patterns. These patterns, largely determined by the extradimensional geometry, constitute the array of possible particle properties observed in the familiar extended dimensions. This means that *extradimensional geometry determines fundamental physical attributes like particle masses and charges that we observe in the usual three large space dimensions of common experience.* (Greene 1999: 206, original emphasis)

The italics are appropriate; it is not every day that a new function for space is discovered.

20.6 Shards

Current formulations of string theory unashamedly presuppose a spacetime manifold. Indeed, as we have just seen, the physical properties of strings are controlled by the manifold's structure. Might this change? Can the spacetime manifold itself be explained in terms of something else? In a speculative vein towards the end of his book, Greene suggests that it might:

> we can ask ourselves whether there is a raw precursor to the fabric of spacetime – a configuration of strings of the cosmic fabric in which they have not yet coalesced into the organized form that we recognize as spacetime. Notice that it is somewhat inaccurate to picture this state as a jumbled mass of individual vibrating strings that have yet to stitch themselves together into an ordered whole because, in our usual way of thinking, this presupposes a notion of both space and time – the space in which a string vibrates and the progression of time that allows us to follow its changes in shape from one moment to the next. But in the raw state, before the strings that makeup the cosmic fabric engage in the orderly, coherent, vibrational dance we are discussing, *there is no realization of space or time.* Even our language is too coarse to handle these ideas, for, in fact, there is no notion of *before.* In a sense, it's as if individual strings are "shards" of space and time, and only when they appropriately undergo sympathetic vibrations do the conventional notions of space and time emerge . . . just as we should allow our artist to work from a blank canvas, we should allow string theory to *create* its own spacetime arena by starting in a spaceless and timeless configuration.
>
> (1999: 378–9)

He goes on to note that a number of leading string theorists are working along these lines.[5]

In the light of the many hard-to-believe-but-possibly-true scientific theories we have encountered thus far it would clearly be a mistake to prejudge this issue. The idea that ordinary material things are constructed from more basic building blocks has proved fruitful; perhaps the same will apply to what underlies material things. Perhaps space is a construct from more basic ingredients, and perhaps these are strings themselves, or related entities such as the "zero-branes", currently viewed as perhaps the most basic ingredient in M-theory (currently viewed as the most promising version of string theory), or alternatively, the constituent parts of the tangled "spin networks" of loop quantum gravity theory, one of string theory's current rivals. However, there is an obstacle that all such approaches must overcome.

When discussing the "connection problem" in §9.3 I suggested that any theory that eliminates space (or spacetime) as a fundamental entity has to provide an account of how the entities that remain can connect with one another. If spacetime is the product of a cooperative string enterprise, the strings doing the work must be able to interact with one another, and a story must be told of how this is possible in the absence of space and time. More generally, no matter what the nature of the pre-spatial shards that are jointly responsible for the physical world as we know it, there must be a connection of *some* sort between them, for in the absence of such a link these entities would – in effect – consist of entirely distinct island universes in their own right. A substantival manifold is one mode of connection, but it is not the only mode. Perhaps there are pre-spatial and pre-temporal shards linked by relations of some sort – if so the relations must themselves be real ingredients of reality in their own right – or perhaps the shards are themselves directly connected to one another (we shall be considering such a case shortly). The mistake to avoid is supposing that these ingredients could start off isolated in the void and then "come together" to constitute an extended spatial network. Nothing can come together in the void; it is not that sort of place. The component parts of a space must come into being together, as a connected (even if growing) whole. Space may not be indispensable, but it is not easily disposed of.

A paradigmatic relational world comprises material objects separated by void (nothingness) but connected by spatial relations of the intrinsic variety (the precise nature of which is usually left unspecified); the paradigmatic substantival world, by contrast, consists of a continuous spatial manifold that may, or may not, contain some material objects. There are possible universes that are spatial, but whose space falls between two stools, being neither relational nor substantival in the usual senses of the terms. One such possibility is the "mesh world" depicted in Figure 20.10 in Flatland form. This space is akin to a grid, with the "fibres" or "threads" of space being separated by void (in accord with relationism), and where the gaps in the grid are smaller than the smallest particles. Since these particles extend beyond the gridlines they "fill" a small region of void (in line with classical relationalism), but their motions are constrained by the components of the mesh (in accord with classical substantivalism).

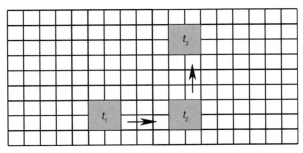

Figure 20.10 Part-void, part-substance: mesh space. An atom is moving through mesh space, constrained by the gridlines and extending through void.

We can further suppose that elementary particles in this world do not move in a smooth continuous fashion, rather they "jump" instantaneously from cell to cell (there are four such jumps between t_1 and t_2, and between t_2 and t_3). Provided the mesh is fine enough (e.g. sub-Planck length) these discontinuities would not be discernible empirically; space would appear continuous, even though it is not.

Although there is certainly a relationalist aspect to this world – the presence of surrounding (or permeating) void – given the role the mesh plays in constraining motion, there is clearly a case for regarding it as a substantival space, albeit of an unusual sort. The "growing mesh" depicted in Figure 20.11 shows how such a space can expand into the non-existent unstructured void, creating new locations and new possibilities for movement as it does so, in just the manner one would expect of a substantival space. This model has more than mere curiosity value: mesh space bears some resemblance to the conception of space (if not objects and motion) to be found in loop quantum gravity theory, the competitor to string theory mentioned above.[6]

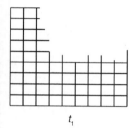 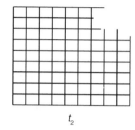

t_1 t_2

Figure 20.11 A growing mesh space.

In one of its forms this theory incorporates a variant of the "spin network" approach originated by Roger Penrose; instead of a spacetime manifold it posits an evolving network of looping intersecting flux-lines. Whereas the spacetime of GTR is continuous, that of the spin network is not, so space and time come in discrete, quantized units. These units are incredibly small. The smallest areas are of the order of 10^{-66} cm², and the smallest volumes about 10^{-99} cm³. The area of a given surface is determined by the number of "edges" of the network it crosses (not the expanse of interior void), and the volume it encloses is given by the number of nodes (or intersections) it contains. Hence circle A in Figure 20.12 encompasses a larger area and volume of the fragment of space shown than the (seemingly larger) circle B. The way these loops and lines intersect and knot together determines the geometry of space at the Planck-scale (the "knotted" aspect of the network would be more apparent in a three-dimensional depiction). Since the network is constantly changing its structure, so, too, is space, albeit at this very (very) small scale. Given their size, these geometrical fluctuations – together with the discrete character of space – are invisible for all practical purposes.

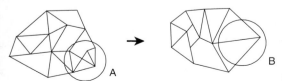

Figure 20.12 An evolving spin network.

Intriguingly (if also optimistically) Smolin suggests that rather than remaining competitors, loop quantum gravity and string theory may one day come together:

> space may be "woven" from a network of loops . . . just as a piece of cloth is woven from a network of threads. The analogy is fairly precise. The properties of the cloth are explicable in terms of the kind of weave, which is to say in terms of how the threads are knotted and linked with one another. Similarly, the geometry of the space is determined only by how the loops link and intersect one another.
>
> We may then imagine a string as a large loop which makes a kind of embroidery of the weave. From a microscopic point of view, the string can be described by how it knots the loops in the weave. But on a larger scale we would see only the loop making up the string. If we cannot see the fine weave that makes up space, the string will appear against a background of some apparently smooth space. This is how the picture of strings against a background space emerges from loop quantum gravity.
>
> If this is right, then string theory will turn out to be an approximation to a more fundamental theory described in terms of spin networks.
>
> (2000a: 186)

Be this as it may, for our purposes what matters is the *type* of world being proposed in these speculations. On various occasions Smolin suggests that not only was the loop quantum gravity approach inspired by Leibnizian-style relationism, but it also vindicates the latter.[7] In one sense it does: what is being proposed is undeniably very different from Newton's absolute space. The structure of the spin network is dynamic, its geometry is continually evolving, and the way one part evolves depends entirely on its relations with other parts; moreover, this evolution does not take place against the backdrop of a separate and independent spatiotemporal framework – the spin network *constitutes* the spatiotemporal framework. But this is only part of the story. Leibniz was also rejecting a substantival conception of space and time, and it is far from clear that space and time as conceived in loop quantum gravity are *not* substantival. Indeed, the loop quantum conception seems *as* substantival as any conception we have encountered thus far. It could even be viewed as vindicating *super*-substantivalism: rather than providing the framework that contains and constrains material objects, space, time and the properties they possess are now being viewed as the sole constituents of physical reality. What could be less Leibnizian?

However, we must not too get carried away by all this. The string and loop theories have fascinating implications, as does the canonical approach, but it is too early to tell whether a viable quantum gravity theory – assuming that physicists succeed in developing one – will resemble anything so far conceived. Nonetheless, since these theories are the best guides we have as to what the future may bring, it may very well be that the next revolution in physics will be as remarkable as any of its predecessors, and its metaphysical implications just as difficult to discern.

Notes

1. Preliminaries

1. Parmenides **B8**, Barnes (1982: 178).
2. The "temporal determinism" entailed by the block view should not be confused with nomological determinism. In a world where the laws of nature are deterministic, how things are at any one time fixes how things will turn out at later times (as well as how things were at earlier times). One can be a nomological determinist without being a temporal determinist (by holding that although the present completely determines how the future will unfold, as of the present time future events are unreal). And one can be a temporal determinist without being a nomological determinist (by holding that although future events are just as real as present events, they are not nomologically determined by them). Since many contemporary physicists believe that our world is (nomologically) indeterministic, the greater threat (if threat it be) may well be posed by determinism of the temporal variety.
3. McTaggart (1908). It has been claimed that the distinction is also to be found in the writings of Iamblichus (died circa. AD325), see Sorabji (1983: Ch. 3).

2. McTaggart on time's unreality

1. Shoemaker (1969) describes a world (quite unlike ours in some respects – there are regular "local freezes") where it would be reasonable to posit the occurrence of occasional universe-wide temporal vacua.
2. It is important to note that the A-theorist cannot seek to avoid the paradox by interpreting "successively" in a B-series way. Suppose we say something along these lines: "No event is past, present and future at the same time. An event E at t is present at t, past at times earlier than t and future at times later than t". This is tantamount to denying the reality of A-series change: all events are equally and timelessly present; an event doesn't *change* its A-attributes – it is *always* true that "E is present at t", likewise it is always true that "before t, E lies in the future" and "after t, E lies in the past".

3. Static time

1. Competing accounts have their adherents, but none of these is *as* popular as the B-theory.
2. See Garrett (1988); for other responses to Prior's argument see Oaklander and Smith (1994: part III), and Gallois (1994).
3. Parfit (1984: 173–7) argues for this view by considering "Timeless", a being who lacks the bias towards the future:

   > Our bias towards the future is bad for us. It would be better for us if we were like Timeless. We would lose in certain ways. Thus we should not be relieved when bad things were in the past. But we should also gain. We should not be sad when good things were in the past . . . There would be other, greater gains. One would be in our attitude to ageing and to death . . . We do not regret our past non-existence. Since this is so, why should we regret our future non-existence? If we regard one with equanimity, should we not extend this attitude to the other? . . . As our life passes, we should have less and less to look forward to, but more and

more to look backward to . . . In giving us this [future] bias, Evolution denies us the best attitude to death.

Einstein was of a similar view. Shortly before his own death, he wrote to the children of one of his oldest and closest friends, Michele Besso, who had just died: "And now he has preceded me briefly in bidding farewell to this strange world. This signifies nothing. For us believing physicists, the distinction between past, present and future is only an illusion, even if a stubborn one" (quoted in Hoffmann 1972: 257–8). Many of us regard death as a misfortune, as something to be feared and dreaded, for we equate it with the onset of complete non-existence. In a block universe this fear is misplaced, for if all times and events are equally real, death does not bring absolute annihilation, lives do not cease to exist, they simply have beginning and ends. If it were not for our deep-seated ("stubborn") bias towards the future, this idea would be more comforting than it actually is.

4. In a four-dimensional space, the distance between two points is given by an extended version of Pythagoras' theorem: $d^2 = w^2 + x^2 + y^2 + z^2$; the *spatiotemporal* distance (or "interval") between two *spacetime* points in special relativity is given by an equation that treats temporal intervals differently from spatial intervals. See §16.5.

5. We will encounter this hypothesis again, in a rather different context, in §6.8.

6. If you find it hard to grasp what having your "belly plump in time" amounts to, it may help to imagine the time axis in Figure 3.3 rotated counter-clockwise so it coincides with (and displaces) the left–right axis. Alice's life now unfolds through space, rather than time (or rather, the dimension that the rest of us call time). In a footnote Williams adds, "I should expect the impact of the environment on such a being to be so wildly queer and out of step with the way he is put together, that his mental life must be a dragged-out monstrous delirium". In this he is surely right, although if he had been writing this passage now he may well have wanted to reconsider the remark about "our comparative indifference to our spatial girth".

4. Asymmetries within time

1. To keep the discussion to a manageable length I have simplified somewhat. A fuller discussion would include the "radiation asymmetry" (we often encounter ripples (e.g. of water or light) moving away from a common source-point, but rarely do we encounter *converging* ripples, even though they are physically possible), and the "counterfactual asymmetry" (true counterfactuals generally concern future rather than past possibilities – e.g. "If I had struck the match, it would have lit" versus "If the match had lit, it would have been struck").

2. In fact, it is far from obvious why a big bang *should* be a low-entropy event; a universe that is very small in spatial size and extremely hot throughout (which is how our universe is envisaged as being just after the initial explosion) would be in a state verging on maximum thermodynamic *disorder*, and hence close to maximum entropy. The "inflationary" big bang model provides a solution; a period of very rapid spatial expansion in the very early universe has the effect of creating a sudden decrease in entropy, hence the subsequent tendency for entropy to increase. Although they are no longer in favour, Boltzmann's cosmological speculations are remarkable on several counts. As well as providing a statistical account of the origin of complex structures such as stars, he also predicted that "time-reversed" regions of the universe exist. Think of the equilibrium condition of the universe as a whole as a straight "baseline" and the occasional eruptions of order as hills or humps along this line; each summit represents a state of maximum local order (minimum entropy). It follows (given the entropic account of the direction of time) that we inhabit a slope of one of these humps, and that the summit of our hump is in our past; ahead of us (in our future) lies disorder and equilibrium. But what lies behind, on the *far side* of the hump? Boltzmann's answer: a region of the universe where entropy is also increasing, and hence where the local arrow of time points in the opposite direction to ours. Hence, for Bolzmann, the arrow of time is only a regional phenomenon, and time-reversed regions of the universe are possible. Of course, this is all based on the assumption that the various observable temporal asymmetries depend on entropic increase, which is precisely what is at issue.

3. Consider the following reasoning: "The processes that are involved in the production of records (such as human memories, or video films, or books) invariably lead to entropy increase, since energy is expended and spread when making the record. Thus the tendency for entropy to increase explains why it is that we have records of the past but not the future." The fact that producing

records is an entropy increasing process doesn't in itself explain why the records produced provide information about the past but not the future.

4. Cf. Mellor (1998: §10.4). Since Mellor takes causation to relate facts rather than events, his formulation differs from that given here.

5. Simultaneous causation poses a similar problem. As noted earlier, Mellor argues that causes cannot be simultaneous with their effects, but many disagree.

6. This was not the first appearance of this notion in science. As noted earlier, Boltzmann deduced the possibility of time-reversed regions of the universe from his own premises.

7. Horwich is far from alone in his view of causation, so his approach is just one implementation of a more general strategy. But few other implementations are so ambitiously wide-ranging.

8. See Price (1996: Ch. 6) for a critical assessment of the strategy of grounding the causal asymmetry on the fork asymmetry. (Augmenting Horwich's fork asymmetry with the more general overdetermination asymmetry – cf. Owens – goes some way towards reducing the force of Price's argument.) Although critical of the fork proposal, Price's view of causation is similar (broadly speaking) to that of Horwich.

5. Tensed time

1. Lowe (1987, 1993). I will draw mainly from Lowe (1998).
2. There are similarities with "deflationary" theories of truth.

6. Dynamic time

1. Tooley argues that the dynamic theorist also needs a more general and non-temporal concept of existence (or actuality) – existence (or actuality) *simpliciter* – so as to be able to make sense of talk of non-temporal entities (such as abstract things), and "the concept of a total, dynamic world, as contrasted with the history of a dynamic world up to some point in time" (1997: 40). It is not clear that this is a necessary move, or a wise one: how can states of affairs that are not actual as of one time *become* actual as of another time if the states of affairs in question are actual *simpliciter*? I prefer not to saddle the growing block model with Tooley's timeless notion of actuality *simpliciter*, and so will ignore it in what follows.

2. The model has evolved over the years. See, among others, McCall (1976, 1984). I rely here on McCall (1994).

3. Or at least they do in universes where indeterminism holds sway. In deterministic universes there is only one nomologically possible future, and so McCall's universe-tree is so slender (and branch free) that it is indistinguishable from a static block universe.

4. It should be noted that the extended case that McCall mounts for his model in *A Model of the Universe* (1994) rests entirely on its explanatory potential (e.g. Chapter 4 provides an interesting interpretation of quantum theory, Chapter 5 is devoted to probability theory). But while the model does have explanatory and problem-solving potential, there are competing solutions to the same problems that bring less surplus baggage.

5. Causal laws, rather than being merely regularities in the history of the world, *control* the course of history; they underlie, and account for, any patterns that the world may exhibit over time. But how is this control to be understood? One way – and, I think, the only satisfactory way – is if causal laws, in conjunction with what is actual as of a given time, determine what states of affairs will ultimately exist, in a tenseless sense.   (Tooley 1997: 111)

6. Le Poidevin (1991) is an exception; see also Zimmerman (1998) and Hinchliff (1998).

7. Any static block theorist who adopts a strict Humean view of causation (and so denies "real connections" between events occurring at distinct moments) may well be endorsing a position that is very similar to that of the many-worlds presentist.

8. A "phase space" is an abstract mathematical representation of a dynamic material system, often of infinite dimensions, whose coordinates are given by the set of independent variables characterizing the state of the system. Each individual point in the phase space corresponds to a possible state of the entire system (in this case, the entire universe), and paths through it correspond to possible histories of the system; or at least they do when time is one of the coordinates. If time is absent, this interpretation is not available.

9. The programme consists of casting general relativity in Hamiltonian form, that is treating it like any other quantum theory. Another approach – string theory – posits basic items that are already quantized. We will be taking a brief look at strings in Chapter 20.
10. For details see Belot & Earman (1999, 2001), Butterfield & Isham (1999), Kuchař (1999).
11. It should be pointed out that not everyone sees the status of the past in such models as unproblematic; see Dummett (1978). Dolev (2000) supplies a succinct introduction to Dummett's position.

7. Time and consciousness

1. In conversation with Carnap, Einstein expressed similar qualms:

> Once Einstein said that the problem of the Now worried him seriously. He explained that the experience of the Now means something special for man, something essentially differ-ent from the past and the future, but that this important difference does not and cannot occur within physics. That this experience cannot be grasped by science seemed to him a matter of painful but inevitable resignation. I remarked that all that occurs objectively can be described in science; on the one hand the temporal sequence of events is described in physics; and, on the other hand, the peculiarities of man's experiences with respect to time, including his different attitude towards past, present and future, can be described and (in principle) explained in psychology. But Einstein thought that these scientific descriptions cannot possibly satisfy our human needs; that there is something essential about the Now which is just outside the realm of science. (Carnap 1963: 37–8)

2. Advocates of this account include Whitehead: "The final facts are all alike, actual entities; and these actual entities are drops of experience" (1929: 25). See also Sprigge (1983: Ch. 1).
3. This, of course, is the problem facing presentism of the non-compound variety, transposed to the phenomenal level.
4. See Broad (1938) and Husserl (1950, 1991).
5. See Dainton (2000: Chs 5–7) for more on the topic of phenomenal time.
6. This example does not establish that the direction of phenomenal flow is independent of the direction of *causality*; the relationship between the two is unclear; as unclear as the notion of the direction of causation. A Humean could define a "personal" or "subjective" arrow of causation in terms of an earlier–later distinction grounded in the direction of phenomenal flow; it seems quite conceivable that this subjective arrow could point in the opposite direction to the "objective" arrow of the inhabitants of a (seemingly) time-reversed planet. Non-Humeans (such as Tooley) might want to say that the phenomenal and causal arrows necessarily point in the same direction, but it is obscure how this claim could be supported, especially since it is far from clear that immanent phenomenal flow depends on "block growth".
7. If this claim seems implausible, consider what it is like to watch televised replays of familiar events: you remember *exactly* what will happen over the next few seconds (the player will miss the penalty, the ball will fly over the crossbar), but is the forward-flow of your experience in any way affected?

8. Time travel

1. This class of machine undermines the "Where are they?" refutation of the possibility of backward time travel popularized (if only semi-seriously) by Stephen Hawking.
2. Some science fiction tales seem to involve a different mode of "time travel which isn't really time *travel*": a powerful device reconfigures the present-day world so that it resembles (with the exception of the presence of whatever "travellers" may be involved) the world as it was at some past time. (I am grateful to Stephen Clark for drawing my attention to this model.) Obviously, in this sort of case there is no objection to "changing the past", since what is being changed is not *the past* at all.
3. See Earman (1995: Ch. 6) for further discussion of the relevant consistency constraints, and how they should be conceived.

4. Or at least it does in worlds where the distinction between earlier and later is sufficiently well defined. In the "Gödel universe" that we shall be encountering later, there is no consistent global time order (moreover, locally speaking, causation along closed time-like curves in this world is usually regarded as always being future directed).

5. Tooley (1997: 63–8) describes a world where backward and forward causation coexist, but are contained within different spatial regions, separated by a barrier that severely restricts interaction in that light (and nothing else) can pass through one way but not the other.

6. What follows is drawn mainly from Price (1984). For more detail see Price (1996: Chs 7–9).

7. In §17.4 we shall see that the matter is far from clear-cut.

8. In Cramer's *transactional interpretation* of quantum mechanics (1980, 1983, 2000), quantum states are determined by wave-functions that propagate both forwards in time and backwards in time, in the manner of the "advanced" and "retarded" electromagnetic waves in the Feynman–Wheeler absorber theory. See Price (1996: Chs 3 & 9) for further details of such proposals, and Maudlin (1994: 196–201) for criticisms of Cramer's theory.

9. See Nahin (1993: 205–17) for a full survey; see Heinlein (1970), Harrison (1979) and MacBeath (1983) for intriguing sexual loops. Some scientists take the "Big Loop" scenario seriously, albeit not in this form. The idea that the universe originated in a causal loop is sympathetically explored in Gott and Li (1998) and Gott (2001), a particularly elegant response to the "What caused the first event?" question, since the universe, in effect, creates itself. Of course, the problem of why the primordial loop exists at all remains unexplained.

10. For a recent survey see Arntzenius and Maudlin (2000) and Nahin (2001).

9. Conceptions of void

1. See also Selby (1982).

2. Think of water flowing over a stone on a riverbed. The water contains the stone, but the water in contact with the stone is constantly changing; consequently, if we were to say that the stone's place is the inner surface of the surrounding water, we would be committing ourselves to the view that the stone was constantly changing place, which is absurd. So Aristotle suggests that we identify the stone's place with the inner surface of the nearest stationary containing body: the river bed. This definition sits uneasily within the rest of Aristotle's cosmology. The first containing body of the Earth and the sub-lunary world as a whole is the Moon's shell, which is rotating; but so is the next (planetary) shell, and so is the *outermost* shell. Consequently, the Earth itself has no place, if we apply Aristotle's own theory rigorously; a fact noted by many commentators.

3. Something akin to Newton's conception of space had been popular in neo-Platonist and mystical circles for some time. According to Giordano Bruno (1548–1600) in *On the Infinite Universe and Worlds*:

> There is a single general space, a single vast immensity which we may freely call VOID; in it are innumerable globes like this one on which we live and grow. This space we declare to be infinite, since neither reason, convenience, possibility, sense-perception nor nature assign to it a limit . . . It diffuseth throughout all, penetrateth all and it envelopeth, toucheth and is closely attached to all, leaving nowhere any vacant space.   (Jammer 1993: 89)

4. Abbott's *Flatland* (first published in 1884 and in print ever since) is the classic literary source (Abbott 1986). See also Dewdney (1984), Rucker (1984) and Stewart (2001).

5. Visit http://casa.colorado.edu/~ajsh/sr/hypercube.html (accessed June 2001) for a simulation of a rotating hypercube.

6. The additional spatial dimensions postulated by string theorists only make their presence felt at a very, very small scale. See Chapter 20.

7. See also Hollis (1967).

8. It is worth noting, however, that some contemporary physicists have floated the idea that our universe is in fact fluid-filled. Grady has devised a model in which the material universe is a giant crystal growing in a five-dimensional fluid. Matter is simply the fluid in frozen form (think of the way thin shards of ice form in water that is nearly frozen). See Chown (1999).

9. See Manders (1982) and Mundy (1983) for some detailed relationalist reconstructions.

10. Space: the classical debate

1. The conversion to Newtonianism was not uniform across national and cultural boundaries. The French in particular were reluctant to abandon Cartesianism, but did so eventually.

2. The Scholium is only a few pages long, and reproduced in full in Thayer (1974), Barbour (1989) and Earman (1989), not to mention Newton (1729). I have given references for Huggett (1999), which also contains useful selections from the relevant writings of Descartes and Leibniz, as well as extracts from Newton's *De Gravitatione*.

3. The idea that space is an infinite, homogeneous and immovable entity, independent of material bodies but capable of containing and constraining them, has a long history, as does the doctrine that space thus conceived is a divine attribute. The latter doctrine dates back at least as far as Palestinian Judaism during the Alexandrian period, and was transmitted through the Jewish esoteric tradition to Italian Renaissance figures such as Mirandola and Campanella, and thence to figures such as Fludd and Gassendi, and subsequently to More and Barrow, both of whom were Cambridge contemporaries of Newton and known to have influenced him. More, who wrestled with the problem of space between 1648 and 1684, sought to synthesize cabalistic doctrines with the "light-metaphysics" of Neo-Platonism, a doctrine that associated space with light, and light with God, and which led Grosseteste to believe that geometrical optics was the key that would unlock the secrets of the universe. Newton's early renown, it is worth noting, rested on his own works in optics.

4. There are passages in Newton's writings in which his mystical-cum-theological bent is evident. In the General Scholium added to the second edition of the *Principia* (1713) we find: "He is not eternity and infinity, but eternal and infinite; he is not duration or space, but he endures and is present. He endures for ever, and is everywhere present; and by existing always and everywhere, he constitutes duration and space". In Query 28 added to the second edition of the *Opticks* in 1706 he writes, "does it not appear from Phaenomena that there is a Being incorporeal, living, intelligent, omnipresent, who in infinite Space, as it were in his Sensory, sees things themselves intimately, and thoroughly perceives them, and comprehends them wholly by their immediate presence to himself". It was this last claim that attracted Leibniz's attentions, and initiated the *Correspondence*. Newton only published these "clarifications" of his theory of absolute space in response to theologically-inspired criticisms of the doctrine that had started to appear in response to the first edition of the *Principia*.

5. Prior to Galileo it was not unknown for anti-Copernicans to argue that since a cannon ball dropped from the mast of a ship in motion does *not* fall straight down to the base of the mast, if the Earth were in motion we could expect the same to apply to objects dropped from towers on land.

6. The situation is rather more complicated: Galileo also subscribed to a principle of straight-line (rectilinear) inertia for things such as bullets shot from guns. The fact that he did not find it necessary to discuss the relationship between circular and rectilinear inertia is another sign of his continued allegiance to Aristotelian doctrines, in this instance the distinction between natural and violent motions.

7. Pondering the reasons why the true laws of dynamics took so long to discover, Barbour observes that the law of inertia would only seem obvious when people became aware of the possibility of free continuation of motion *for ever* once commenced:

> and from what activities could man have gained such an awareness? Skating perhaps, but that was scarcely a sport practised by the ancients in the Mediterranean. (One wonders if it is entirely fortuitous that the first reasonably clear statement of the law of inertia in its modern form was made by a man – Descartes – who spent most of his adult working life in *Holland*, at a time moreover in which skating had become a great vogue.)
>
> (1989: 35, original emphasis)

8. It follows from this that in the strict sense there is no motion occurring in the case of the earth or even the other planets, since they are not transferred from the vicinity of those parts of the heaven with which they are in immediate contact, in so far as those parts are considered as being at rest. (*Principles* II, 28)

9. Although Leibniz was a long-standing critic of Newton (in 1705 he had argued that the latter's gravitational force was an unwelcome reintroduction of "occult qualities" into physics), they rarely entered into direct correspondence; relations between them had been strained beyond breaking point by an acrimonious priority dispute over the invention of the calculus.

10. I leave the notion of "genuine" property vague, but there are certain properties for which PII is certainly not a valid principle. For example, if the property "is believed by Lois Lane to be a superhero" is a genuine property, then PII entails that the names "Clark Kent" and "Superman" cannot refer to the same person.

11. There is a well known (but disputed) counterexample to PII (due to Max Black). Couldn't there be a universe whose sole contents are two iron spheres, of exactly the same size and shape, sitting at a distance of six feet away from one another? It seems so. But since the two spheres have exactly the same properties, including the relational property of "being six feet away from an iron sphere of such and such shape, size, etc.", PII entails that there aren't in fact two spheres at all, only one. Note also that PII is not the only principle which is sometimes called "Leibniz's Law", there is also the *principle of the indiscernibility of identicals*: if a and b name the same object, then a and b have exactly the same properties. This is far less controversial, although it has to be indexed to times to accommodate the fact that persisting objects can change their properties.

12. As Maudlin notes, "Leibniz, in a classic modus ponens/modus tollens reversal, argues that since only those relations can be observed, the supposed shifts create no real change at all" (1993: 188).

11. Absolute motion

1. For a discerning analysis of just what Newton was trying to achieve in the Scholium see Rynasiewicz (1995).

2. The Newtonian account of inertial forces requires absolute time as well as absolute space. Accelerated motion gives rise to effects that nonaccelerated motion does not, and this acceleration is absolute. It isn't a matter of moving with respect to other bodies; acceleration is rate of change of velocity, i.e. how much space is covered during a certain interval of *time*. Now, the relationist holds that there is no absolute measure of periods of time, just more or less convenient measures as recorded by different clocks. If I select as my clock one that reveals a body to be accelerating, but the body suffers no force, we can conclude that the body isn't really accelerating, so the clock is wrong. Hence there is an "absolute" time scale, and good clocks accurately reflect this time scale, bad clocks don't.

3. In recent years work has been done to improve on Mach. Barbour and Bertotti (1977, 1982) have formulated precise and rigorous Machian-style theories of gravitation and inertia. Their proposals are serious and well regarded, and although far from complete (they have not as yet incorporated electromagnetism or quantum mechanics into their Machian schemes), do show that it is possible to go far further in the direction that Mach advocated than many have supposed. Still, it remains the case that the viability of a Machian approach is as yet unproven. For evaluations, see Earman (1989: 92–6), Belot (2000) and Misner *et al.* (1970: §21.12).

12. Motion in spacetime

1. For an early and influential statement of this view, see Howard Stein's "Newtonian space-time" (1970).

2. The following succinct definition of neo-Newtonian spacetime is due to Huggett:

> Between any two events p and q, (i) let there be a definite temporal interval, $T(p, q)$; (ii) if they are simultaneous ($T(p, q) = 0$), let there be a definite spatial distance $D(p, q)$; (iii) given any curve c through p, let c have a definite curvature at p, $S(c, p)$ (i.e. a curve c is straight if $S(c, p) = 0$ for all points p on c).   (1999: 194)

3. Largely drawn from Robert Geroch (1978: Ch.3).

4. If the relationist recognizes absolute directions there is no choice of orienting M in S in different ways; this additional degree of freedom becomes available if only relative angular differences are recognized.

5. This way of characterizing the difference between relationists and substantivalists is now standard: see Friedman (1983: Ch. 6) and Mundy (1983).

6. While this notion may be unproblematic in the context of the world*tubes* of non-punctiform objects, defining a notion of curvature for the strictly one-dimensional world*lines* of point-particles is another matter. Perhaps the curvature of a given worldline can be defined as the limit of series of nested worldtubes, all with the same (or minimally different) curvature along the time-axis, but diminishing radii (in a manner reminiscent of Whitehead's method of extensive abstraction). In any

event, there is no reason why the relationist need shy away from introducing new geometrical primitives, even if working out the detailed mathematics is a non-trivial task.
7. Maudlin's *Quantum Non-Locality and Relativity* (1994) is entirely devoted to exposing the difficulties of reconciling special relativity and quantum theory.

13. Curved space

1. A perfected Euclidean system was finally achieved at the end of the nineteenth century by Hilbert, who used a total of twenty axioms.
2. This is the "Playfair" version. Euclid's version is "If a straight line falling across two straight lines makes the sum of the interior angles on the same side less than two right angles, then the two straight lines intersect, if sufficiently extended on that side."
3. In a letter to Bessel, quoted in Jammer (1993: 148).
4. See Norton (1992: Pt II) for a good discussion of this point in a simple context. There is clear and succinct account of the basics of Riemannian geometry in Reichenbach (1958: §39); see Egan (1998, II) for a somewhat more technical (but not too advanced) explanation of what is involved mathematically. As Sklar notes (1975: 45), in characterizing the intrinsic structure of a Riemannian space it is not necessary to start off with a particular coordinatization and specification of the *g*-function; there are "coordinate free" methods, and "much of modern differential geometry is devoted to the problem of characterizing the intrinsic structure of general curved spaces without reference to an imposed coordinatization of the space". See Egan (1998) and Baez (1996) for comparatively accessible routes to the curvature tensor via parallel transport of vectors and connection coefficients. For an idea as to what "parallel vector transport" involves, imagine walking around the four sides of a football field holding a walking stick out in front of you, making sure you keep the stick pointing in the same direction at all times (this may take some doing when you reach the corners, but you are allowed to change hands). Assuming that the field is flat, when you complete the trip and arrive back where you started, the stick (which has undergone a "parallel transport") will still be pointing in the same direction. If you conduct a similar round-trip on a curved surface, the stick will *not* arrive back pointing in the same direction; the difference in direction constitutes a measure of the curvature of the relevant surface – the greater the curvature, the greater the direction change after parallel-transport. It is easy to verify this effect of curvature by spending a few minutes experimenting with an apple and pencil.
5. See Callender and Weingard (1996) for an informal exposition of some of the fundamentals.
6. In the two-dimensional case, five types of space are locally Euclidean: the plane, the cylinder, the Möbius strip, the torus and the Klein bottle. Of these, only the plane is simply-connected. There are two spaces of constant positive curvature – elliptical (or spherical) space and the "projective plane" – and the latter is multi-connected, the former simply-connected. As for two-dimensional spaces of constant negative curvature, the Lobachevskian space that we encountered earlier is simply-connected, but the number of multi-connected spaces of this type is infinite. The same pattern extends to the three-dimensional case, but direct comparisons become trickier as the numbers on both sides are infinite.
7. For more on life in topologically unusual spaces see Reichenbach (1958: Ch. 1, §12), Nerlich (1994b: Ch. 5).
8. In fact, in travelling between two locations light follows the path that takes the least time – which even *in vacuo* isn't quite the same as travelling in a Euclidean straight line, as we shall see.

14. Tangible space

1. This is oversimplifying considerably. Although an object entering the hole will acquire stress energy, and (given the principle of the conservation of energy) this energy must come from somewhere; in some cases it may be supplied from the outside (a push), but it may also be supplied (partly or wholly) from within: the moving object will lose some of its kinetic energy, and so change either its velocity or momentum. But as Nerlich notes:

> The interaction of the tensor field of stress with the vector field of momentum is so complex that we can say very little in general about how they will sum over a volume. Furthermore,

in certain cases, kinetic energy may be gained at the expense of stress energy, and the elastic volume will accelerate without the intervention of an outside force.   (1994a: §5, 178–9)

Even so, having one's middle regions undergo unforced accelerations would be likely to induce an unusual sensation.
2. See Van Cleve (1998) for a different response to the "lesson" from Flatland.
3. As most will recall from early school days, you can make one by taking a longish strip of paper, giving it a half-twist, and then sticking the two ends together.

15. Spatial anti-realism

1. The odd goings-on across the boundaries of R_1 and R_2 in Foster's scenario are due to place-specific laws of nature, which, in effect, teleport objects between the regions concerned. There may be another way in which the same effect could be produced. In §13.5 we encountered a space with a toroidal topology. When a plane "exits" from the space on the left, it instantly re-emerges on the right; when it "exits" on the right, it re-emerges from the left, and so on. These effects were not due to strange laws of nature, but rather to the fact that the points on opposite sides of the space are numerically identical. If, in similar fashion, the points on the boundaries of R_1 and R_2 in the Oxford–Cambridge case were numerically identical, on entering R_1 from the left you would find yourself inside R_2, continuing on in the same direction. Would this "re-location" be detectable?
2. To be specific, the bulk of Chapters 8–10.
3. Foster points out (1993: 305) that for his purposes he could take any aspect of the geometrical structure of a space to be essential to it (e.g. metrical or affine structure), since the functional and intrinsic geometries of S could differ in any of these other ways too. But since "topological essentialism" is the weakest version of "geometrical essentialism" it is also the least contentious.
4. Here is how Foster himself formulates the modal argument:

> We know that the physically relevant geometry (topology) of S logically depends upon its nomological organization. And we also know that S possesses this nomological organization (including those aspects relevant to the physical geometry) only contingently. From these premises, it immediately follows that S possesses its physically relevant geometry (topology) only contingently. But we also know that, as a genuine space, P possesses its physical geo-metrical (topological) structure *essentially*. So, standing as they do in different modal rela-tions to the physical (or physically relevant) geometry, P and S must be numerically distinct. Moreover, since the physical geometry (topology) which is essential to P is essential to the identities of its points, we can conclude, more strongly, that the points of P are numerically distinct from the points of S. And, points being the smallest components of a space, this means that there are no divisions of S and P relative to which the components of one can be identified with those of the other.   (1993: 306).

To simplify, I have replaced Foster's F (in the original) with S.
5. Rejecting Foster's essentialist doctrine concerning points would permit a response of this sort:

> If geometry (or topology) is not essential to the identity of points, spaces with different geometries can contain the same points; consequently, the fact that S and P can differ geo-metrically is no obstacle to holding that they are the same space, i.e. the same collection of points.

Among those who question the doctrine of the necessity of identities, in its traditional guise, are Lewis (1971, 1986), Noonan (1991, 1993) and Gallois (1998).

16. Special relativity

1. Davies (1995: 69) provides a nice example of a case in which these effects are observable in everyday life:

> Most metals have a silvery appearance, but not gold. Its distinctive and attractive glitter can be traced to the effects of relativity on the motions of the electrons inside the metal that are responsible for reflecting light. So it is no exaggeration to say that this precious metal is

precious – and financially valuable – partly as a result of time dilation operating within the gold atoms.

2. It should be pointed out that Einstein's theory has implications of a very practical kind. Einstein published two papers on STR in 1905; the first lays down the basic principles of the theory, the second (only three pages long) establishes the equivalence of mass and energy via the famous equation $E = mc^2$. Matter and energy are interconvertible, and the amount of energy contained in a body at rest is huge by ordinary standards (the energy "stored" in a 1 kg mass would keep around 30 million hundred watt light bulbs going for a year). While the full implications of this did not emerge at once, it was only another forty years before the first atomic bombs exploded.

3. STR gives rise to several intriguing puzzle cases (often referred to as "paradoxes"). For more on these see entries for The Twin Paradox, The Barn and the Pole, The Superluminal Sissors and (more advanced) Bell's Spaceship Paradox in the Usenet Relativity FAQ at www.math.ucr.edu/home/baez/physics/relativity.html (accessed September 2001).

17. Relativity and reality

1. To keep things as simple as possible I will restrict my attention to *growing* block theories, but the same considerations will apply to dynamic block models that do not posit a preferred temporal direction.

2. Stein states his theorem thus:

> If R is a reflexive, transitive relation on a Minkowsi space (of any number of dimensions – of course at least two), invariant under automorphisms that preserve the time-orientation, and if Rab holds for some pair of points (a, b) such that ab is a past-pointing (time-like or null) non-zero vector, then for any pair of points (x, y), Rxy holds if and only if xy is a past-pointing vector.
>
> (1991: 149)

3. For example, Dorato (1995: Ch. 11), Clifton and Hogarth (1995) and Tooley (1997: 337).

4. Hinchliff (1998), but see also Godfrey-Smith (1979), who seems to have been the first to propose this version of presentism. Compound presentists will extend reality to sheaves of neighbouring light-cones. Another option for presentists would be to hold that the present of a given point consists of the space-like separated regions of spacetime (the point's "absolute elsewhere").

5. The cone presentist is confronted with another awkward question: why confine the present only to the surfaces of past light-cones? Why not include future light-cones too? After all, events on the latter are also light-like related. Hinchliff rejects this "double-cone presentism", appealing to an asymmetrical causal relation: "The difference between the cones is due to the asymmetry built into the nature of a light *ray* or *signal*. And that asymmetry arises from the asymmetric nature of causation itself, which is a non-arbitrary foundation on which to rest the distinction between cone and double-cone presentism" (1998: 582). Presentists who reject this view of causation may well be obliged to embrace double-cone presentism, and all that it entails.

6. For a different line of response see Smith (1993, 1998).

7. "[We] cannot maintain a Lorentz invariant interpretation of the quantum nonlocal connection of distant systems . . . there has to be a unique frame in which these nonlocal connections are instantaneous" Bohm and Hiley (1993: 271). See also Bohm (1952), and Albert (1992: Ch. 7). The theory (also known as the "De Broglie-Bohm Theory" since it is a development of a line of thinking that dates back to De Broglie in the 1930s) posits a form of non-locality over and above the EPR variety. The global wave function, which according to the theory simultaneously fixes the motions of all particles, is itself influenced by the condition of individual particles everywhere, so it "turns out that in determining the trajectory of a particle, one may have to take into account the wave functions of particles in other galaxies!" (Chalmers 1996: 344).

18. General relativity

1. In a letter to Sommerfeld, quoted in Greene (1999: 62).

2. This example is drawn from Maudlin (1994: 225–7).

3. What Newton could have done (had he had the requisite mathematics) was formulate an explanation of gravity in term of mass-induced curvature in *Newtonian spacetime* – and the precise form

such a theory would take has been worked out. See Friedman (1983: 95–104) for details. Friedman comments, "In an important sense, then, general relativity is just Newtonian gravitation theory plus special relativity" (1983: 184). But since the empirical predictions of this theory differ from those of Einstein's theory, and the latter's predictions have (so far) proved to be correct, there are no takers for the modified Newtonian theory.

4. The conventionalism stance is far from dead, especially among scientists who regard theories as primarily tools for prediction: cf. Thorne (1994: Ch. 11).

5. The twenty components of the Riemann curvature tensor – which generate the twenty numbers needed to completely describe curvature in four dimensions – can be divided between the "Ricci tensor" and the "Weyl tensor". The Ricci tensor describes changes in *volume*, whereas the Weyl tensor describes changes in *shape* (objects can change their shape without changing their volume). The Einstein tensor is defined via the Ricci tensor, and since it only has ten components (likewise the stress-energy tensor) it does not completely determine the geometry of spacetime. The remaining aspects of curvature are characterized by the Weyl tensor. In regions where there is no mass-energy the stress-energy tensor is zero, and likewise the Einstein tensor and the Ricci curvature. But the Weyl curvature in such regions need not be zero: there is thus an aspect of curvature that is independent of mass-energy. Gravitational waves and tidal forces involve variations in the Weyl curvature. See Baez (2001) and Penrose (1989: 210–11). Note also that since GTR is a spacetime theory, it is formulated in terms of a *pseudo-* or *semi*-Riemannian manifold, not the spatially four-dimensional Riemannian manifold we encountered in §13.4.

6. Torretti writes:

> The [stress-energy] tensor is something of a joke, for we are quite incapable of accurately representing the distribution of matter and energy in the universe and, if we could do it, we would be unable to solve the [Einstein Field Equations] EFE for a matter-energy tensor field of the required complexity. In actual practice, the EFE are solved exactly only for ideally simple fields. Thus the Schwarzschild solution, used in celestial mechanics, concerns the spherically symmetric field of a central body, and the Friedmann solutions, used in cosmology, assume a distribution of matter which is perfectly isotropic about every point in space.
>
> (1999a: 77)

7. Newtonian gravitational theory leads to the same result. In 1785, John Michell, working on the assumption that light is corpuscular, showed that mass, sufficiently concentrated, would have a gravitational pull so that strong light could not escape (assuming that light is affected by gravity in the way of other particles). Since nothing ruled out the possibility of matter existing in so condensed a state, he speculated that space might contain a huge number of "dark stars", all invisible to us.

8. For more on black holes and their strange properties see Penrose (1989: 330–45) and Thorne (1994). For a discussion (often advanced) of the problems posed by singularities, see Earman (1995).

9. These properties are distinct: an infinite field of exactly similar (and similarly spaced) blades of grass, all blown in precisely the same direction is homogeneous (the same everywhere) without being isotropic (there is a preferred direction).

10. "Dynamic" in this context should not be confused with "dynamic time"(of the sort we looked at in Chapter 6). A universe is dynamic in the current sense if it is of different *spatial* sizes at different times; universes in which *time* is dynamic are of different *temporal* sizes as of different times. It is perfectly possible for a static-time universe to be spatially dynamic, just as it is possible for a dynamic-time universe to be spatially static (or dynamic).

11. A significant minority of cosmologists remained hostile to the big bang: the "steady state theory", proposed by Bondi, Gold, Hoyle and Narlikar (among others) in the late 1940s, remained a serious competitor for at least two decades, before falling out of favour due to its incompatibility with observational data. Underlying the theory is the so-called "perfect cosmological principle", according to which the universe is homogeneous and isotropic in *time* as well as space. Since the steady state theorists accepted that the universe is expanding, they had to come up with an explanation for how the average density of matter could remain constant over time (other things being equal, it would decrease with the expansion of space). Their solution was the C-*field*, which generates a steady flow of new matter (by a process of "continuous creation") to compensate for the

drop in density brought about by expansion. Continuous creation has never been observed in the laboratory, but given the tiny quantities required by the theory – only one hydrogen atom per cubic metre over the entire age of the universe – this was not regarded as problematic. The C-field, along with the notion of continuous creation, could be regarded as a spatial (albeit small-scale) analogue of the continual diachronic *ex nihilo* creation that growing block models of time require. See Coles (2001: 330–31) for more on the steady state theory.

12. Current estimates place the critical value between 1 and 2.10^{-29} g cm^{-3}, which works out at about three hydrogen atoms per cubic metre.

13. Cf. §13.5.

14. For a full review see Lachièze-Rey & Luminet (1995); also Luminet (1998, 2001), and Luminet *et al.* (1999). The fundamental domains for (three-dimensional) multi-connected spaces are polyhedra; the idea that these geometrical objects could have cosmological significance has definite Platonic overtones.

19. Spacetime metaphysics

1. These "holes", it will be recalled, are not cavities in (or absences of) substantival space, but rather regions of extreme local curvature *in* substantival space.

2. It should be noted that gravity waves have yet to be detected under laboratory conditions, but there is astronomical evidence for their existence, e.g. the observed energy loss of binary pulsar systems match the energy that GTR predicts will be transported away by gravitational ripples. See Will (1988: 201–6) and Thorne (1994: Ch. 10).

3. Wheeler combines the two: "The geometry of spacetime, and therefore the inertial properties of every infinitesimal test particle are determined by the distribution of energy and energy-flow throughout all space" (quoted in Ray 1991: 134).

4. For more detail see Sklar (1977: 210–44), Raine (1981), Earman (1989: 105–9) and Ray (1991: Ch. 7).

5. The diffeomorphism is of the "active" rather than "passive" variety; i.e. it involves a switching about of geometrical objects – points – rather than merely a relabelling of the same objects via a different system of coordinates or names.

6. See, for example, Butterfield (1989a,b), Brighouse (1994) and and Rynasiewicz (1994). Some might be tempted simply to reject the hole argument on the grounds that it is foisting upon GTR superfluous metaphysical commitments concerning the numerical identities of spacetime points. To take a simpler case, suppose we have a theory that predicts the forces that exist within systems of interconnected load-bearing beams at any given time. We would have grounds for complaint if our theory delivered alternative models of a given beam structure, some of which predict collapse, some of which don't. But we would scarcely complain if the differences between the alternative models involved nothing more than the numerical identity of the beams themselves (e.g. there are transformations that switch the locations of intrinsically similar beams within a structure at a time). All our theory purports to offer is an account of synchronic forces within structures of certain *types* of beam, so its "failure" to be sensitive to the identities of *token* beams is not really a failure at all. Might not precisely the same be said of the alleged failures of determinism in GTR? Proponents of the hole argument would say not: unlike our imagined theory of beams, the mathematical formalism used in GTR is sensitive to token-identities, and so is *not* neutral on the question of the numerical identities of spacetime points; consequently, the alternative models are genuinely distinct by the lights of the theory itself.

7. I say a little more about these issues in Chapter 20.

8. See also Figure 18.6. The uniform microwave background radiation left over from the big bang (measured with great accuracy by COBE) provides a valuable guide here:

> Although the view from Earth is of a slightly skewed cosmic heat bath, there must exist a motion, a frame of reference, which would make the bath appear *exactly* the same in every direction. It would in fact seem perfectly uniform from an imaginary spacecraft travelling at 350 km per second in a direction away from Leo (towards Pisces, as it happens) . . . We can use this special clock to define a *cosmic* time . . . Fortunately, the Earth is moving at only 350 km per second relative to this hypothetical special clock. This is about 0.1 per cent of the speed of light, and the time-dilation factor is only about one part in a million. Thus to an

excellent approximation, Earth's historical time coincides with cosmic time, so we can recount the history of the universe contemporaneously with the history of the Earth, in spite of the relativity of time. (Davies 1995: 128–9)

See also Dorato (1995: Ch. 13).

9. This analogy should not be extended too far. If our spacetime is, say, negatively curved and open, what "grows" is already spatially infinite.
10. Thorne (1994: Ch. 14); see Al Khalili (1999: 225–39) for a more recent survey.
11. Savitt explains it thus:

any way we extend the Gödelians' local time function to the whole spacetime will have an odd consequence. Pick any spacetime point x somewhat distant from the Gödelians but simultaneous (according to this extended time function) with some point of their history. A possible observer, Kurt, can be born at x, live an entire lifetime experiencing time just as they (and we) do, yet no part of his life is assigned a time different from that assigned to x. . . . Thus Kurt's "whole existence objectively would be simultaneous" if one tried to identify objectively lapsing time with the Gödelians time. (1994: 467–8)

12. For further discussion of Gödel's argument see Yourgrau (1991) and Earman (1995: Ch. 6, appendix).

20. Strings

1. Strong interactions bind protons and neutrons (and their component quarks) together in atomic nuclei, which is roughly the range of the strong force (approximately 10^{-13} cm). The weak force only operates over distances a hundred times smaller than this, but nonetheless plays an important role in particle physics: neutrinos feel the weak force but not the strong, and are produced in bulk by the nuclear reactions that power the sun.
2. Needless to say, there is rather more to it. See Greene (1999: 152–65) for a fuller account of the way string theory overcomes the problems associated with point-interactions.
3. It is possible to produce two-dimensional representations of what three-dimensional "sections" of such a manifold would look like. There is a beautiful example in Greene (1999: 207), which resembles a cross between a tangle of worms and an impossibly intricate piece of origami.
4. See Greene (1999: Ch. 13) for more on string theory and black holes. In the mid-1990s the five main competing versions of string theory were brought into a single unified framework known as "M-theory". Once again, the unification was brought about by the "discovery" of another spatial dimension: in M-theory there are ten spatial dimensions, and one of time. Thanks to the extra dimension, strings can possess length *and* depth. M-theory posits not just one-dimensional strings, but two-dimensional membranes (two-branes), along with pulsating blobs (three-branes) and other exotic Planck-scale entities. See Greene (1999: Ch. 12).
5. In his "Reflections on the fate of spacetime" (1996), Witten suggests "'spacetime' seems destined to turn out to be only an approximate, derived notion, much as classical concepts such as the position and velocity of a particle are understood as approximate concepts in the light of quantum mechanics" (Callender and Huggett 2001: 134).
6. Or at least it does when one variant of the theory is construed both literally and realistically, rather than being viewed as merely a useful mathematical model.
7. This theme runs throughout Smolin's book (e.g. 2000a: 119–20). The theoretical underpinning is diffeomorphism invariance, a feature of GTR that is often taken to have Leibnizian implications (see §19.3 & §19.4), and which is preserved in loop quantum theory; see Gaul and Rovelli (1999). Whereas in the context of classical GTR it is questionable whether the classical point-manifold really is dispensable, there is no reason for such misgivings in the loop quantum case, since a substitute for the manifold (the spin lattice) is supplied. For more on loop quantum gravity follow the links at www.qgravity.org

Glossary

Many of the technical terms used in the main text can be found below, sometimes explained in less detail, occasionally more. I have taken this opportunity to introduce some standard philosophical terminology that is used but *not* explained in the main text.

absolute: quantities or relationships that are independent of **frames of reference** and **coordinate systems**.

absolute becoming: the coming-into-existence of objects, events and times posited by dynamic theories of time, such as the **growing block model** and **presentism** in some of its guises. "Absolute annihilation" is the ceasing-to-exist of objects, times and events.

absolute motion: motion with respect to substantival space. See **substantivalism, relationism.**

absolute simultaneity: two events are simultaneous if the temporal interval between them is zero; if the temporal interval between them is zero in all frames of reference, they are absolutely simultaneous. Since absolute simultaneity is an equivalence **relation**, it partitions points in space and time into mutually exclusive and exhaustive "planes of simultaneity". The notion is abandoned in Einstein's special theory of **relativity**, a fact that has led many to suppose that the notion of a single universe-wide "now" cannot be physically real.

absolute space: the sort of space Newton believed in: substantival and with a structure that is independent of, and unaffected by, any material bodies that occupy it. Subsequent developments (such as Einstein's general theory of relativity) have made it clear that the structure of a substantival space need not be independent of its occupants.

acceleration: rate of change of velocity. "Absolute acceleration" is rate of change of velocity with respect to a substantival space.

action-at-a-distance: a force or influence that acts between two objects directly, without passing through the intervening space. Prime example: Newtonian gravity. In recent time, theories that posit action-at-a-distance have been regarded as suspect, or even "spooky" (see **field**), but developments in quantum theory have brought the notion to the fore once again (see **Bell's inequality, non-local**).

affine geometry: the study of properties that remain constant under "affine transformations", i.e. those that preserve parallelism and collinearity (points are collinear if they lie on the same straight line), but not angles or distances. The slogan "In affine geometry, all triangles are the same" is a useful reminder of what is involved.

analytic–synthetic: another way of categorizing truths and falsehoods, also the subject of dispute. Analytic truths are true in virtue of the meanings of the words used to express them ("Bachelors are unmarried"), whereas synthetic truths are not ("Jupiter is the largest planet in the solar system").

anti-realism: the term is used in a variety of ways in contemporary philosophy, but, generally speaking, anti-realism is concerned with the rejection (or downgrading) of existence and/or truth. An anti-realist about unicorns (say) maintains that although we talk about unicorns in much the same way as we talk about cats or oranges, in reality unicorns do not exist, and so statements about unicorns are either false, or neither true nor false; a realist about unicorns maintains that unicorns exist, and that statements about them can be true. So far as scientific theories are concerned, some anti-realists hold that we have no reason to believe that any of our theories are true, others say that we have no reason to believe that the unobservable entities posited by certain theories exist (these claims can, of course, be combined). Some anti-realists (or "reductionists") about the past hold that, although statements about the past can be true or false, what makes these statements true are not past states of affairs, but presently available evidence, and, consequently, that statements about the past that cannot be verified or falsified by presently existing evidence are neither true nor false – a position sympathetically explored by Michael Dummett.

a priori–a posteriori: a statement or claim is a priori if it is possible to determine whether it is true (or false) without consulting the world via observation or experiment. To verify (or falsify) an a posteriori claim one has to consult the world. The application of the distinction is much disputed, but most would agree that "1 + 1 = 2" and "widows are women" are a priori, whereas "some cats are black all over, others are not" and "radiation causes cancer" are a posteriori. See also **analytic–synthetic, necessary, contingent.**

A-properties/B-properties: A-properties include *present, past, future*; B-properties include *earlier than, simultaneous with, later than*. The A-properties of an event are different as of different times, B-properties are unchanging. A-properties are also called "tensed" and B-properties "tenseless". "X happened three days ago" (i.e., in the past) is an A-sentence, and "X happened three days earlier than Y" is a B-sentence.

arrow of time: used in connection with the idea that time has a direction, in a way that space does not. Temporal **passage**, if it were real, would provide time with an arrow, but so too might asymmetries among the contents of time. There are numerous apparent content-asymmetries: the causal, electromagnetic and thermodynamic, the "psychological" (intention, belief), the "phenomenal" (involving experience), together with the knowledge-asymmetry (we know more about the past than the future). How these various arrows are related to one another is one of the central problems in the philosophy of time. See also **time-reversibility, entropy.**

A-theory/B-theory: prior to the 1980s, an A-theorist was usually someone who believed that B-sentences could be translated (without change of meaning) by A-sentences, and a B-theorist someone who believed the opposite. As it became clear that such translations are impossible, usage diversified. The **new (tenseless) theory of time** is often called the "B-theory", but the term is also used simply to refer to the **block view**, irrespective of other commitments. Consequently, an "A-theorist" might be someone who either (i) believes that there are genuine A-properties and facts, and that it is these that make A-statements true, and/or (ii) believes that an adequate account of time cannot be formulated in entirely tenseless terms, and/or (iii) rejects the block view, and holds that time is (in some way) dynamic rather than static.

atoms: the tiny constituents of complex material objects and substances. A classical (or metaphysical) atom is simple (has no parts) and indivisible. The atom of contemporary science consists of a nucleus (made up of protons and neutrons), and a surrounding swarm of electrons. The atoms of different elements have different numbers of protons, neutrons and electrons.

backward causation: normally, a cause c occurs before its effect e; in backward causation, e occurs before c.

Bell's inequality (or theorem): if the inequality holds, quantum mechanics is **non-local**. Experiments have confirmed that the inequality holds, but the precise manner in which quantum mechanics is non-local remains a matter of controversy.

big bang model: the hypothesis that the universe began somewhere between 10 and 15 billion years ago in a tiny primeval fireball of extreme density, and has been expanding and cooling ever since. Among the key pieces of evidence for the model are the observed red-shift of light from distant galaxies (which means the distance between us and them is increasing) and **microwave background radiation.**

block view: the doctrine that past, present and future are all equally real (sometimes known as "eternalism"). Although block theorists subscribe to a **static** conception of time – they reject temporal passage and the moving present – they reject the allegation that a block *universe* is "static" in the sense of containing no change. For the block theorist, a universe (or object) changes if it has different properties at different times, and our universe does. One occasionally (not here) finds "the block universe" being used to refer to universes where **determinism** (of the **nomological**) variety obtains.

calculus: the branch of mathematics concerned with the study of the rates of change of continuously varying quantities; used in most branches of science. "Differential" calculus provides the means for finding the slope of a tangent to a curve at a given point, and the maximum and minimum values of functions; it can be used to work out the precise rate at which one variable quantity is changing relative to another (e.g. velocity is rate of change of distance with respect to time, acceleration is rate of change of velocity with respect to time). "Integral" calculus is used to calculate the lengths of curves, the areas of flat and curved surfaces, and hence magnitudes such as the total force acting on a body.

causal theory of time: a programme of analysing temporal concepts in causal terms.

centre of gravity: the point at which the total weight of a body can be considered to act.

classical physics: Newtonian and post-Newtonian physics, prior to the development of **relativity** and **quantum mechanics** in the early years of the twentieth century.

congruence: two objects or shapes are congruent if they have exactly the same size and shape.

connection: any sort of link or relation, but in spacetime theories a specification of points which fall on straight (or straightest) lines (see **geodesic**).

consciousness: your consciousness as you read this sentence consists of everything you are experiencing: your visual, auditory and tactile experiences, your thoughts, mental images, bodily sensations, and so forth. We are not "conscious of" (in the sense of "thinking about") most of our consciousness for most of the time. Consciousness (in this semi-technical sense) does not require wakefulness: we are conscious (i.e. have conscious states) when dreaming.

contingent: something that is not **necessary**. A true statement is contingent if it could have been false, an entity is contingent if it might not have existed, an object's condition is contingent if it could have been different. The laws of nature are usually taken to be contingent – since it seems reasonable to suppose there are logically **possible worlds** where the laws of nature are different.

continuant: a persisting object (or **substance**), specifically an object that lacks temporal parts (see **event**).

continuum: in discussions of space and time "continuous" usually means infinitely divisible (so there is an infinite number of points between any two points, no matter how close together they are); a "continuum" is thus a time, space or spacetime which is infinitely divisible.

coordinate system: a systematic assignment of numbers to spatial, temporal (or spatiotemporal) points, such that each point receives a different number. For three-dimensional spaces, the *Cartesian* coordinate system (familiar from school geometry, and invented by Descartes) employs three axes, x, y, z, centred on a given point O, all at right angles to each other, and pointing in fixed directions; by convention, the x- and y-axes are represented as horizontal, the z-axis as vertical; each point can then be identified by three numbers, one for each axis. In a two-dimensional space, two numbers suffice to identify each point; in a four-dimensional spacetime, four numbers are required, and so on up – hence the connection between coordinates and **dimension**. Other methods of coordinatization are available: e.g. *polar* coordinates use combinations of distance and angle. Points in **curved** spaces can also be assigned coordinates, but somewhat different procedures are required, and in the case of variably curved spaces it will often be the case that several different coordinate systems (or "charts") are required to cover the whole space.

cosmic time: if we assume that the universe is effectively homogeneous and **isotropic** it is possible to define a single timescale that is valid across the entire universe; observers who move with the cosmic expansion can synchronize their clocks by reference to the local matter-density. In homogeneous universes equipped with a cosmic time coordinate, it is possible (mathematically) to ignore the intermixing of space and time that relativity theory usually brings, and treat space as an independent three-dimensional continuum.

counterfactual: a statement that makes a claim that is contrary to fact, describing a possible but non-actual state of affairs. A "counterfactual conditional" is an if–then claim with a counterfactual antecedent ("If it were sunny then I would be happier.")

covariance: the property possessed by quantities that do not depend on their method of representation in a given theory (e.g. the unit of measurement, or system of coordinates); only features that are *not* artifacts of particular methods of representation are candidates for being physically real. A **field** theory is said to be "generally covariant" if it is insensitive to the points underlying the field being swapped around in certain specified ways (the relevant transformations are called "diffeomorphisms").

curved space: spaces that conform to Euclidean geometry are said to be "flat", and consequently spaces that are non-Euclidean are said to be "curved".

determinism: the **nomological** (or causal) determinist believes that the laws of nature are such that how things are at one moment of time precisely fixes how things will be at all subsequent times, and how things were at all earlier times. The *temporal* determinist believes that all times and events are equally real, and hence that we can no more change the future than we can change the past (which is not to say that what we do in the present will not contribute to the future being as it is). The two determinisms are independent: one can be a nomological determinist without subscribing to the **block view** (and hence temporal determinism), and the laws of nature in a block universe need not be (nomologically) deterministic. See also **fatalism**.

differential geometry: the branch of mathematics that involves the application of **calculus** to curved surfaces and spaces, of any number of dimensions, hence the study of "differential (or differentiable) **manifolds**".

dimension: when used loosely (as in "I believe in other dimensions") the term simply means **universe**, in the narrow sense. In geometry, a point is said to be zero dimensional, a line one-dimensional, a plane two-dimensional, and a cube (or sphere) three-dimensional, likewise the space in which we appear to live. Although we cannot imagine them clearly, spaces of higher dimensions are **logically possible**, and their properties have been explored by mathematicians, and exploited by physicists: the **spacetime** of relativity theory is four-dimensional, and the idea that there are additional (but concealed) spatial dimensions has proved useful in particle physics in recent years. Formulating a precise mathematical definition of dimension is a non-trivial task (most current definitions appeal to concepts in **topology**). The intuitively appealing idea that a space's dimensionality is given by counting up how many numbers are needed to provide a unique "address" for each point in the space using a Cartesian-style **coordinate system** was undermined by the nineteenth-century discovery that a single continuous line could completely fill (and hence be used to coordinatize) a two-dimensional square.

dynamic models of time: dynamic theorists accord metaphysical significance to the passage of time, and hence reject the **static** or **block view**. There are different accounts of what passage involves: some dynamists equate it with the acquisition and loss of the A-properties *pastness*, *presentness* and *futurity* (which are construed as making a real difference to what possesses them), other dynamists posit variations in the sum total of reality.

dynamics: the branch of physics concerned with bodies moving under the influence of forces (cf. **kinematics**)

electromagnetic spectrum: the range of frequencies at which electromagnetic waves (also known as "radiation") can exist. It is customary to divide the spectrum into several broad bands, consisting of radio waves (starting at about 3×10^4 Hz), microwaves, infrared, visible light (a narrow band between 10^{14} and 10^{15} Hz), ultraviolet, X-rays and gamma rays (starting at about 3×10^{21} Hz). Lower frequency waves have longer wavelengths: radio waves can have wavelengths of several hundred metres, whereas gamma waves are at the pico-metre scale (10^{-12} metres); the wavelength of visible light ranges between 700 nanometres (red) and 420 nanometres (violet).

electromagnetic wave: a wave that consists of an electrical field and a magnetic field, oscillating at the same rate at right-angles to each other – in short, a ripple in the electromagnetic field. Electromagnetic waves travel at the speed of light in a vacuum, and, in effect, are self-perpetuating and self-propelling, and so unlike other waves (such as sound), do not require a medium in which to propagate. The existence (and speed) of such waves was predicted by Maxwell in the nineteenth century; he also drew the (correct) conclusion that light was itself a form of electromagnetic radiation. Maxwell's theory of electromagnetism is a **classical** theory; the quantum theory of electromagnetism – quantum electrodynamics (QED) – developed by Feynman – is regarded by some as being the most successful current theory of matter.

empirical: concerned with what can be discovered using observation and experiment, as opposed to what can be discovered using reason alone (see **a priori**).

entropy: a measure of the internal disorder (or randomness) of a closed system; the greater the disorder, the greater the entropy. The second law of **thermodynamics** says that the entropy of a closed system can never decrease, and so remains constant or increases. The second law is now seen as probabilistic rather than absolute, and neutral with respect to time: if a system exists in a state of highly improbable order, it is likely to exist in a condition of greater (and more probable) disorder at later *and* earlier times. This temporal symmetry leaves us with a puzzle: why does our universe have a low-entropy past? See also **time-reversibility**.

epistemology: concerned with knowledge and what can be known.

essential v. non-essential: an entity's essential properties are those that it cannot exist without. A triangle is essentially three-sided; I am not essentially two-legged (I could lose one and still survive).

event: a happening or process, typically extended over both time and space; standard examples include battles, concert performances and arguments; events such as these have both spatial and temporal parts (e.g. the first-half and second-half of a performance). Some maintain an event requires a change of some sort to occur; others, drawing a distinction between "boring" and "interesting" events, disagree. My usage in the main text reflects the latter view: I often use "event" to refer to what happens at a single moment of time – events of this duration cannot involve change. In spacetime theories this practice is taken a step further: individual spacetime *points*, occupied or unoccupied, are called events. According to traditional **ontology**, although entities such as cats and planets are spatially extended and endure through time, they do not have temporal parts, and so fall into an ontological category of their own: they are "objects" or "things" rather than events (see **substance, continuant**). The distinction between object and event has come under pressure since the advent of the four-dimensional (spacetime) way of thinking; in this context, everyday objects are often regarded as extended events possessing both spatial and temporal parts.

fatalism: the doctrine that what will be will be. In the philosophical literature the term is often used to refer to the doctrine that statements about **contingent** future happenings ("Jones will win the lottery next week.") have a definite truth value as of the present. Thus construed, fatalism is entailed by the **block view** of time, at least in its standard guise. The doctrine of temporal **determinism** is also called "fatalism" by some writers.

Feynman–Wheeler absorber theory: an elegant but controversial account of electromagnetic radiation. Electromagnetic waves that propagate in the earlier-to-later direction are called "retarded" – they arrive after their departure. These are the only sort of electromagnetic waves we are familiar with: we never receive television broadcasts from the future, we receive them only from the past (even "live" broadcasts suffer a slight time-delay, since electromagnetic radiation travels at the speed of light, i.e. about one foot per nanosecond.) However, Maxwell's equations for electromagnetism also predict "advanced" waves, which travel backwards in time and arrive before they were emitted. This was generally regarded as an insignificant mathematical oddity, but in the 1940s Feynman and Wheeler developed an account of electromagnetic radiation which involved both retarded and advanced waves. According to this account, when energy is converted into electromagnetic radiation, only half of the resulting radiation takes the form of retarded waves, the remainder is sent back into the past, in the form of advanced waves. The retarded waves that do get emitted are absorbed by electrons in the future, which in turn emit both advanced and retarded waves; these advanced waves travel back in time to stimulate the original emitter – so doubling its output, in accord with what we observe – as well as cancelling out all the advanced waves it emits – which explains why we never seem to receive emissions from the future. One thing the theory does not explain is why the overall earlier–later asymmetry exists in the first place.

field: in physics, not a patch of grass but a force or influence that extends through space. A particular field is represented by an assignment of numbers (or, if the relevant quantities have direction, **vectors**) to each point of space at each moment of time. The field-concept came to the fore in physics with Maxwell's theory of electromagnetism in the nineteenth century. Field theories are *local*, in that physical (or causal) influence is always transmitted from one small region of space to another, hence there is no **action-at-a-distance**. Einstein's theory of gravity is a field theory, Newton's is not.

frame of reference: a **coordinate system** centred on a particular point O in space or spacetime, and assumed to be at rest. In classical (Newtonian) physics, the standard reference frame is Cartesian (three axes at right angles to each other). Different frames of reference can be created by moving the original axes to a different point in space, retaining their directions (a "translation"), or by rotating the axes around O, or by combining translation and rotation.

function: hair-dryers have a function – to dry hair. The term is used differently in mathematics: a function is akin to an input-output device, where what comes out depends upon the input and the function. A simple example is the function "... + ... = ...": insert 2 either side of the "+" and 4 is delivered as output, insert a couple of 3s and 6 is delivered.

Galilean relativity (or "Galilean invariance" or "Galilean equivalence"): physical laws have this property if they remain valid in reference frames in uniform motion relative to one another. Newton's laws of motion are Galilean invariant. Suppose we start with a **frame of reference** centred on O, which we assume to be at rest in absolute space, and a system of bodies S; we now introduce a second frame, M, moving uniformly relative to O: the axes of O and M remain parallel to one another, but the distance between O and M is increasing at a steady rate. The bodies in S which are moving inertially relative to O will also be moving inertially relative to M. Any non-inertial motions (accelerations) will have the same magnitudes in both frames, and if Newton's laws are used to calculate the forces acting on these accelerating bodies, precisely the same results will be obtained in both frames. As a consequence, the fact (if we assume it is one) that O is at rest and M is moving has no experimental or observational consequences in the circumstances envisaged.

geodesic: the path of least distance connecting two points in a flat or curved space. Geodesics are also paths of "least curvature": although in curved regions of a space geodesics are themselves curves, their curvature is limited to what the space itself imposes; they have no "additional curvature", as it were. (These notions can be made mathematically precise.)

Gold universe: a universe of finite duration that is (largely if not completely) symmetrical about its mid-point, at least in respect of discernible physical processes. If the inhabitants of one side of a Gold universe could travel to the "other half", it would seem as if everything were running backwards.

gravity: what keeps us anchored to the Earth's surface and the planets in their orbits. According to Newton's universal law of gravitation, each material body exerts an attractive force on every other body in the universe, a force that is proportional to its mass, and that decreases with the square of the intervening distance. According to Einstein, gravitational effects are due not to a force, but to spacetime curvature. See also **inertial mass vs. gravitational mass**.

growing block model: the universe is a spacetime block that is gradually increasing in size; past and present are real, but the future is not.

hole (in space): there are different types of spatial "hole" referred to in the literature on space and spacetime. A hole in a substantival space is a "region" within a **manifold** that contains no spatial points, where the spatial substance, so to speak, is absent. A "non-Euclidean hole" is a region of strong local spatial curvature. In the "hole argument", the hole is a region in representations of a spacetime where the points are switched about via a mathematical transformation.

idealism: the doctrine that our universe is entirely mental (i.e. composed of minds and mental states).

identity: two senses must be distinguished: *numerical* and *qualitative*. To say that "x is identical (or the same as) y" could mean "x and y are one and the same object" – this is numerical identity – or it could mean "x and y are two distinct objects which are exactly alike" – this is qualitative identity.

identity-conditions: an object's *synchronic* (or at-a-time) identity conditions are the way an object must be in order to qualify as an object of a particular kind. Spelling out precisely what these are is not easy, but it is clear that the conditions an object must meet (at a given moment) to be a brick are quite different from those it must meet to be a horse. Specifying an object's *diachronic* (across-time) identity conditions involves spelling out the sorts of changes an object of a given kind can (and cannot) undergo while remaining in existence. Different sorts of object have different diachronic identity conditions.

identity of indiscernibles: the (dubious) claim that, for any objects x and y, if x and y are exactly alike then they are also numerically identical (see **identity**). Not to be confused with the *indiscernibility of identicals*, the far less contentious claim that if x and y are numerically identical (i.e. "x" and "y" are simply different names of the same object), then x and y are exactly alike.

iff: an abbreviation for "if and only if".

indexicals: words such as "here", "now", "I", "her", whose reference (e.g. a particular place, time, person) depends upon the context of utterance, i.e. where a particular **token** of the indexical is used, when, by whom, etc.

inertia: the tendency to resist acceleration; bodies with greater mass have greater inertia, and so require greater force to accelerate them.

inertial forces: the forces which are suffered by a body when – due to the impact of forces upon it – it moves non-inertially (i.e. when it accelerates or decelerates).

inertial frame (of reference): any **frame of reference** in which Newton's laws of motion hold; more generally, a non-accelerating frame.

inertial mass v. gravitational mass In Newtonian physics, an object's "inertial mass" is the property responsible for its resistance to acceleration; the greater an object's inertial mass, the greater the force needed to accelerate that object by a given amount. "Gravitational mass" is the property which features in the universal law of **gravity**: the greater an object's gravitational mass, the more it attracts (and is attracted by) other objects. Inertial and gravitational mass are exactly equal – which explains the observed fact that bodies of different masses starting from the same point undergo the same gravity-induced acceleration when dropped. However, there is no obvious reason *why* such distinct properties should be exactly equal. (Electrical charge, which – like gravitational mass – generates a force, is not equivalent to inertial mass.) In developing his alternative account of gravity, Einstein sought an explanation for this mysterious coincidence; his solution involved the abolition of gravity as a *force* operating between bodies.

inertial motion: the motion of a body which is not subject to any forces; sometimes called "free-fall" motion. In Newtonian physics, bodies in inertial motion move in straight lines at constant speeds; in general relativity "straight line" is replaced by **geodesic**.

inflationary universe: not a universe-wide tendency for prices to rise, but a universe that undergoes a period of very rapid growth shortly after the **big bang**. The big bang models of the 1960s and early 1970s faced a number of problems. One worry was the "flatness problem" (why the universe at the present time doesn't have a higher degree of curvature, whether positive or negative, than observations suggest it actually possesses), another was the "smoothness problem" (why the average density of matter is so close to being homogeneous), another was the source of the "primordial density fluctuations" responsible for galaxy formation. It turned out that all these problems could be solved by positing a short period of exponential expansion: "inflation", as it become known. The "inflationary period" starts when the universe is about 10^{-43} seconds old, and comes to an end shortly afterwards, well before the first second has ended. This short span of time is sufficient for very considerable growth. One version of the inflationary theory has it that the universe doubles in size every 10^{-34} seconds, a rate that amounts to 100 doublings in 10^{-32} seconds. Several different quantum-theoretical explanations of where the energy that fuels inflation comes from have been developed.

intentional: as well as using the term to talk about actions that are performed on purpose, philosophers also use it in connection with *representation*. Mental states (such as thoughts, beliefs, desires), along with signs, pictures and sentences (whether written or spoken) can be *about* things other than themselves; such items are said to possess the property of "intentionality". There are also "intentional objects" – the intentional object of "Texas is big" is Texas. Since we can think (and have beliefs) about entities such as unicorns and elves, intentional objects need not exist.

interval: in special **relativity**, the spatial and temporal distances between events are not **absolute**, they vary between reference frames. However, one quantity *is* absolute: the square of the separation in space minus the square of the separation in time. This quantity – known as the spacetime "interval" – does not vary between reference frames. Events whose spatial separation is greater than their temporal separation have a positive interval, and are said to be *space-like* connected; events whose temporal separation is greater than their spatial separation have a negative interval, and are said to be *time-like* connected; events whose interval is zero are said to be *light-like* connected.

intrinsic curvature: the curvature of a space (or surface) which is determined solely by features of the space itself, without reference to any higher-order "embedding space". Since the **universe** as a whole cannot be embedded in a higher-dimensional space the idea that the space of our universe could be curved only became a serious option when mathematicians (starting with Gauss and Riemann) worked out ways of characterizing curvature intrinsically.

intrinsic nature: the way an object is in itself. Foster uses the expression in a more restricted way, to refer to the qualitative characteristics a thing has over and above its causal and "structural" properties (intrinsic properties such as size and shape).

invariants: laws or quantities which remain unchanged by certain specified transformations (usually with respect to position or coordinates). Shape and size are invariants of translations, rotations and reflections in Euclidean space; Newton's laws of motion are invariant under translations and rotations (sudden, once and for all) of reference frames. The speed of light is an invariant in special relativity.

is: often used with different meanings, and hence a potential source of confusion. "John is washing up" features the so-called "tensed *is*"; the utterance means that John is *now* doing the washing up. "Paris is the capital of France" features the so-called "tenseless *is*": there is no reference to any specific time; the same holds for the present tensed sentence "Water boils at 100 degrees centigrade". "Superman is Clark Kent" features the "is of **identity**", whereas "Superman is strong" employs the "*is* of predication".

isotropic: a universe in which no particular direction is distinguished by the laws of nature, or the behaviour and distribution of the contents of the space. A universe where this is not the case is *anisotropic*.

kinematics: the branch of physics concerned with unforced motion.

logical possibility: the only things that are *not* logically possible are those which involve an internal incoherence. Five-headed flying pigs are logically possible, round squares are not. See **possible worlds**.

manifold: used loosely, another term for a spatial, temporal or spatiotemporal **dimension**. The term has a more technical usage: in **differential geometry** the term refers to a **continuous** space (or collection of points); a manifold in this sense has no spatiotemporal properties until one or more spacetime structures are imposed (e.g. **affine** structure, a **metric**).

measurement problem: see **quantum mechanics**.

metaphysics: the investigation into the most basic and/or general features of either reality itself, or the conceptual schemes with which we operate. P.F. Strawson called the latter "descriptive metaphysics" and the former "revisionary metaphysics" – the rationale being that revisionary metaphysicians are usually open to the possibility that reality may not conform to our ordinary ways of thinking. So-called "metaphysical systems" (many of the best-known philosophers have produced one) are full-scale attempts to bring all aspects of reality under a unified conceptual system. Although physics is also concerned with the most basic and general aspects of reality, physicists confine themselves to trying to explain the observable features of the universe (albeit often by positing unobservable entities and forces); for better or for worse, metaphysics (of the revisionary sort) is not so constrained.

metric: a measure of the distance between points in a space or spacetime, the numerical value of which will depend on the **coordinate system** employed. In a flat space equipped with Cartesian coordinates, the distance between two points is given by Pythagoras' Theorem: $d^2 = x^2 + y^2 + z^2$, where x, y and z are the distances separating the points along the x-, y- and z-axes respectively. In a **curved space**, the axes of any coordinate system will themselves be curved, and a more complex formula is required, one employing correction factors (or "curvature coefficients"), the precise numerical value of which will depend upon the coordinates chosen and the curvature of the space in the relevant regions. In discussions of general relativity, "the metric" often refers to the *metric field tensor* – g (or g_{ik}) – which determines the inertial structure of spacetime (the **affine connection**), and thus which motions are accelerated and which are not, as well as the **space-like/time-like** distinction, and the distances between points along all paths connecting them.

microwave background radiation: the slowly fading microwave "glow" left by the **big bang**, measured to be around 2.7 degrees above absolute zero, a result which fits the predictions made by most versions of the big bang model.

Minkowski spacetime: the flat spacetime of special **relativity** theory; see **interval**.

modal: concerning necessity and possibility. An object's "modal properties" concern the sorts of change it is possible for it to undergo while remaining in existence (see **identity conditions**).

monadic: a non-relational property or predicate, such as "... *is running*" or "... *is sad*"; not to be confused with "monads": the spatially non-extended non-physical minds which feature prominently in Leibniz's metaphysical system.

multiverse: sometimes used to refer to all the myriad branches of the universe posited by many-worlds interpretations of **quantum mechanics**; more generally, any **universe** (in the broad sense) which contains more than one spatiotemporal system can be said to constitute a "multiverse".

necessary: a necessary truth is one which could not be otherwise than true, a necessary entity is one which could not fail to exist. See also **contingent** and **possible worlds**.

necessary condition: see **sufficient condition**.

new (tenseless) theory of time: the central claim is that although tensed sentences cannot be translated by tenseless sentences, the conditions under which tensed sentences (as used on particular occasions) are true can be specified in tenseless terms. There can thus be "tensed truth" in a world lacking genuine tensed properties.

nomological: concerning the laws of nature. Something is "nomologically necessary" (or "physically necessary") if it is required by the laws of nature; something is "nomologically possible" if it is permitted by the laws of nature. In nomologically **possible worlds** the laws of nature are the same as in the actual universe, but history unfolds differently, often very differently.

non-local: theories which posit direct interactions between objects that are independent of distance (see **action-at-a-distance**) are said to be "non-local", whereas "local theories" only recognize interactions which are transmitted in a continuous fashion across space, usually through **fields**.

noumenal: Kant distinguished the "phenomenal world" – the realm of appearances, the world we can know something about – from the "noumenal world" – the world as it really is, about which (Kant claimed) we can know nothing at all. More generally, something is "noumenal" if it has no **empirical** consequences.

Occam's razor: the methodological precept of not multiplying entities beyond necessity; entities that are not needed for explanatory purposes should be expelled from one's **ontology** – i.e. not taken to exist.

ontology: the branch of metaphysics concerned specifically with what exists. The "ontology of a theory" (or its "ontological commitments") are the entities (objects, properties, etc.) that you will have to believe exist if you accept the theory as true.

overdetermination: an **event** is "causally overdetermined" if it has two (or more) causes, each of which would have sufficed to bring the event in question about. Overdetermination of a different sort occurs when an object or event is ascribed combinations of properties which cannot be possessed together.

passage of time: the future-directed "movement" of the present, which results in present times becoming past, past times becoming more past, and future times becoming less future, until they finally become present and then past. The dispute between **static** and **dynamic** views of time is centred on the question of whether time really does pass.

persistence: entities that endure through time are said to "persist"; there are competing accounts of what persistence involves. See also **substance, event.**

phenomenal property: those properties that we are directly aware of in our experience, e.g. the smell and colour of a rose, the sound of a car-horn, the taste of nutmeg, pains, aches, feelings of joy or nausea, and so forth. See **secondary property, consciousness.**

phenomenology: the enterprise of describing the character of **consciousness**, in all its aspects, in as clear and systematic way as is possible. Phenomenology is concerned only with describing how things seem to the experiencing subject, and ventures no claims about the causes of experience, or what lies outside or beyond experience.

photon: a particle (or basic quantum-unit) of light.

positivism: see **verificationism.**

possible worlds: a (logically) possible **world** is a way the universe as a whole might have been, but isn't. There is a possible world which consists of nothing but a single elementary particle; there is a possible world which is exactly like ours save for a single trivial detail; there are untold numbers of possible worlds which are richer and more complex than the actual universe, and possible worlds where the laws of nature are very different. A possible world is a consistent ensemble of logically possible (non-contradictory) states of affairs; every consistent set of sentences describes a possible world – but no doubt there are possible worlds which cannot be described by any human language. The concept of a possible world has proved useful in **modal** logic, and provides a convenient way of talking about possibility and necessity; for example, a statement that expresses a **necessary** truth is true in every possible world, whereas a **contingent** truth is true in at least one possible world. Most philosophers are of the opinion that only one possible world is real or actual – this one. But a minority – "modal realists" – take the view that all possible worlds are equally real. Also see **nomological.**

predicate: any linguistic expression used to ascribe a **property** to something.

presentism: the doctrine that only what is present is real. Despite the simplicity of the basic claim, there are very different variants of presentism.

proper time: time as measured by a single observer (or any other system which changes in a regular way); in special **relativity**, the time between two **events** as measured by an observer whose **worldline** passes through both events.

property: a feature or characteristic, usually of a thing – though some believe ordinary things are themselves nothing but collections of interrelated properties. An object's "intrinsic properties" depend on how the object is in itself (". . . is square", ". . . has a hole"), whereas an object's "relational properties" depend on how the object is related to other objects ("... is bigger than . . .", ". . . is between . . . and . . ."). Since ". . . is a round-square" seems a perfectly respectable **predicate**, it should not be assumed that every predicate refers to a genuine property, i.e. something which could be a real feature of a world. The question of whether A-predicates such as ". . . is past" refer to genuine properties is much disputed (see **A-theory/B-theory**).

proposition: often used to refer to the content of an utterance or belief. (See also **sentence v. statement**.)

quantum mechanics: our current best theory of the realm of the very small, which underlies a good deal of our modern technology (e.g. lasers, computer chips). Quantum mechanics originated with the discovery that electromagnetic radiation is packaged into discrete units (or "quanta"); in the 1920s and 1930s detailed mathematical formalisms were developed (by Schrödinger, Heisenberg and Dirac, among others) which proved highly successful in delivering accurate predictions; later developments rendered quantum theory compatible with special (but not general) relativity. Despite its empirical power, quantum theory has several highly puzzling features, and debate continues to rage over how the quantum formalisms should be interpreted. The "measurement problem" is particularly significant in this respect. The fundamental dynamical law of quantum theory is Schrödinger's wave equation, which assigns specific probabilities to all the different ways a given system of particles might evolve, but does not specify the way the system *will* in fact evolve – this can only be discovered by performing a test or observation. When a test is conducted, the wave equation is said to "collapse", since once a system is observed as being in a given state, the probability of finding it in all the other possible states reduces to zero. Prior to an observation being made, the system is said to exist in a "superposition" of states, though quite what this means for the state in question is unclear. According to the "Copenhagen interpretation", from which the preceding terminology derives, the question is simply illegitimate; a system only acquires definite properties as and when these properties are observed or measured. However, this response leaves many questions unanswered, and many theorists unsatisfied. Why should conducting a measurement have such profound consequences? What happens to a system if no measurement is ever conducted, perhaps because no conscious observers exist at the time? Various alternatives to the Copenhagen interpretation have been proposed. "Hidden variables" theorists argue that the system in question *is* in a definite state, and since quantum theory does not tell us what that state is, the theory is incomplete. "Many worlds" theorists argue that each of the possibilities which feature in the Schrödinger equation are realized: the wave equation never collapses; rather, when certain interactions occur (including measurements) the universe splits into separate branches, one for each element of the superposition. Other theorists argue that the wave equation quickly collapses of its own accord due to interactions with other systems at the micro-level, which is why we never encounter strange "superposed" objects in everyday life. Quantum theory has other notable and surprising features, including wave-particle duality and the Heisenberg "uncertainty principle", according to which the more we know about one quantum quantity (e.g. position) the less we can know about certain others (e.g. momentum). See also **Bell's inequality**.

realism: see **anti-realism.**

reductionism: see **anti-realism**

relations: objects can stand in many different relations to one another, and different relations themselves have different properties. I will only mention a few of the most important. Let x, y and z refer to objects, and ". . . is R-related to . . ." mean ". . . bears the relation R to . . .". If a relation is *symmetrical*, then if x is R-related to y, then y will be R-related to x, e.g. ". . . is next to . . .". If a relation is *transitive*, then if x is R-related to y, and y is R-related to z, then x is R-related to z, e.g. ". . . is larger than . . .". A relation is *reflexive* if anything which possesses it bears the relation to itself, e.g. "is the same size as" (we are all the same size as ourselves). A relation which is reflexive, symmetrical and transitive is known as an *equivalence relation*. Equivalence relations partition all their relata into distinct mutually exclusive groups, in such a way that each member of a group bears the relation in question to every other member of the group. "Having the same age" is an equivalence relation: it divides a collection of people into non-overlapping groups, and each person belongs to only one group. Not all relations involve genuine *connections* between objects, though some do: compare ". . . is tied to . . ." with ". . . is the same shape as . . .". Objects with the relational property "same shape" need not be in the same universe, but objects which are tied together (by a piece of string, say) are linked by a real physical connection.

relationism: the view that space, time or spacetime do not exist as entities in themselves, over and above material objects and their spatiotemporal relations.

relative motion: motion relative to other objects, rather than space itself.

relativity theory: a shorthand way of referring to Einstein's special theory of relativity, his general theory of relativity, or both. The special theory posits a flat spacetime, and is based on the idea that the laws of physics (and the speed of light) are the same in all **inertial frames**; it has several counterintuitive consequences, such as length contraction, time dilation and the abolition of **absolute simultaneity**. The general theory extends the special theory by explaining gravitational effects in terms of spacetime curvature.

secondary properties (or **qualities**): a race of intelligent aliens with very different perceptual systems from our own might well have no idea what we mean when we say that a certain flower is "red and tastes bitter", or than an object "makes a whining sound", but they would have no trouble with properties (and concepts) such as "square", "mass" and "size". The latter sort of property are called "primary", the former "secondary". More generally, secondary properties are perceived characteristics of objects that depend heavily upon the peculiarities of the perceptual system of whoever is doing the perceiving; by contrast, "primary properties" are independent of the peculiarities of perceptual systems. According to one influential account (Locke's), although secondary properties such as colour seem to be intrinsic features of objects, in just the same way as the primary properties, this is an illusion; in reality, secondary qualities are causal properties: a red object is one that causes certain types of experience (red-as-we-experience-it) in certain sorts of perceiver (typical human) in certain sorts of circumstance (ordinary lighting conditions). As a consequence, there is no reason to suppose red-as-we-experience it is an intrinsic feature of red objects themselves; if so, physical reality, as it is in itself, is very different from how it appears to us.

sentence/statement: a sentence is a grammatical sequence of words (or more generally, meaningful signs), a statement is what a particular sentence expresses on a particular occasion of use. Different sentences can express the same statement ("It's snowing", "Il neige"), and the same sentence can express different statements (if you say "I'm hungry" and I say "I'm hungry", we are saying different things using the same words). The term "content" is often used to refer to what statements and beliefs (and other mental states) have in common when they involve the same assignment of properties to the same objects; e.g. I say "Mary is unhappy", you believe that Mary is unhappy, and someone else hopes that Mary is unhappy. See **proposition**.

singularity: roughly speaking, a point or region of spacetime where curvature (or some other quantity) becomes infinite, spacetime structure breaks down, and the standard laws of physics cannot be applied. There are different types of singularity.

solipsism: the doctrine that only one thing exists, me. "Solipsism of the present moment" is the doctrine that only one moment of time exists, this one.

space-like (and time-like) separation: see **interval**.

spacetime: often used to refer to the four-dimensional spatiotemporal frameworks employed in **relativity theory**, according to which space and time are no longer independent dimensions, but "intermingled" in a distinctive way: the spatial and temporal distances between the same two events can vary from one **frame of reference** to another, and so are no longer **absolute**. However, the concept has wider applicability, and is used to refer to any collection of individual points which jointly constitute a single spatiotemporal system. Contemporary geometrical treatments of spacetime begin with a **manifold** of points and add one or more geometrical structures (such as a **metric**).

static time: a time which does not pass; see **dynamic models of time**.

substance: in ordinary language the term can be used to refer to just about anything ("What's that strange green sticky substance on your shirt?"), but in metaphysics the term is generally used exclusively to refer to enduring *objects* (dogs, bicycles) rather than *stuffs* (patches of green sticky material, a litre of beer), or **events**. Another term sometimes used for substance in this sense is **continuant**. Further restrictions are sometimes imposed (e.g. "only an object which does not depend on any other object for its existence is a *substance*"), but not here.

substantivalism: the doctrine that space, time or spacetime are entities in their own right; denied by **relationism**.

sufficient condition: if X is a sufficient condition of Y, then if X obtains, so too does Y. If X is a **necessary** condition of Y, then Y cannot exist in the absence of X. If X is necessary *and* sufficient for Y, then Y obtains **iff** X does, and vice-versa.

supervenience: an asymmetrical relationship of dependency, which can come in different forms, depending on the precise mode of dependency, and the items so related. A (comparatively) uncontroversial example of a supervenience claim: "The large-scale shape of physical objects supervenes upon their micro-structure." At the very least, this means that two objects with similar micro-constituents arranged in similar ways will have similar macro-shapes, and that changes in macro-shape depend on, and are brought about by, changes at the micro-level.

symmetry: in physics, used in connection with properties of a system which are unaffected by specified transformations.

tangent: a line which touches a curve at only one point.

temporal part: see **event.**

temporal vacuum: the temporal counterpart of the "**hole** in space", which like the latter, can come in different forms. A "mild" temporal vacuum consists of a period of time during which nothing moves or changes – the universe, as it were, undergoes a temporary *total freeze.* A stronger form of temporal vacuum consists of a period of time during which nothing *exists* other than time and space – the universe is completely empty (of matter, minds) for a while. Relationists will generally reject the possibility of (at least) the stronger form of temporal vacuum, whereas spacetime substantivalists will be inclined to accept the possibility of both.

tensor: a mathematical object, a generalization of the **vector**-concept, specified with respect to a given coordinate system and able to undergo transformation to other coordinate systems; there are various types (or "ranks") of tensor (a vector is a first-rank tensor); generally employed to deal with vectors in complex dependence relationships representing quantities that are not parallel. A tensor in a spacetime is a **function** at a given point which takes vectors (or functions of vectors) as inputs and delivers as output a number or vector, where the output depends on the inputs in a linear way. A tensor **field** is an assignment of tensors to each point in a spacetime. Einstein's general theory of relativity is formulated in terms of tensor-equations.

thermodynamics: the branch of physics concerned with the transformation of energy by heat and work.

tidal forces: forces suffered by extended bodies in a gravitational field. If you were to fall feet-first into a black hole your body would be torn apart by tidal forces before you reached the **singularity**, since a stronger "pull" would be exerted on your feet than on the upper parts of your body (your feet being nearer to the source of the gravitational field).

time-reversibility: if the laws of nature are fully time-reversible (or "time-symmetric"), any physical process would be able to unfold in either temporal direction. The laws of **classical physics** are time-reversible, likewise **relativity** and the basic equations of **quantum mechanics** (or so it is generally held). However, some *interpretations* of quantum theory do seem committed to a deep asymmetry between past and future: those which both posit and attribute ontological significance to the "collapse of the wave-equation". On this view, the past and present are fixed and determinate, whereas the future is unfixed and indeterminate. Whether this interpretation of quantum theory is correct is another matter. One piece of evidence from particle physics suggests the laws of nature can in fact distinguish between the two directions of time: the decay of the neutral K-meson (or *kaon*) is not perfectly time-symmetrical. The significance of this rare and subtle effect is not yet known, but there is no obvious connection between it and the various macroscopic asymmetries. See **arrow of time, entropy.**

tokens v. types: a token is a particular instance of a general kind, or type. In "dddd" there are four tokens of a single type (in this case, a letter of the alphabet).

topic-neutral: a theory is topic-neutral if it restricts itself to making claims about causal and/or structural features, and says nothing about the **intrinsic natures** of the items which possess these features.

topology: the branch of geometry concerned with the properties of spaces (and spacetimes) that are independent of **metric**: i.e. with how the different parts of a space are connected, rather than how far apart they are. Squeezing and stretching a space (or shape) do not affect its topological properties, provided the space isn't torn and the local "betweeness" relation between nearby points is not altered.

transcendental: pertaining to reality as it really is.

universe: in the broad sense of the term, the universe is all that exists; in its narrow sense, a universe is a single spatiotemporal system and all that it contains. It may well be that the universe (in the broad sense) consists of more than one spatiotemporal system. If so, then these different systems will not be spatially or temporally related to one another, so to call them "parallel universes" is misleading.

vector: a quantity which has both a magnitude and a direction (e.g. velocity, acceleration); "scalar" quantities have magnitude but lack direction (e.g. weight).

velocity: speed in a given direction.

verificationism: the doctrine (associated with logical positivism) that unless a (non-**analytic**) statement is empirically verifiable or falsifiable it is meaningless. Hence claims about the world which go beyond what is scientifically verifiable (or falsifiable) are literally meaningless (or "cognitively empty"); many traditional metaphysical claims fall into this category – or would do, if verificationism were true. In the opinion of many, the doctrine applies to itself and so is self-refuting.

void: nothingness.

wave function: see **quantum mechanics.**

worlds: in philosophical writings, "worlds" are often entire universes, rather than individual planets; see **possible worlds.**

worldline: the career of an object in spacetime, from beginning to end, represented as an extended four-dimensional entity. Strictly speaking, only points have world*lines*; spatially extended objects – such as your body – occupy four-dimensional volumes. In spacetime diagrams, a curved worldline indicates acceleration, a straight worldline indicates uniform velocity, and an object at rest has a worldline that is straight and vertical.

Web resources

The web changes quickly, and links come and go. For a regularly updated listing of materials relevant to the issues discussed in the book, see the dedicated site:

http://liv.ac.uk/~bdainton/time&space.html

Listed below are a few of the more useful science-oriented sites:

1. The Los Alamos e-print archive (formerly xxx.lanl.gov) is a fully automated electronic archive and distribution server for research papers in physics, astronomy and other sciences. The search engine is efficient, and downloads are available in multiple formats

   http://arXiv. org (accessed June 2001)

2. "Relativity on the world wide web", maintained by John Baez.

   http://math. ucr. edu/home/baez/relativity. html (accessed June 2001)

   A wealth of information is available at this site, including links to the physics and relativity frequently asked questions (FAQs), guides and tutorials on relativity pitched at beginner, intermediate and advanced levels.

3. "Foundations", Greg Egan, has useful introductions to STR, GTR and quantum theory.

   www.netspace.net.au/~gregegan/FOUNDATIONS/index.html
   (accessed June 2001)

4. "Lecture notes on general relativity", Sean M. Carroll (other useful links).

   http://pancake.uchicago.edu/~carroll/notes/ (accessed June 2001)

5. Jeff Weeks's topology games (do crossword puzzles in multi-connected spaces):

   http://www.northnet.org/weeks/ (accessed June 2001)

6. Andrew Hamilton's homepage, with videos of what you would see as you fall into a black hole singularity. Also pay a visit to the STR "tour", which features a "Guide to relativistic flight simulators" (web-based); the simulations of four-dimensional geometrical objects is also worth a visit.

   http://casa.colorado.edu/~ajsh/home.html (accessed June 2001)

7. "'Warp' allows you to visualize relativistic phenomena that are generally unobservable in the everyday world. These are the changes in appearance that an observer will see as an object reaches very high speed relative to the observer." Available to download at:

> www.barneyhawes.com/~warp/downloads.html (accessed September 2001)

8. For introductions to quantum theory, a useful starting point is the "Thinkquest" site "Quantum mechanics made simple":

> http://library.thinkquest.org/C005775/frameset.html
> (accessed June 2001)

The *Stanford Encyclopedia* has several useful (some quite technical) entries, e.g. "Quantum theory":

> http://plato.stanford.edu/entries/qm/ (accessed June 2001)

and "Holism and nonseparability in physics" by Richard Healey:

> http://cd1.library.usyd.edu.au/stanford/entries/physics-holism/
> (accessed June 2001)

See also Healey's "The meaning of quantum theory":

> http://w3.arizona.edu/~phil/faculty/healeytx.html (accessed June 2001)

9. Steve Preston's "Time travel" site has useful information and links:

> http://freespace.virgin.net/steve.preston/Time.html
> (accessed June 2001)

See also www.pbs.org/wgbh/nova/time/ (accessed June 2001)

10. For striking Hubble Space Telescope images of the far-reaches of space and time:

> http://oposite.stsci.edu/pubinfo/pictures.html (accessed June 2001)

Bibliography

Abbot, E. A. (1986) *Flatland: A Romance of Many Dimensions by a Square*. New York: Penguin.

Albert, D. Z. (1992) *Quantum Mechanics and Experience*. Cambridge, MA: Harvard University Press.

Alexander, H. G. (1956) *The Leibniz-Clarke Correspondence*. Manchester: Manchester University Press.

Al-Khalili, J. (1999) *Black Holes, Wormholes and Time Machines*. Bristol: Institute of Physics.

Arntzenius, F. & Maudlin, T. (2000) "Time travel and modern physics", in *Stanford Encyclopedia of Philosophy*, http://plato.stanford.edu/ (accessed June 2001).

Arthur, R. (1994) "Space and relativity in Newton and Leibniz", *British Journal for the Philosophy of Science* 45, 219–40.

Atkins, P. W. (1986) "Time and dispersal: the second law". See Flood & Lockwood (1986), 80–98.

Baez, J. C. (1996) "General relativity: long course outline", http://math.ucr.edu/home/baez/gr/outline2.html (accessed June 2001).

Baez, J. C. (2001) "The meaning of Einstein's equation", www.math.ucr.edu/home/baez/einstein/ (accessed September 2001).

Barbour, J. (1982) "Relational conceptions of space and time". See Butterfield *et al.* (1996), 141–64.

Barbour, J. (1989) *Absolute and Relative Motion*. Cambridge: Cambridge University Press.

Barbour, J. (1999) *The End of Time*. London: Weidenfeld & Nicholson.

Barbour, J. & Bertotti, B. (1977) "Gravity and Inertia in a Machian Framework", *Nuovo Cimento* 38B, 1–27.

Barbour, J. & Bertotti, B. (1982) "Mach's Principle and the Structure of Dynamical Theories", *Proceedings of the Royal Society of London* 382, 295–306.

Barnes, J. (1982) *The Presocratic Philosophers*. London: Routledge.

Belot, G. (2000) "Geometry and motion", *British Journal for the Philosophy of Science* 51, 561–95.

Belot, G. & Earman, J. (1999) "From physics to metaphysics", in *From Physics to Philosophy*, J. Butterfield & C. Pagonis (eds), 166–86. Cambridge: Cambridge University Press.

Belot, G. & Earman, J. (2001) "Pre-Socratic quantum gravity". See Callender & Huggett (2001), 213–55.

Bigelow, J. (1991) "World's enough for time", *Nous* 25, 1–19.

Bigelow, J. (1996) "Presentism and Properties", *Philosophical Perspectives* 10, 35–52.

Bohm, D. (1952) "A suggested interpretation of quantum theory in terms of 'hidden variables', parts I and II" *Physical Review* 85, 166–93.

Bohm, D. & Hiley, B. J. (1993) *The Undivided Universe*. London: Routledge.

Bricker, P. (1993) "The fabric of space", *Midwest Studies in Philosophy XVIII*, P. A. French, T. E. Uehling & H. K. Wettstein, (eds). Notre Dame, IN: Notre Dame University Press.

Brighouse, C. (1994) "Spacetime and holes", in D. Hull, M. Forbes & R. M. Burian (eds) *PSA 1994* 1, 117–25.

Broad, C. D. (1923) *Scientific Thought*. London: Kegan Paul.

Broad, C. D. (1938) *An Examination of McTaggart's Philosophy*. Cambridge: Cambridge University Press.

Broad C. D. (1998) "McTaggart's arguments against the reality of time: an excerpt from *Examination of McTaggart's Philosophy*". See Van Inwagen & Zimmerman (1998), 74–9.

Butterfield, J. (1989a) "The hole truth", *British Journal for the Philosophy of Science* 40, 1–28.

Butterfield, J. (1989b) "Albert Einstein meets David Lewis", in A. Fine & J. Leplin (eds) *Proceedings of the 1988 Biennial Meeting of the Philosophy of Science Assocation, Volume 2*. East Lansing, MI: Philosophy of Science Association; reprinted in Butterfield *et al.* (1996), 295–311.

Butterfield, J. (1999) *The Arguments of Time*. Oxford: The British Academy.

Butterfield J. & Isham, C. (1999) "On the emergence of time in quantum gravity". See Butterfield (1999), 111–68.

Butterfield, J. & Pagonis, C. (eds) (1999) *From Physics to Philosophy*. Cambridge: Cambridge University Press.

Butterfield, J., Hogarth, M. & Belot, G. (eds) (1996) *Spacetime*. Aldershot: Dartmouth.

Callender, C. (1998a) "The view from nowhen", *British Journal for the Philosophy of Science* 49, 135–59.

Callender, C. (1998b) "Shedding light on time", *Proceedings of the 1998 Biennial Meeting of the Philosophy of Science Association* 67(3), part II, supplement.

Callender, C. (2000) "Is time 'handed' in a quantum world?", *Proceedings of the Aristotelian Society*, 15 May, 247–69.

Callender, C. & Huggett, N. (eds) (2001) *Physics Meets Philosophy at the Planck Scale*. Cambridge: Cambridge University Press.

Callender, C. & Weingard, R. (1996) "An introduction to topology", *The Monist* 79, 21–33.

Carnap, R. (1963) "Carnap's intellectual biography", in *The Philosophy of Rudolf Carnap*, P. A. Schilpp (ed.). La Salle, IL: Open Court.

Cau, T. C. (1998) *Conceptual Developments of 20th Century Field Theories*. Cambridge: Cambridge University Press.

Chalmers, D. (1996) *The Conscious Mind*. Oxford: Oxford University Press.

Chown, M. (1999) "Cosmic Crystal", *New Scientist*, 13 February, 161(2173), 42.

Chown, M. (2000) "Backwards to the future", *New Scientist*, 6 February, 165(2224): 26.

Clifton, R. & Hogarth, M. (1995) "The definability of objective becoming in Minkowski spacetime", *Synthese* 103, 355–87.

Coles, P. (2001) (ed.) *The New Cosmology*. London: Routledge.

Cramer, J. G. (1980) "Generalized absorber theory and the Einstein–Podolsky–Rosen paradox", *Physical Review D*, 22, 177–89.

Cramer, J. G. (1983) "The arrow of electromagnetic time and the generalized absorber theory", *Foundations of Physics* 13, 887–902.

Cramer J. G. (1986) "The transactional interpretation of quantum mechanics", *International Journal of Theoretical Physics* 27, 227–36.

Cramer, J. G. (2000) "The plane of the present and the new transactional paradigm of time", in *Time and the Instant*, R. Durie (ed.), 177–89. Manchester: Clinamen Press.

Dainton, B. (1992) "Time and division", *Ratio* 5, 102–28.

Dainton, B. (2000) *Stream of Consciousness*. London: Routledge.

Davies, P. (1986) "Time asymmetry and quantum mechanics". See Flood & Lockwood (1986), 98–124.

Davies, P. (1995) *About Time: Einstein's Unfinished Revolution*. Harmondsworth: Penguin.

Davies, P. C. W. and Brown, J. (eds) (1988) *Superstrings: A Theory of Everything?*, Cambridge: Cambridge University Press.

Deutsch, D. (1997) *The Fabric of Reality*. Harmondsworth: Penguin.

Deutsch, D. & Lockwood, M. (1994) "The quantum physics of time travel', *Scientific American*, March, 50–56.

Dewdney, A. K. (1984) *The Planiverse*. London: Picador.

Dolev, Y. (2000) "Dummett's antirealism and time, *European Journal of Philosophy* 8, 253–76.

Dorato, M. (1995) *Time and Reality: Spacetime Physics and the Objectivity of Temporal Becoming*. Bologna: CLUEB.

Dorato, M. (1998) "Becoming and the arrow of causation", *Proceedings of the 1998 Biennial Meeting of the Philosophy of Science Association* 67(3), part II, supplement.

Dummett, M. (1964) "Bringing about the past", *Philosophical Review* 73, 338–59; reprinted in Le Poidevin & MacBeath (1993), 117–33.

Dummett, M. (1978) "On the reality of the past", in *Truth and Other Enigmas*. London: Duckworth.

Dummett, M. (1986) "Causal loops". See Flood & Lockwood (1986), 135–69.

Dummett, M. & Flew, A. (1954) "Can an effect precede its cause?", *Aristotelian Society Supplement* 28, 27–62.

Durie, R. (ed.) (2000) *Time and the Instant*. Manchester: Clinamen Press.

Earman, J. (1971) "Kant, incongruous counterparts and the nature of space and space-time", *Ratio* **13**, 1–18.

Earman, J. (1974) "An attempt to add a little direction to 'The problem of the direction of time'", *Philosophy of Science* **41**, 15–47.

Earman, J. (1989) *World Enough and Spacetime*. Cambridge, MA: MIT Press.

Earman, J. (1995) *Bangs, Crunches, Whimpers and Shrieks*. Oxford: Oxford University Press.

Earman, J. & Norton, J. (1987) "What price space-time substantivalism? The hole story", *British Journal for the Philosophy of Science* **38**, 515–25.

Earman, J., Janis, A. J., Massey, J. & Rescher, N. (eds) (1993) *Philosophical Problems of the Internal and External Worlds*. Pittsburgh, PA: University of Pittsburgh Press.

Einstein, A. (1922) *Sidelights on Relativity*. London: Methuen.

Einstein, A. (1961) *Relativity: The Special and the General Theory*. New York: Three Rivers Press.

Egan, G. (1995) "The hundred light year diary", in *Axiomatic*. London: Orion/Millenium.

Egan, G. (1997) *Diaspora*. London: Orion/Millenium.

Egan, G. (1998) *Foundations 1-4*, www.netspace.net.au/~gregegan (accessed 17 June 2001).

Ellis, G. F. R. (1978) "Is the universe expanding?", *General Relativity and Gravitation* **8**, 87–94.

Fitzgerald, P. (1985) "Four kinds of temporal becoming", *Philosophical Topics* **13**, 145–77.

Flew, A. (1954) "Can an effect precede its cause?", *Proceedings of the Aristotelian Society*, Supplementary Volume **28**, 45–62.

Flew, A. (1956) "Effects before their causes? – Addenda and corrigenda", *Analysis* **16**, 104–10.

Flew, A. (1957) "Causal disorder again", *Analysis* **17**, 81–6.

Flood, R. & Lockwood, M. (eds) (1986) *The Nature of Time*. Oxford: Blackwell.

Foster, J. (1979) "In *self*-defence", in *Perception and Identity*, G. F. Macdonald (ed.). London: Macmillan.

Foster, J. (1982) *The Case for Idealism*. London: Routlege & Kegan Paul.

Foster, J. (1985) *Ayer*. London: Routledge & Kegan Paul.

Foster, J. (1991) *The Immaterial Self*. London: Routledge.

Foster, J. (1993) "The succinct case for idealism", in *Objections to Physicalism*, H. Robinson (ed.). Oxford: Clarendon.

Foster, J. (2000) *The Nature of Perception*. Oxford: Oxford University Press.

Friedman, M. (1983) *Foundations of Space-Time Theories*. Princeton, NJ: Princeton University Press.

Gallois, A. (1994) "Asymmetry in attitudes and the nature of time", *Philosophical Studies* **76**, 51–69.

Gallois, A. (1998) *Occasions of Identity*. Oxford: Clarendon.

Garrett, B. J. (1988) "'Thank goodness that's over' revisited", *Philosophical Quarterly* **39**, 201–5; also in L. N. Oaklander & Q. Smith (eds) (1994) *The New Theory of Time*, 316–21. New York: St Martin's Press.

Gaul, M. & Rovelli, C. (1999) "Loop quantum gravity and the meaning of diffeomorphism invariance", http://arXiv.org/abs/gr-qc/9910079 (accessed June 2001).

Geroch, R. (1978) *General Relativity from A to B*. Chicago, IL: Chicago University Press.

Gibbard, A. (1975) "Contingent identity", *Journal of Philosophical Logic* **4**, 187–222.

Gödel, K. (1949) "A remark about the relationship between relativity theory and idealistic philosophy", in *Albert Einstein: Philosopher-Scientist*, P. A. Schilpp (ed.). La Salle, IL: Open Court.

Godfrey-Smith, W. (1979) "Special relativity and the present", *Philosophical Studies* **36**, 233–44.

Gold, T. (1962) "The arrow of time", *American Journal of Physics* **30**, 403–10.

Gott, R. J. (2001) *Time Travel in Einstein's Universe*. London: Orion.

Gott, R. J. and Li, L.-X. (1998) "Can the universe create itself?", *Physical Review D*, **58**, 023501.

Greene, B. (1999) *The Elegant Universe*. London: Jonathan Cape.

Grey, W. (1999) "Troubles with time travel", *Philosophy* **74**, 55–70.

Guth, A. H. (1997) *The Inflationary Universe*. London: Jonathan Cape.

Harrison, J. (1979) "Jocasta's crime", *Analysis* **39**, 65.

Harrison, J. (1980) "Report on *Analysis* problem no. 18", *Analysis* **40**, 65–9.

Heinlein, R. A. (1970) "All you zombies", in *The Mirror of Infinity*, R. Silverberg (ed.). San Francisco, CA Canfield.

Hinchliff, M. (1998) "A defense of presentism in a relativistic setting", *Proceedings of the 1998 Biennial Meeting of the Philosophy of Science Association* **67**(3), part II, supplement.

Hoefer, C. (1996) "The metaphysics of spacetime substantivalism", *The Journal of Philosophy* **93**, 5–27.

Hoefer, C. (1998) "Absolute versus relational spacetime: for better or worse, the debate goes on", *British Journal for the Philosophy of Science* 49, 451–67.

Hoefer, C. & Cartwright, N. (1993) "Substantivalism and the hole argument", in *Philosophical Problems of the Internal and External Worlds*, J. Earman, A. J. Janis, J. Massey & N. Rescher (eds), 23–43. Pittsburgh, PA: University of Pittsburgh Press.

Hoffman, B. (1972) *Albert Einstein: Creator & Rebel*. New York: New American Library.

Hollis, M. (1967) "Time and spaces", *Mind* 76, 524–36.

Horwich, P. (1975) "On some alleged paradoxes of time travel", *Journal of Philosophy* 72, 432–44.

Horwich, P. (1978) "'On the existence of time, space and space-time", *Nous* 12, 396–419.

Horwich, P. (1987) *Asymmetries in Time*. Cambridge, MA: MIT Press.

Hospers, J. (1997) *An Introduction to Philosphical Analysis*, 4th edn. London: Routledge.

Huggett, N. (1999) *Space from Zeno to Einstein*. Cambridge, MA: MIT Press.

Husserl, E. (1950) *The Phenomenology of Internal Time-Consciousness*, J. S. Churchill (trans.). Bloomington, IN: Indiana University Press.

Husserl, E. (1991) *On the Phenomenology of the Consciousness of Internal Time* (1893–1917), J. B. Brough (ed. and trans.). Dordrecht: Kluwer.

James, W. (1952) *The Principles of Psychology*. Chicago, IL: Encyclopedia Britannica.

Jammer, M. (1993) *Concepts of Space*, 3rd edn. London: Dover.

Kant, I. (1996) *Prolegomena to Any Future Metaphysics*, B. Logan (ed.). London: Routledge.

Kuchař, K. (1999) "The problem of time in quantum geometrodynamics". See Butterfield (1999), 169–95.

Lachièze-Rey, M. & Luminet, J.-P. (1995) "Cosmic topology", *Physics Reports* 254, 136; also http://arXiv.org/abs/gr-qc/9605010 (accessed June 2001).

Le Poidevin, R. (1991) *Change, Cause and Contradiction*. Basingstoke: Macmillan.

Le Poidevin, R. (ed.) (1998a) *Questions of Time and Tense*. Oxford: Oxford University Press.

Le Poidevin, R. (1998b) "The past, present and future of the debate about tense". See Le Poidevin (1998a), 13–42.

Le Poidevin, R. & MacBeath, M. (eds) (1993) *The Philosophy of Time*. Oxford: Oxford University Press.

Lewis, D. (1971) "Counterparts of persons and their bodies", *Journal of Philosophy* 68, 203–11.

Lewis, D. (1976) "The paradoxes of time travel". See Le Poidevin & MacBeath (1993), 134–46; reprinted from *American Philosophical Quarterly*.

Lewis, D. (1986) *On the Plurality of Worlds*. Oxford: Blackwell.

Lockwood, M. (1989) *Mind, Brain and the Quantum: The Compound "I"*. Oxford: Blackwell.

Loemker, L. E. (ed.) (1970) *Leibniz: Philosophical Papers and Letters*. Dordrecht: Reidel.

Lowe, E. J. (1987) "The indexical fallacy in McTaggart's proof of the unreality of time", *Mind* 96, 62–70.

Lowe, E. J. (1993) "McTaggart's paradox revisited", *Mind* 101, 323–6.

Lowe, E. J. (1998) "Tense and persistence". See Le Poidevin (1998), 43–60.

Lucas, J. R. (1984) *Space, Time and Causality*. Oxford: Clarendon.

Lucas, J. R. (1986) "The open future". See Flood & Lockwood (1986), 125–34.

Lucas, J. R. (1999) "A century of time". See Butterfield (1999), 1–20.

Lukasiewicz, J. (1967) "Determinism", in *Polish Logic 1920-1939*, S. McCall (ed.). Oxford: Oxford University Press.

Luminet, J.-P. (1998) "Past and future of cosmic topology", arXiv:gr-qc/9804006 (accessed September 2001).

Luminet, J.-P. (2001) *L'Univers Chiffonné*. Paris: Fayard.

Luminet, J.-P. & Roukema, B. F. (1999) "Topology of the universe: theory and observation", http://arXiv.org/abs/astro-ph/9901364 (accessed June 2001).

Luminet, J.-P., Starkman, G. D., Weeks, J. R. (1999) "Is space finite?", *Scientific American Online*, April, www.sciam.com/1999/0499issue/0499weeks.html (accessed June 2001).

Manders, K. L. (1982) "On the space-time ontology of physical theories", *Philosophy of Science* 49, 575–90.

MacBeath, M. (1982) "Who was Dr Who's father?", *Synthese* 51, 397–430.

MacBeath, M. (1983) "Communication and time reversal", *Synthese* 56, 26–46.

Markosian, N. (1993) "How fast does time pass?", *Philosophy and Phenomenological Research* 53, 829–44.

Maudlin, T. (1989) "The essence of spacetime", *PSA 1988* **2**, 82–99; reprinted in Butterfield *et al.* (1996), 313–22.

Maudlin, T. (1993) "Buckets of water and waves of space: why spacetime is probably a substance", *Philosophy of Science* **60**, 183–203; also in Butterfield *et al.* (1996), 263–83.

Maudlin, T. (1994) *Quantum Non-Locality and Relativity*. Oxford: Blackwell.

McCall, S. (1976) "Objective time flow", *Philosophy of Science* **43**, 337–62.

McCall, S. (1984) "A dynamic model of temporal becoming", *Analysis* **44**, 172–6.

McCall, S. (1994) *A Model of the Universe*. Oxford: Oxford University Press.

McTaggart, J. E. (1908) "The unreality of time", *Mind* **17**, 457–74.

McTaggart, J. E. (1927) *The Nature of Existence*, vol. 2. Cambridge: Cambridge University Press.

Meiland, J. W. (1974) "A two-dimensional passage model of time for time travel", *Philosophical Studies* **26**, 153–73.

Mellor, D. H. (1981) *Real Time*. Cambridge: Cambridge University Press.

Mellor, D. H. (1994) "Thank goodness that's over", in *The New Theory of Time*, L. N. Oaklander & Q. Smith (eds), 293–304. New York: St Martin's Press

Mellor, D. H. (1998) *Real Time II*. London: Routledge.

Minkowski, H. (1908) "Space and time", in *Problems of Space and Time*, J. J. C. Smart (ed.) (1964), 297–312. London: Macmillan.

Misner, C. W., Thorne, K. S., Wheeler, A. (1970) *Gravitation*. San Francisco, CA: Freeman.

Mundy, B. (1983) "Relational theories of Euclidean space and Minkowski space-time", *British Journal for the Philosophy of Science* **50**, 205–26.

Nahin, P. J. (1993) *Time Machines*. New York: American Institute of Physics.

Nahin, P. J. (2001) *Time Machines* (2nd edn). New York: Springer-Verlag.

Nerlich, G. (1979) "What can geometry explain?", *British Journal for the Philosophy of Science* **30**, 69–83.

Nerlich, G. (1991) "How Euclidean geometry has misled metaphysics", *The Journal of Philosophy* **88**, 169–89.

Nerlich, G. (1994a) *What Spacetime Explains*. Cambridge: Cambridge University Press.

Nerlich, G. (1994b) *The Shape of Space*. Cambridge: Cambridge University Press.

Newton, I. (1729) *Mathematical Principles of Natural Philosophy*, A. Motte and F. Cajori (trans.). Berkeley, CA: University of California Press (1962).

Newton-Smith, W. H. (1980) *The Structure of Time*. London: Routledge & Kegan Paul.

Newton-Smith, W. H. (1986) "Space, time and space-time: a philosopher's view". See Flood & Lockwood (1986), 22–35.

Noonan, H. (1991) "Indeterminate identity, contingent identity and Abelardian predicates", *Philosophical Quarterly* **41**, 183–93.

Noonan, H. (1993) "Constitution is identity", *Mind* **102**, 133–46.

Norton, J. (1989) "The hole argument", in *Proceedings of the 1988 Biennial Meeting of the Philosophy of Science Assocation, Volume 2*, A. Fine & J. Leplin (eds). East Lansing, MI: Philosophy of Science Association; reprinted in Butterfield *et al.* (1996), 285–93.

Norton, J. (1992) "Philosophy of space and time", in *Introduction to the Philosophy of Science*, M. Salmon (ed.), 179–232. New Jersey: Prentice Hall; reprinted in Butterfield *et al.* (1996), 3–56.

Norton, J. (1999) "The hole argument", *Stanford Encyclopedia of Philosophy*. http://plato.stanford. edu

Oaklander, L. N. & Smith, Q. (eds) (1994) *The New Theory of Time*. New York: St Martin's Press.

Owens, D. (1992) *Causes and Coincidences*. Cambridge: Cambridge University Press.

Parfit, D. (1984) *Reasons and Persons*. Oxford: Oxford University Press.

Penrose, R. (1989) *The Emperor's New Mind*. Oxford: Oxford University Press.

Poincaré, H. (1952) *Science and Hypothesis*. New York: Dover.

Price, H. (1984) "The philosophy and physics of affecting the past", *Synthese* **16**, 299–323.

Price, H. (1996) *Time's Arrow and Archimedes' Point*. Oxford: Oxford University Press.

Prior, A. N. (1959) "Thank goodness that's over", *Philosophy* **34**, 12–17.

Putnam, H. (1967) "Time and physical geometry", *Journal of Philosophy* **64**, 240–47.

Putnam, H. (1979) "A philosopher looks at quantum mechanics", *Philosphical Papers Vol. I*, 2nd edn. Cambridge: Cambridge University Press.

Quinton, A. (1962) "Spaces and times", *Philosophy* **37**, 130–47; reprinted in Le Poidevin & MacBeath (1993), 203–20.

Ray, C. (1991) *Time, Space and Philosophy*. London: Routledge.

Raine, D. J. (1981) "Mach's principle and space-time structure", *Reports on Progress in Physics* **44**, 1151–95.

Reichenbach, H. (1956) *The Direction of Time*. Berkeley, CA: University of California Press.

Reichenbach, H. (1958) *The Philosophy of Space and Time*. London: Dover.

Rietdijk, C. W. (1966) "A rigorous proof of determinism derived from the special theory of relativity", *Philosophy of Science* **33**, 341–4.

Rovelli, C. (1999) "Quantum spacetime: what do we know?", http://arXiv/abs/gr-qc/9903045 (accessed June 2001); also in Callender & Huggett (2001), 101–22.

Rucker, R. (1984) *The Fourth Dimension*. Boston, MA: Houghton Mifflin.

Russell, B. (1915) "On the experience of time", *Monist* **25**, 212–33.

Rynasiewicz, R. (1994) "The lessons of the hole argument", *British Journal for the Philosophy of Science* **45**, 407–36.

Rynasiewicz, R. (1995) "By their properties, causes and effects: Newton's scholium on time, space, place and motion", *Studies in History and Philosophy of Science* **26**, 133–53, 295–321.

Rynasiewicz, R. (1996) "Absolute versus relational space-time: an outmoded debate?", *Journal of Philosophy* **93**, 279–306.

Rynasiewicz, R. (2000) "On the distinction between absolute and relative motion", *Philosophy of Science* **67**, 71–93.

Savitt, S. (1994) "The replacement of time", *Australasian Journal of Philosophy* **72**, 463–74.

Savitt, S. F. (ed.) (1995) *Time's Arrow Today*. Cambridge: Cambridge University Press.

Savitt, S. F. (1996) "The direction of time", *British Journal for the Philosophy of Science* **47**, 347–70.

Savitt, S. F. (1998) "There's no time like the present (in Minkowski spacetime)", *Proceedings of the 1998 Biennial Meeting of the Philosophy of Science Association*, **67**(3), part II, supplement.

Savitt, S. (2001) "Being and becoming in modern physics", *Stanford Encyclopedia of Philosophy*, http://plato.stanford.edu (accessed August 2001).

Schilpp, P. A. (ed.) (1959) *Philosophy of C. D. Broad*. New York: Tudor Publishing Company.

Schlesinger, G. N. (1991) "E pur si muove", *The Philosophical Quarterly* **41**, 427–41.

Selby, D. (1982) "Two conceptions of vacuum", *Phronesis* **27**(2), 175–93.

Shoemaker, S. (1969) "Time without change", *Journal of Philosophy* **66**, 363–81; reprinted in Le Poidevin & MacBeath (1993), 63–79.

Sklar, L. (1974) "Incongruous counterparts, intrinsic features and the substantivality of space", *Journal of Philosophy* **71**, 277–90.

Sklar, L. (1977) *Space, Time and Spacetime*. Berkeley, CA: University of California Press.

Sklar, L. (1981) "Up and down, left and right, past and future", *Noûs* **15**, 111–29; reprinted in LePoidevin and MacBeath (1993), 99–116.

Sklar, L. (1992) *Philosophy of Physics*. Oxford: Oxford University Press.

Sklar, L. (1985) *Philosophy and Spacetime Physics*. Berkeley, CA: University of California Press.

Smart, J. J. C. (1964) *Problems of Space and Time*. London: Macmillan.

Smart, J. C. C. (1980) "Time and becoming", in *Time and Cause*, P. van Inwagen (ed.), 3–15. Dordrecht: Reidel.

Smith, Q. (1993) *Language and Time*. New York: Oxford University Press.

Smith, Q. (1998) "Absolute simultaneity and the infinity of time". See Le Poidevin (1998), 135–84.

Smolin, L. (2000a) *Three Roads to Quantum Gravity*. London: Weidenfeld & Nicolson.

Smolin, L. (2000b) "The present moment in quantum cosmology: challenges to the arguments for the elimination of time", in *Time and the Instant*, R. Durie (ed.), 112–43. Manchester: Clinamen Press.

Sorabji, R. (1983) *Time, Creation and the Continuum*. London: Duckworth.

Sprigge, T. (1983) *The Vindication of Absolute Idealism*. Edinburgh: Edinburgh University Press.

Stachel, J. (1993) "The meaning of general covariance", in *Philosophical Problems of the Internal and External Worlds*, J. Earman, A. J. Janis, J. Massey & N. Rescher (eds), 129–60. Pittsburgh, PA: University of Pittsburgh Press.

Stein, H. (1968) "On Einstein–Minkowski space-time", *Journal of Philosophy* **65**, 5–23.

Stein, H. (1970) "Newtonian spacetime", *Texas Quarterly* **10**, 174–200; also in Butterfield *et al.* (1996), 79–105.

Stein, H. (1991) "On relativity theory and the openness of the future", *Philosophy of Science* **58**, 147–67; reprinted in Butterfield *et al.* (1996), 239–59.

Stewart, I. (2001) *Flatterland*. London: Macmillan.

Teller, P. (1991) "Substance, relations, and arguments about the nature of space-time", *The Philosophical Review* **100**, 353–97.

Thayer, H. S. (ed.) (1974) *Newton's Philosophy of Nature: Selections from his Writings*. New York: Macmillan, Hafner Press.

Thorne, K. (1994) *Black Holes and Time Warps: Einstein's Outrageous Legacy*. London: Picador.

Tipler, F. (1974) "Rotating cylinders and the possibility of global causality violation", *Physical Review* **D9**, 2203–6.

Tooley, M. (1997) *Time, Tense and Causation*. Oxford: Oxford University Press.

Torretti R. (1998) "Space", in *Routledge Encylopedia of Philosophy*, CD-ROM. London: Routledge.

Torretti, R. (1999a) "On relativity, time reckoning and the topology of time series". See Butterfield (1999), 65–82.

Torretti, R. (1999b) *The Philosophy of Physics*. Cambridge: Cambridge University Press.

Van Cleve, J. (1998) "Incongruent counterparts and higher dimensions". See Van Inwagen & Zimmerman (1998), 111–20.

Van Cleve, J. & Frederick, R. (eds) (1991) *The Philosophy of Left and Right*. Dordrecht: Kluwer.

Van Inwagen, P. & Zimmerman, D. W. (eds) (1998) *Metaphysics: The Big Questions*. Oxford: Blackwell.

Whitehead, A. N. (1925) *Process and Reality*, Cambridge: Cambridge University Press.

Will, C. M. (1988) *Was Einstein Right?* Oxford: Oxford University Press.

Williams, D. C. (1951) "The myth of passage", *Journal of Philosophy* **48**, 457–72.

Winnie, J. A. (1970) "Special relativity without one-way velocity assumption", *Philosophy of Science* **37**, 81–99, 223–38.

Witten, E. (1996) "Reflections on the fate of spacetime", *Physics Today* **96**(4), 24–30; reprinted in Callender & Huggett (2001), 125–37.

Yourgrau, P. (1991) *The Disappearance of Time: Kurt Gödel and the Idealistic Tradition in Philosophy*. Cambridge: Cambridge University Press.

Zelicovici, D. (1989) "Temporal becoming minus the moving-now", *Nous* **23**, 505–24.

Zimmerman, D. W. (1998) "Presentism and temporary intrinsics". See Van Inwagen & Zimmerman (1998), 206–19.

Index